VD-125a

THE CONTROL OF EUTROPHICATION OF LAKES AND RESERVOIRS

MAN AND THE BIOSPHERE SERIES

Series Editor:
JOHN JEFFERS
The University, Newcastle-upon-Tyne, NE1 7RU, United Kingdom and Ellerhow, Lindale, Grange-over-Sands, Cumbria LA11 6N1, United Kingdom

Editorial Advisory Board

EDOUARD G. BONKOUNGOU
Centre de Recherche en Biologie et Ecologie Tropicale, B.P. 7047, Ouagadougou, Burkina Faso

GONZALO HALFFTER
Insituto de Ecología, A.C., Apatado Postal 18-8-45, Delegación Miguel Hidalgo, 11800 Mexico, D.F., Mexico

OTTO LANGE
Lehrstuhl für Botanik II, Universität Würzburg, Mittlerer Dallenbergweg 64, D-8700 Würzburg, Federal Republic of Germany

LI WENHUA
Commission for Integrated Survey of Natural Resources, Chinese Academy of Sciences, P.O. Box 787, Beijing, People's Republic of China

GILBERT LONG
Centre d'Ecologie Fonctionnelle et Evolutive Louis Emberger (CNRS-CEPE), Route de Mende, B.P. 5051, 34033 Montpellier Cedex, France

IAN NOBLE
Research School of Biological Sciences, Australian National University, P.O. Box 475 Canberra City A.C.T. 2601, Australia

P.S. RAMAKRISHNAN
G.B. Pant Institute of Himalayan Environment and Development, Kosi, Almora, U.P. 263643, India

VLADIMIR SOKOLOV
Institute of Evolutionary Morphology and Animal Ecology USSR Academy of Sciences, 33 Lininsky Prospect, 117071 Moscow, USSR

ANNE WHYTE
Division of Social Sciences, International Development Research Centre (IDRC), P.O. Box 8500, Ottawa K1G 2119, Canada

Ex. officio:
BERND VON DROSTE
Division of Ecological Sciences, Unesco, 7, place de Fontenoy, 75700 Paris, France

MALCOLM HADLEY
Division of Ecological Sciences, Unesco, 7, place de Fontenoy, 75700 Paris, France

MaB
MAN AND THE BIOSPHERE SERIES

Series Editor: J.N.R. Jeffers

VOLUME I

THE CONTROL OF EUTROPHICATION OF LAKES AND RESERVOIRS

Edited by
Sven-Olof Ryding and Walter Rast

with the assistance of
Dietrich Uhlmann — Jürgen Clasen — Laszlo Somlyódy

PUBLISHED BY

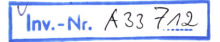

PARIS

AND

The Parthenon Publishing Group
International Publishers in Science, Technology & Education

Published in 1989 by the United Nations Educational Scientific and Cultural Organization,
7 Place de Fontenoy, 75700 Paris, France — Unesco ISBN 92-3-102550-3
and
The Parthenon Publishing Group Limited
Casterton Hall, Carnforth
Lancs LA6 2LA, UK — ISBN 1-85070-257-8
and
The Parthenon Publishing Group Inc.
120 Mill Road,
Park Ridge
New Jersey 07656, USA — ISBN 0-929858-13-1

© Copyright **Unesco 1989**

The designations employed and the presentation of the material in this publication do not imply the expression of any opinion whatsoever on the part of the publishers concerning the legal status of any country or territory, or of its authorities, or concerning the frontiers of any country or territory.

The authors are responsible for the choice and the presentation of the facts contained in this book and for the opinions expressed therein, which are not necessarily those of Unesco and do not commit the organization.

No part of this book may be reproduced
in any form without permission from the
publishers except for the quotation of
brief passages for the purpose of review

Printed and Bound in Great Britain by
Butler and Tanner Ltd., Frome and London

British Library Cataloguing in Publication Data

The control of eutrophication of lakes and reservoirs.
 1. Lakes. Management 2. Natural resources: water. Reservoirs. Management
 I. Ryding, Sven-Olof II. Rast, Walter III. Series 333.91'63

ISBN 1-85070-257-8

Library of Congress Cataloging-in-Publication-Data

The Control of eutrophication of lakes and reservoirs/edited by Sven-Olof Ryding and Walter Rast, with the assistance of Dietrich Uhlmann, Jürgen Clasen, Laszlo Somlyody.
    p.   cm. — (MAB; v. 1)
  Bibliography: p.
  Includes index.
  ISBN 0-929858-13-1: $48.00 (U.S.)
   1. Eutrophication—Control—Handbooks, manuals, etc. I. Ryding, Sven-Olof. II. Rast, Walter. III. Unesco. IV. Series: MAB (Series); 1.
QH96.8.E9C66 1989   89-2992
628.1'68—dc19   CIP

SERIES PREFACE

Unesco's Man and the Biosphere Programme

Improving scientific understanding of natural and social processes relating to man's interactions with his environment, providing information useful to decision-making on resource use, promoting the conservation of genetic diversity as an integral part of land management, enjoining the efforts of scientists, policymakers and local people in problem-solving ventures, mobilizing resources for field activities, strengthening of regional cooperative frameworks. These are some of the generic characteristics of Unesco's Man and Biosphere Programme.

Unesco has a long history of concern with environmental matters, dating back to the fledgling days of the organization. Its first Director General was biologist Julian Huxley, and among the earliest accomplishments was a collaborative venture with the French Government which led to the creation in 1948 of the International Union for the Conservation of Nature and Natural Resources. About the same time, the Arid Zone Research Programme was launched, and throughout the 1950s and 1960 this programme promoted an integrated approach to natural resources management in the arid and semi-arid regions of the world. There followed a number of other environmental science programmes in such fields as hydrology, marine sciences, earth sciences and the natural heritage, and these continue to provide a solid focus for Unesco's concern with the human environment and its natural resources.

The Man and Biosphere (MAB) Programme was launched by Unesco in the early 1970s. It is a nationally based, international programme of research, training, demonstration and information diffusion. The overall aim is to contribute to efforts for providing the scientific basis and trained personnel needed to deal with problems of rational utilization and conservation of settlements. MAB emphasizes research for solving problems: it thus involves research by interdisciplinary teams on the interactions between ecological and social systems; field training; and applying a systems

Control of eutrophication

approach to understanding the relationships between the natural and human components of development and environmental management.

MAB is a decentralized programme with field projects and training activities in all regions of the world. These are carried out by scientists and technicians from universities, academies of sciences, national research laboratories and other research and development institutions, under the auspices of more than a hundred MAB National Committees. Activities are undertaken in cooperation with a range of international governmental and non-governmental organizations.

Further information on the MAB Programme is contained in *A Practical Guide to MAB*, *Man Belongs to the Earth*, a biennial report, a twice-yearly newsletter *InfoMAB*, MAB technical notes, and various other publications. All are available from the MAB Secretariat in Paris.

Man and the Biosphere Book Series

The Man and the Biosphere Series has been launched with the aim of communicating some of the results generated by the MAB Programme to a wider audience than the existing Unesco series of technical notes and state-of-knowledge reports. The series is aimed primarily at upper level university students, scientists and resource managers, who are not necessarily specialists in ecology. The books will not normally be suitable for undergraduate text books but rather will provide additional resource material in the form of case studies based on primary data collection and written by the researchers involved; global and regional syntheses of comparative research conducted in several sites or countries; and state-of-the-art assessments of knowledge or methodological approaches based on scientific meetings, commissioned reports or panels of experts.

The series will span a range of environmental and natural resource issues. Currently available in press or in preparation are reviews on such topics as control of eutrophication in lakes and reservoirs, sustainable development and environmental management in small islands, reproductive ecology of tropical forest plants, the role of land/inland water ecotones in landscape management and restoration, ecological research and management in alpine regions, structure and function of a nutrient-stressed Amazonian ecosystem, non-conventional conservation and the role of biosphere reserves in the quest for alternatives, assessment and control of non-point source pollution, research for improved land use in arid northern Kenya, ecological and social effects of large-scale logging of tropical forest in the Gogol Valley (Papua New Guinea), changing land-use in the European Alps.

The Editor-in-Chief of the series is John Jeffers, until recently Director

Series preface

of the Institute of Terrestrial Ecology in the United Kingdom, who has been associated with MAB since its inception. He is supported by an Editorial Advisory Board of internationally-renowned scientists from different regions of the world and from different disciplinary backgrounds: E.G. Bonkoungou (Burkina Faso), Gonzalo Halffter (Mexico), Otto Lange (Federal Republic of Germany), Li Wenhau (China), Gilbert Long (France), Ian Noble (Australia), P.S. Ramakrishnan (India), Vladimir Sokolov (USSR) and Anne Whyte (Canada).

A publishing rhythm of three to four books per year is envisaged. Books in the series will be published initially in English, but special arrangements will be sought with different publishers for other language versions on a case-by-case basis.

CONTENTS

1. **INTRODUCTION**   1
   The aging process of lakes and reservoirs   1
   The need for concern in regard to eutrophication   3

2. **PURPOSE OF BOOK AND HOW IT CAN BEST BE USED**   5
   Intended audience for the book   6
   Focus of the book   7

3. **A MANAGEMENT FRAMEWORK FOR THE POLICYMAKER**   11
   The role of the policymaker   11
      The policymaker and environmental issues   11
      Development of a eutrophication management framework   12

   Identify eutrophication problems and establish management goals   14
      Why is eutrophication a problem?   14
      Establishment of eutrophication management goals   15
      Who should be involved in addressing the problem?   16

   Assess the extent of information available about the waterbody   18

   Identify available options for management of eutrophication   19
      Should one treat the causes or the symptoms?   19
      Consider full range of available control options   20

   Analyze all costs and expected benefits of alternative management strategies   22
      Consider relative costs of control options   22
      Compare management goals, available resources and expected benefits   23
      Cost–benefit analysis   25

Analyze adequacy of existing legislative and regulatory
framework for implementing eutrophication control programmes 30
    Institutional concerns 30
    Regulatory concerns 31

Select eutrophication control strategy and disseminate
summary to affected parties prior to implementation 32
    Select and summarize desired control strategy 32
    Other practical considerations 33

Provide periodic progress reports on the control programme to the
public and other interested parties 34
    The role of public awareness 34
    The value of public feedback 35

4. CHARACTERISTICS OF EUTROPHICATION 37

The concept of eutrophication as related to trophic status 37

Symptoms of eutrophication 38
    General observations 38
    Differences between lakes and reservoirs 42
    Differences between temperate and tropical lake systems 46

The limiting nutrient concept 49
    Some assumptions in using the limiting nutrient concept 51
    The use of nitrogen and phosphorus concentrations and
        ratios to assess nutrient limitation 54
    The use of algal bioassays to assess nutrient limitation 58
    The use of physiological indicators to assess nutrient
        limitation 60

5. FACTORS AND PROCESSES AFFECTING THE DEGREE OF EUTROPHICATION 65

Factors related to the drainage basin 65
    Natural factors 65
    Anthropogenic factors 71

Factors related to the waterbody 74
    Morphology of the lake basin 74
    In-lake nutrient sources 77
    Flushing rate 79
    Light intensity 80
    Biological controls 80
    Macrophyte growths 81
    Cage fish farming 83

6. THE USE OF MODELS 85

General types of models 85

Watershed models 87
Empirical watershed models 88
Simulation watershed models 90

Waterbody models 94
Empirical waterbody models 94
Simulation waterbody models 107

7. ESTIMATING THE NUTRIENT LOAD TO A WATERBODY 115

Major phosphorus and nitrogen sources 115
Point sources 115
Diffuse (non-point) external sources 116
Diffuse internal sources 118
Biological availability of nutrients 119

Quantifying the nutrient load 120
Point source loads 121
Diffuse (non-point) external loads 122
Diffuse internal loads 131
Summary comments on nutrient load quantification 135

A simplified approach for estimating the annual nutrient load to a lake or reservoir 135
Specific steps in use of simple watershed model 138

Nutrient estimates for waterbodies with rapid flushing rates 143

Reliability of nutrient load estimates 146

8. GUIDELINES FOR SAMPLING A WATERBODY 147

What to sample 147

Necessary temporal and spatial resolution for data 150
Where to sample 150
When to sample 154
Sampling strategies in waterbodies with longitudinal water quality gradients 155

Calculating the costs of sample collection 158

Compilation and presentation of data 164

9. AVAILABLE TECHNIQUES FOR TREATING EUTROPHICATION 169

General considerations 169

Control of the external phosphorus load 169
  Direct reduction of phosphorus at the source 170
  Treatment of tributary influent waters 174
  Canalization/diversion of wastewaters 179

In-lake eutrophication control methods 181

Control of non-point source nutrients in the drainage basin 191
  Major non-point nutrient sources and possible remedial measures 191

Assessment of costs of achieving phosphorus control goals 195
  Relative costs of point and non-point source phosphorus control measures 199
  Use of optimization models in selection of phosphorus control options for lakes and reservoirs 210

10. THE POSSIBILITIES FOR THE REUSE OF NUTRIENTS 213

Use of algae and macrophytes for nutrient removal 213
  Phytoplankton 213
  Harvesting of macrophytes and filamentous algae 215
  Use of macrophyte biomass as animal food 216
  Use of plant biomass as raw materials for commercial purposes 217

Aquaculture for the production of fish and other aquatic food organisms 218
  Aquaculture for the control of nutrients 219
  Estimation of fish yield 222
  Considerations for the management of fish ponds and small waterbodies 226

Other nutrient reuse possibilities 229

Eutrophication and acidification of lakes 230

11. SELECTION OF EFFECTIVE STRATEGIES FOR THE MANAGEMENT OF EUTROPHICATION 231

General considerations 231

Water quality as related to desired water use 232

A simple approach for defining and selecting an effective eutrophication control programme 234
  General overview 234
  Specific steps to follow 238
  What if the waterbody does not respond as expected? 245

Detailed scenario analysis as an aid in selecting eutrophication
control measures — 251

Post-treatment monitoring — 254

ANNEX 1: CLASSIFICATION OF WATERBODIES IN RELATION TO THEIR DESIRED USES — 257

A simple lake/reservoir classification system — 257

A detailed classification system — 262
  Hydrographic and territorial criteria — 264
  Trophic criteria — 268
  Dissolved solids, special and hygienically-relevant criteria — 269
  Lake classification in regard to possible water uses — 269
  How to classify a waterbody — 269

ANNEX 2: CASE STUDIES OF EUTROPHICATION CONTROL MEASURES — 277

LITERATURE CITED — 283

INDEX — 309

LISTING OF CORE EDITORIAL GROUP, CONTRIBUTING AUTHORS AND REVIEWERS
(listed alphabetically)

CORE EDITORIAL GROUP

S.-O. Ryding, (CHAIRMAN), Federation of Swedish Industries, Stockholm, Sweden.

J. Clasen, Wahnbachtalsperrenverband, Siegburg, Federal Republic of Germany.

W. Rast, (Current address: Water Resources Division, US Geological Survey, Austin, Texas, USA).

L. Somlyódy, Research Center for Water Resources Development (VITUKI), Budapest, Hungary.

D. Uhlmann, Sektion Wasserwesen, Technische Universität, Dresden, German Democratic Republic.

SCIENTIFIC ADVISORY GROUP

C. Forsberg, Institute of Limnology, University of Uppsala, Sweden.

H.L. Golterman, Station Biologique, De La Tour du Valat, Camargue, France.

R.A. Vollenweider, Canada Centre for Inland Waters, Burlington, Ontario, Canada.

CONTRIBUTING AUTHORS

Chapter 1. *Introduction*:
W. Rast
S.-O. Ryding

Chapter 2. *Purpose of book and how it can best be used*:
S.-O. Ryding

Chapter 3. *A management framework for the policymaker*:
M.M. Holland, Ecological Society of America, Washington, DC, USA
J.G. Moore, Jr, Environmental Sciences Program, University of Texas at Dallas, Richardson, Texas, USA.
W. Rast
G. Thornburn, International Joint Commission, Ottawa, Ontario, Canada.
J.A. Thornton, Town Planning Branch, c/o 87 Shore Drive, Auburn, Massachusetts, USA.

Chapter 4. *Characteristics of eutrophication*:
W. Rast
V.H. Smith, Department of Biology, University of North Carolina, Durham, North Carolina, USA.
J.A. Thornton

Chapter 5. *Factors and processes affecting the degree of eutrophication*:
D.E. Canfield, Center for Aquatic Weeds, University of Florida, Gainesville, Florida, USA.
J.R. Jones, School of Forestry, Fisheries and Wildlife, University of Missouri, Columbia, Missouri, USA.
S.-O. Ryding
D. Uhlmann

Chapter 6. *The use of models*:
D.A. Haith, Department of Agricultural Engineering, Cornell University, Ithaca, New York, USA.
W. Rast
K.H. Reckhow, School of Forestry and Environmental Sciences, Duke University, Durham, North Carolina, USA
L. Somlyódy
G. van Straten, Department of Chemical Technology, Twente University of Technology, Enschede, The Netherlands.

Chapter 7. *Estimating the nutrient load to a waterbody*:
R.C. Loehr, Department of Civil Engineering, University of Texas, Austin, Texas, USA.
S.-O. Ryding
W.C. Sonzogni, Department of Environmental Engineering, University of Wisconsin, Madison, Wisconsin, USA.

Listing of core editorial group

Chapter 8. *Guidelines for sampling a waterbody*:
H. Klapper, Institut für Wasserwirtschaft, Abt. Oberflächengewässer, Magdeburg, German Democratic Republic.
W. Rast
D. Uhlmann

Chapter 9. *Available techniques for treating eutrophication*:
J. Clasen
W. Rast
S.-O. Ryding

Chapter 10. *The possibilities for the reuse of nutrients*:
D. Barthelmes, Institut für Binnenfischerei, Berlin, German Democratic Republic.
J.M. Melack, Department of Biological Sciences, University of California, Santa Barbara, California, USA.
R.T. Oglesby, Cornell University, Ithaca, New York, USA.
D.W. Smith, Department of Biological Sciences, University of California, Santa Barbara, California, USA.
D. Uhlmann

Chapter 11. *Selection of effective strategies for the management of eutrophication*:
W. Rast
S.-O. Ryding

Annex 1. *Classification of waterbodies in relation to their desired uses*:
H. Klapper
W. Rast
D. Uhlmann

Annex 2. *Case studies of eutrophication control measures*:
W. Rast
S.-O. Ryding

LIST OF REVIEWERS

V. Coehlo, Companhia de Tecnologia de Saneamento Ambiental (CETESB), Sao Paulo, Brazil.
D.G. Cooke, Kent State University, Kent, Ohio, USA.
J. Derisio, Companhia de Tecnologia de Saneamento Ambiental (CETESB), Sao Paulo, Brazil.
J.I. Furtado, University Malaya, Kuala Lumpur, Malaysia
D.J. Gregor, Environmental Canada, Regina, Saskatchewan, Canada.
G. Jolánkai, Research Centre for Water Reources Development (VITUKI), Budapest, Hungary.
O. Jóo, Western Transdanubian Water Authority, Szombathely, Hungary
R.H. Kennedy, US Army Corps of Engineers, Vicksburg, Mississippi, USA.
H.L. Ludwig, SEATEC International, Bangkok, Thailand.
J.G. Limón M., Grupo Hidrosanitec Ingeniería Ambiental, University of Guadalajara, Jalisco, Mexico.
F. Liscum, U.S. Geological Survey, Houston, Texas, USA.
A. Muñoz, Centro Panamericano de Ingeniería Sanitaría y Ciencias del Ambiente (CEPIS), Pan American Health Organization, Lima, Peru.
V. Pantulu, Mekong Basin Commission, Bangkok, Thailand.
E. Pieczynska, Zoological Institute, University of Warsaw, Warsaw, Poland.
K. Pütz, Sektion Wasserwesen, Technische Universität, Dresden, German Democratic Republic.
O. Ravera, Commission of European Communities, Euratom, Ispra, Italy.
T.C. Rey, Laguna Lake Development Authority, Manila, Philippines.
H. Salas, Centro Panamericano de Ingeniería Sanitaria y Ciencias del Ambiente (CEPIS), Pan American Health Organization, Lima, Peru.
J. Santos, Centro de Estudios Hidrograficos, Ministerio de Obras Publicas y Urbanismo , Lisbon, Portugal.
J.C. Stephens, Electronic Data Systems, Dallas, Texas, USA.
M. Straškraba, Institute of Landscape Ecology, Czechoslovak Academy of Science, České Budějovice, Czechoslovakia.
I. Toms, Thames Water Authority, London, United Kingdom.
E. Welch, University of Washington, Seattle, Washington, USA.
P. Woods, US Geological Survey, Boise, Idaho, USA.
D.P. Zutshi, University of Kashmir, Srinagar, India.

In addition to the above list of reviewers, the individual contributing authors provided a review of the chapters in which they were involved.

FOREWORD

Human settlement of a drainage basin, and the associated clearing of forests, building of farms and towns, etc. can dramatically accelerate the aging process (eutrophication) of a natural lake or man-made reservoir, and significantly reduce its life span. This process, often called cultural eutrophication, usually is accompanied by detrimental changes in water quality, which can interfere with human use of the water resource for many purposes (e.g. drinking water, recreation, industrial uses).

On a global scale, eutrophication of lakes and reservoirs continues to rank as one of the most pervasive water pollution problems. Because of this problem, an International Workshop on the Control of Eutrophication was held at the International Institute for Applied Systems Analysis (IIASA) in Laxenburg, Austria, in October, 1981. This workshop was co-sponsored by the United Nations Educational, Scientific and Cultural Organization (Unesco), the Organization for Economic Cooperation and Development (OECD) and IIASA. A strong recommendation made at this workshop was that a book or manual for the assessment and control of eutrophication should be developed. This recommendation resulted in the formation of a multinational, Core Editorial Group by Unesco. This group was given the mandate of preparing such a document.

To aid in this task, the Core Editorial Group also recruited a number of contributing authors from several countries around the world, to assist in providing materials and expertise for specific portions of the book. Furthermore, the document has been reviewed by a number of scientists, engineers and water managers around the world. All persons involved in the preparation or review of this book are identified individually on pages xv–xix.

This book focuses on the *practical* control of eutrophication, as contrasted with a strictly academic treatment of the subject. It is not meant to be an extensive treatise on the subject of eutrophication, but rather a simplified handbook, which attempts to bring existing knowledge and experience together in a form for widespread use by both a technical and non-technical audience.

In contrast to previous efforts, this document attempts to discuss practical control of eutrophication. It specifically emphasizes a number of topics not considered in previous publications dealing with the control of eutrophication. As examples, this book contains discussions of comparable and contrasting features of tropical/sub-tropical, temperate and sub-arctic waterbodies, as well as the similarities and differences between natural lakes and man-made reservoirs. It also discusses aquaculture and fisheries enhancement for the purposes of food production and other uses, as one of the positive aspects of eutrophication. Furthermore, this book presents the minimum requirements for an in-lake monitoring programme for obtaining suitable data inputs for existing eutrophication models, and for assessing the trophic conditions and water quality of a waterbody relative to its water use. In addition, the book attempts to deal with the special needs of the decision maker and manager (in contrast to the more technical concerns) in regard to development of sound eutrophication-related water policy. As this is intended to be a practical document, this book attempts to cover most of the lake and reservoir situations likely to be encountered. Hopefully, many persons dealing with eutrophication problems around the world will find it useful in their work.

Since this book is intended to bring together existing knowledge and experience in a form useful for the practical control of eutrophication, its contents necessarily reflect our varying understanding of the eutrophication process and the major factors which affects its timing and intensity in lakes and reservoirs. It is hoped that further experiences gained by those using this document in many different limnological situations will provide the basis for revising and augmenting the practical guidance contained in it.

This book is meant to be an integrated document with the same basic theme uniting all the chapters. It is emphasized that this document does not necessarily represent an official policy statement either of Unesco, or of the home organizations of the individuals involved in its preparation. Rather, it is an integration, by the Core Editorial Group, of the personal and professional expertise of all the involved authors.

I wish to acknowledge Leo Teller of Unesco for his efforts in initiating this interesting project, and the members of the Core Editorial Group for their efforts and assistance in making this document become a reality. I also wish to acknowledge the excellent written contributions, technical advice and reviews of this book from the many individuals involved in its preparation, and to extend my sincere appreciation. These many individuals are identified in the following pages. Special thanks are given to J.A. Thornton for providing both his technical expertise and meticulous reviews of the book through its multiple stages of development. The assistance and patience of the Ecological Sciences Division of Unesco, including individuals involved in the Programme on Man and the Biosphere (MAB), is gratefully appreciated.

Sven-Olof Ryding, Chairman
Unesco Core Editorial Group
Stockholm, Sweden

CHAPTER 1

INTRODUCTION

THE AGING PROCESS OF LAKES AND RESERVOIRS

During the last 20 years, the word 'eutrophication' has been used more and more to denote the artificial and undesirable addition of plant nutrients, mainly phosphorus and nitrogen, to waterbodies. In some situations (see Chapter 10), this view can be misleading, since what is an undesirable addition to one waterbody may be harmless, or even beneficial, in another waterbody. Nevertheless, eutrophication is most commonly known as the state of a waterbody which is manifested by an intense proliferation of algae and higher aquatic plants, and their accumulation in the waterbody in excessive quantities. These accumulations can result in detrimental changes in the water quality and in the biological populations of a waterbody, which can interfere significantly with man's use of the water resource. The definition of eutrophication developed by the Organization for Economic Co-operation and Development (OECD, 1982) was adopted as the primary working definition for the materials discussed in this report. OECD defined eutrophication as 'the nutrient enrichment of waters which results in the stimulation of an array of symptomatic changes, among which increased production of algae and macrophytes, deterioration of water quality and other symptomatic changes, are found to be undesirable and interfere with water uses'.

Eutrophication, in the original sense, represents the natural aging process of a lake. A lake receives inflows of water from its surrounding drainage basin, along with materials carried in the water from the land surface, e.g. following a rain storm or from irrigation drainage. Materials associated with rain, snow and wind-blown substances, as well as ground water inflows (sub-surface flow), can also directly enter a lake. The observed water quality and biological communities in a lake, therefore, reflect the cumulative impacts of all the water and material inflows into the lake.

Over time, a lake will be slowly filled in with soil and other materials carried by inflowing waters, and eventually become a marsh and, ultimately, a terrestrial system. This process usually takes many hundreds of thousands of years to occur and is largely irreversible. Lakes undergoing such natural eutrophication generally have good water quality and exhibit a diverse biological community throughout much of their existence.

Where man has not settled a drainage basin, the growth of algae and other aquatic plants in a lake in the drainage basin is usually minimal, and generally in balance with the input of plant nutrients. However, human settlement of a drainage basin, and the associated clearing of forests, development of farms and cities, etc., usually changes the natural eutrophication process in a dramatic way. The runoff of most materials from the land surface to the waterbody is greatly accelerated. An increased input of plant nutrients (mainly phosphorus and nitrogen) to a lake can stimulate algal and aquatic plant growths which, in turn, can stimulate the growth of fish and other higher trophic level organisms in the aquatic food chain. This latter phenomenon is often termed 'cultural eutrophication' to distinguish it from the natural process (the terms 'artificial', 'anthropogenic' or 'man-made' are also often used to describe the same phenomenon). A waterbody undergoing cultural eutrophication can be treated so that it will again exhibit an 'aging' rate more characteristic of natural eutrophication. However, for waterbodies (lakes and reservoirs) undergoing extensive cultural eutrophication, the necessary control measures can be quite expensive and difficult to administer.

The effects of eutrophication are considered negative in many places around the world, and often reflect human perceptions of good versus bad water quality. Excessive algae and aquatic plant growths are highly visible and can interfere significantly with the uses and aesthetic quality of a waterbody. One consequence of such growths can be the production of taste and odor problems in drinking water drawn from a waterbody, even though the water may be treated and filtered prior to use. The water treatment process itself can become more expensive and time-consuming for eutrophic waters. The water transparency may be greatly reduced.

There are also significant ecological consequences related to cultural eutrophication. As algal populations die and sink to the bottom of a waterbody, their decay by bacteria can reduce oxygen concentrations in bottom waters to levels which are too low to support fish life, resulting in fish kills. Such oxygen-deficient conditions can also result in excessive levels of iron and manganese in the water, which can interfere with drinking water treatment. There are also potential health effects, especially in tropical regions, related to such parasitic diseases as schistosomiasis, onchocerchiasis and malaria, all of which can be aggravated by cultural eutrophication, and which can enhance the appropriate habitats for these organisms.

Eutrophication does have its positive aspects to consider. If the eutrophication process is used to enhance the production of fish, or in other forms of aquaculture, for the purpose of producing protein as a food supply, then the management goal is to maximize or optimize such productivity at minimal cost and effort. It is pointed out, however, that both the types and quality of fish in a lake or reservoir can change as a result of increasing eutrophication. This latter aspect is discussed further in Chapters 4 and 10.

It is emphasized that this general description of the eutrophication process, and its accompanying symptoms, also applies to man-made lakes. As used in this document, man-made lakes refer to waterbodies which have been created artificially by the construction of a dam across a flowing river or stream. Such waterbodies are usually called reservoirs or impoundments to distinguish them from natural lakes. As discussed in more detail in Chapter 4, there are some differences in the characteristics of reservoirs and natural lakes. These differences should be considered when assessing the degree of eutrophication in a man-made reservoir as compared to a natural lake system. Nevertheless, the factors to be considered in selecting eutrophication control measures are sufficiently similar in both types of waterbodies that the terms 'lake' and 'reservoir' are used throughout this document (usually interchangeably). Where the differences between lakes and reservoirs can influence the selection of specific eutrophication control measures, they are discussed in the text of the document.

THE NEED FOR CONCERN IN REGARD TO EUTROPHICATION

On a global scale, the demand for surface waters for many purposes is increasing. There are a number of established water usages, including industrial use, drinking water supply, transportation, recreation, irrigation, aquaculture and habitat preservation. The increasing demands for surface waters of good quality for such uses can easily lead to conflicts if there are shortages in the supply. Many developed and developing countries experience this problem. This is often the case in areas where evaporation exceeds precipitation, thereby minimizing runoff or ground water recharge from rains. Furthermore, even in countries with a net surplus of water on an annual basis, severe shortages of water during limited periods can still occur. Climatic variations can make it difficult to predict the amount of water that will be available for various purposes from one year to the next.

The eutrophication problem often becomes apparent to the public in many countries as a function of the progressive growth of densely populated areas. In many situations, the discharge of raw or inadequately treated sewage (a major source of aquatic plant nutrients) to the nearest body of water has increased several fold over the years. Because of the nature of municipal or industrial nutrient sources, it is often easy to demonstrate the usefulness of efforts to reduce nutrient pollution from such sources, based on the expected beneficial changes in water quality. As a result, in many countries around the world, large investments have been made in the construction of treatment facilities for municipal wastewaters (such facilities constitute the majority of so-called 'point' sources of nutrients). However, in spite of efforts to reduce nutrient pollution from municipal or industrial sources by the introduction of advanced wastewater treatment, water quality has continued to deteriorate in many streams, lakes, reservoirs and coastal areas. This is due in many cases to a continuing increase in pollutant runoff from the land surface or in materials deposited directly into a waterbody from the atmosphere (these sources constitute a major portion of so-called 'non-point' or diffuse sources, see Chapter 7). In principle, diffuse nutrient pollution can occur from both external sources, such as atmospheric precipitation and surface runoff from urban and agricultural areas in drainage basins, and internal sources, such as regeneration from sediments.

In view of the large number of pollution sources to be considered, and our need to increase our understanding of the complexity of maintaining an adequate water supply, there is an urgent need for an increased understanding of the behavior of aquatic ecosystems. This is especially true with regard to the extent of man's activities in the ecosystem and the increasing need for water on a global scale for different purposes. As a consequence, a suitable approach to addressing the problem is an integrated view of land and water resources, including consideration of the simultaneous impacts on a lake or reservoir of all the significant nutrient inputs. In addition, it is likely to be an even more difficult task to combat eutrophication in the future, due to the increasing influence of diffuse pollution from land use practices, etc., in the drainage basin.

The need for accurate knowledge of the relevance and magnitude of different pollutant sources to the state of the aquatic ecosystem is essential. Such knowledge will provide a better basis for planning future water management efforts, both for the control of eutrophication and for other types of lake and reservoir pollution.

CHAPTER 2

PURPOSE OF BOOK AND HOW IT CAN BEST BE USED

Past scientific studies have provided ample evidence that, technically, effective control of eutrophication of lakes and reservoirs is possible. Whether or not effective control also is economically feasible, however, depends on the dominant factors in a specific situation (see Chapters 3 and 11). During the last several decades, a number of studies focusing on the eutrophication problem, and the relationship between nutrient load and responses in waterbodies of different character, have been undertaken. Many types of eutrophication relationships and mathematical models, both simple and complex, have been presented in the scientific literature.

One example of a study which focuses on the use of simple eutrophication models is the Cooperative Programme for Monitoring of Inland Waters, organized and coordinated by the Organization for Economic Co-operation and Development (OECD). This international, cooperative programme synthesized the results from four different regional eutrophication projects into a comprehensive summary document (OECD, 1982). Furthermore, the Pan American Center for Sanitary Engineering and Environmental Sciences (CEPIS) is coordinating a project for the development of simplified methodologies for the evaluation of eutrophication of warm water (i.e. tropical/sub-tropical) lakes (Salas, 1982). When completed, this study will provide a general quantitative framework similar to that developed in the OECD study. Another study, emphasizing a more detailed modeling approach to assess the eutrophication of shallow lakes, was conducted under the auspices of the International Institute for Applied Systems Analysis (Somlyódy & van Straten, 1986). Because eutrophication will continue to be a problem in many countries, there is an urgent need to make practical use of such studies, both for the accurate assessment and the effective long-term control of lake and reservoir eutrophication.

Many scientific and technical eutrophication studies are referenced in this book as sources of relevant eutrophication information. However, it is emphasized that primary attention in this document is given to methodologies which have been shown, over time, to be useful and practical in the assessment and control of eutrophication. Accordingly, the main purpose of this book is to bring together current knowledge and experience on this complex topic in a form which is both usable and practical for the control of eutrophication. Thus, some recent or controversial studies may receive only minor attention in this document, in favor of emphasizing those items and topics that have received adequate technical review and have been shown to be useful over the long term.

This book focuses primarily on lakes and reservoirs, both natural and impounded. It does not directly address other water systems (e.g. flowing waters or estuarine systems). These latter systems are no less important than lakes and reservoirs. However, even though the same basic principles would apply to a large degree, the available knowledge for dealing effectively with flowing and estuarine waters on a practical level is less extensive than that pertaining to lakes and reservoirs.

It is emphasized that this document focuses primarily on the average conditions in the whole waterbody, rather than on seasonal dynamics or specific regions in the waterbody. This focus may not be adequate in some cases. Examples are situations in which the specific timing and/or location of algal blooms constitute the primary concern. However, present knowledge regarding the factors that control the seasonal and temporal dynamics of the eutrophication process is still rudimentary. In most cases, accurate knowledge of the the annual, average conditions in a lake or reservoir is sufficient for the purpose of eutrophication management. Consequently, the practical nature of this book dictates the focus on average, lake-wide conditions.

This book also does not address conditions in the nearshore regions or littoral zone of waterbodies as a primary concern. However, Chapters 5, 6 and 10 do provide discussions of the problems related to rooted and floating aquatic plants growing in the shallow waters near the shoreline.

INTENDED AUDIENCE FOR THE BOOK

It is the intention that virtually anyone concerned with the inherent problems and/or control of eutrophication of lakes and reservoirs will find this document to be useful. It should be especially useful to individuals in local water authorities, private companies, state agencies, etc., working on the control of eutrophication of lakes and reservoirs (e.g. development of monitoring programmes, collection and evaluation of data, etc.). It discusses available techniques and strategies for attempting to control eutrophication, and contains practical guidance for assessing the potential outcome of various eutrophication control alternatives.

Purpose of book and how it can best be used

An important audience for this book is individuals with responsibility for making management decisions regarding the control of eutrophication. This group of individuals includes *politicians, legislators, managers, senior administrators, company presidents*, etc. For them, the book concentrates on those items important or unique to the decision making process, including guidelines for decision making related to eutrophication control, some indications of the costs associated with different control strategies, and a bibliography of documented case studies.

Following the chapter dealing with the primary concerns of the decision maker (Chapter 3), the book is roughly divided into two broad categories. The first major part (Chapters 4 to 6) is relatively descriptive in nature, and provides details regarding the processes and factors affecting eutrophication. This portion of the document should be useful to *scientists, engineers* and *educators*.

The second major part of the book (Chapters 7 to 11) deals with the practical aspects for control or management of eutrophication. In addition, Chapter 10 provides a discussion of the positive aspects of eutrophication, as they apply to various forms of aquaculture and to the reuse of nutrients. This latter portion of the book is intended to be especially useful for the work of project managers and technicians, as well as of scientists and engineers.

FOCUS OF THE BOOK

A schematic guide to the contents of this document is presented in Figure 2.1. The main topics addressed in each chapter are listed in a logical sequence for eutrophication management considerations. The specific topics and related materials are discussed in the referenced chapters. *Why cultural eutrophication is a problem* is discussed in *Chapters 1 and 4*. These chapters (especially Chapter 4) describe the causes and symptoms of eutrophication. They also discuss the differences and similarities of the eutrophication process in various regions of the world. Information on water quality, as related to intended water uses, is also provided. Since successful management of eutrophication is often dependent on proper identification of the most likely growth-limiting nutrient for algae and other aquatic plants, special emphasis is given to the limiting nutrient concept.

Topics that are of special importance to the decisionmaker are dealt with in *Chapter 3*. The role of the decision maker is discussed in terms of identifying eutrophication management goals, determining the extent of available information, identifying different management options and cost–benefit considerations, assessing the adequacy of the existing legislative/regulatory framework, and selecting specific eutrophication control strategies.

Factors that can affect the extent of eutrophication are presented in

Control of eutrophication

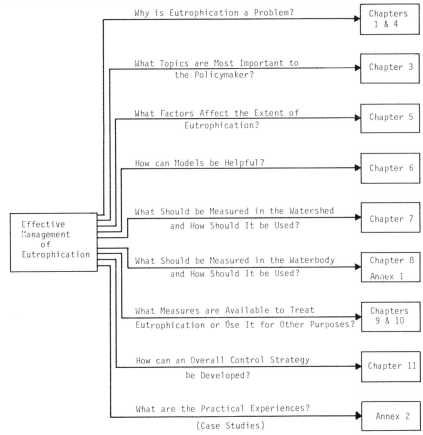

Figure 2.1 Guide to contents of manual

Chapter 5. These factors are related both to the catchment area and to the waterbody. In the first case, both natural and man-made factors are considered. In the latter case, various in-lake characteristics are identified. These factors include physical (e.g. morphology of lake basin, light intensity), chemical (e.g. in-lake nutrient sources) and biological (e.g. plankton and macrophyte production) components.

The use of mathematical models in assessment and management of eutrophic lakes and reservoirs is considered in *Chapter 6*. This chapter presents an overview of relevant eutrophication models (both watershed and waterbody models). It also presents guidelines for using such models to assess the trophic conditions and associated water quality conditions of lakes and reservoirs in a range of limnological settings, including tropical/sub-tropical, temperate and sub-arctic regions. The book stresses the use of simple, practical relationships and models.

Guidelines for assessing nutrient inputs from different point and non-point sources in a drainage basin (i.e. *what should be measured in the watershed and how should it be used*) are presented in *Chapter 7*. This chapter discusses all relevant point and non-point nutrient sources, and their potential impacts on the eutrophication process. This chapter also presents a simple approach to determining the total component nutrient sources (point and non-point) in the drainage basin, as well as the general reliability of nutrient load estimates.

What should be measured in the waterbody and how should it be used is discussed in *Chapter 8 and Annex 1*. Guidelines for development of an effective monitoring programme are provided. Annex 1 presents both a simple and a more detailed approach for classifying a lake or reservoir with regard to its trophic state, and relating it to intended or desired water uses.

Available methods for the treatment of eutrophication are identified and discussed in *Chapter 9*. The methods discussed include the control of the external nutrient supply, as well as different in-lake control methods. This chapter also provides a basis for developing sound land use practices which focus on the management of nutrient inputs to lakes and reservoirs. Different water quality problems treatable by various in-lake nutrient control measures are also identified.

Positive features associated with cultural eutrophication are discussed in *Chapter 10*. This chapter focuses on optimizing the reuse of aquatic plant nutrients. One topic of discussion is nutrient removal from nutrient-rich waterbodies, in the form of algae and macrophytes. The use of such materials for commercial purposes is discussed. Chapter 10 also deals with increasing the food production of a lake or reservoir by the aquaculture of different types of aquatic organisms, especially fish. Other nutrient reuse possibilities, such as the use of sludge from fish ponds and dredged sediments as fertilizers and the use of hypolimnetic waters for irrigation, are discussed.

The development of a eutrophication control strategy is discussed in *Chapter 11*. This chapter focuses on general considerations, and presents an overall approach for the selection of a specific eutrophication control programme. It considers both the control of external point and non-point nutrient sources, and the possible use of in-lake control measures. It also discusses special limnological conditions and situations where such measures should, and should not, be applied.

Information on the *past experiences of different control measures* is very useful, prior to making specific decisions on the development and implementation of eutrophication control measures. *Annex 2* provides a list of documented case studies regarding various eutrophication control programmes around the world.

This book represents a multinational approach to the practical application of our present knowledge of the eutrophication process, and generally accepted nutrient load–lake and reservoir response relationships for eutrophication management. It is recognized that other publications on the subject of eutrophication control do exist. Examples include the reports of Walmsley & Butty (1980), Rast (1981a), United States Environmental Protection Agency (1980; 1988), OECD (1982), and V.H. Smith *et al.* (1984). However, the existing publications do not appear to cover the full range of experiences documented thus far. In addition, these latter publications generally focus only on certain types of waterbodies, or on specific geographic areas in the world. In contrast to the above, this book attempts to present a broader view of the practical control of eutrophication around the world, and in most of the lake and reservoir situations likely to be encountered.

CHAPTER 3

A MANAGEMENT FRAMEWORK FOR THE POLICYMAKER

THE ROLE OF THE POLICYMAKER

Detailed discussions of the scientific and technical aspects of the eutrophication process, and its potential impacts on the biological, chemical and physical properties of lakes and reservoirs are provided in Chapters 4 to 11 of this book. In contrast to these strictly scientific concerns, however, this particular chapter focuses on those aspects of eutrophication control which are of primary importance to the policymaker or 'decisionmaker' (e.g. executive, legislator, senior administrator, director, etc.).

As a preface, it is difficult to present a characterization of the 'policymaker' which is accurate in all cases. Furthermore, because of the differing (and sometimes conflicting) concerns and perspectives within a country, and on a global scale, the elements discussed in this chapter will have different priorities in different countries. Thus, the circumstances in a given situation may require the policymaker to interject issues and concerns other than those discussed here in the formulation of the 'solution' to a given problem. This possibility should be kept in mind while considering the contents of this chapter.

The policymaker and environmental issues

Both the public and responsible officials are becoming increasingly concerned with changes in environmental quality resulting from human activities. These changes are related to a number of environmental stresses, including eutrophication, toxic substances and acid rain. An example of this concern is reflected in the United Nations' designation of the 1980's as the 'Water Decade'. However, this increased environmental consciousness is occurring coincidently with significant socio-economic concerns such as depressed economies, unemployment and other financial constraints.

Recognizing these sometimes conflicting pressures, this book was developed to provide both a management and technical framework for addressing one of these important issues; namely, the eutrophication of lakes and reservoirs. This long-term problem continues to be of environmental, economic and social concern around the world. The primary exception is the use of the eutrophication process to enhance the production of fish or other types of aquaculture for the purpose of producing food supplies (see Chapter 10). In this latter case, particularly in developing countries, eutrophication also has a beneficial aspect to be considered.

The role of the policymaker in the protection of the natural environment often focuses on the development and implementation of management strategies and/or control programmes for dealing effectively with environmental issues. Furthermore, effective environmental management usually requires substantial recognition that an environmental problem exists, as well as sufficient support to formulate and implement a corrective policy. Yet, because of both public and political pressures, decisions regarding environmental management or control programmes sometimes have to be made in a relatively short time frame, regardless of the state of scientific knowledge on a specific item of concern. In some cases, the problem may not even be lack of information, but rather the resolution of 'an array of more or less persuasive fact or opinion on both sides' of an issue. In such a setting, the policymaker may be called upon to resolve conflicting claims, sometimes with 'little real knowledge of the facts and less knowledge about the consequences' (Gonzalez, 1984). Indeed, the policymaker will often be confronted simultaneously by advisors and supporters who argue that immediate action is vital, as well as by those who contend that corrective efforts should be delayed until more is known about the extent and severity of the problem and the most desirable measures to treat it. Therefore, responsible officials must attempt to balance the need and desire for immediate action against the need for further study.

Development of a eutrophication management framework

Because of the type of concerns identified above, this book attempts to incorporate the need for balance in evaluating the potential impacts of human activities and pollution control programmes on the environment. Recognizing that the specific needs of policymakers and administrators are usually different from those of the strictly technical audience, the primary purpose of this document is to:

(1) provide quantitative tools for assessing the state of eutrophication of lakes and reservoirs;

(2) provide a framework for developing cost-effective eutrophication management strategies;

(3) provide a basis upon which strategies can be tailored for each specific case according to the physical, social, institutional, regulatory and economic characteristics of the local area or region; and

(4) provide specific technical guidance and studies of the effective management of eutrophication.

The approach presented in this chapter is also sufficiently general that it can be applied, with minimal modification, to the assessment of other environmental problems and to the development of effective management strategies for such problems.

A basic approach to achieving the basic objectives stated above consists of the following components, applied approximately in the order presented:

1. Identify eutrophication problem and establish management goals
   * What is cultural eutrophication and why is it a problem?
   * How can management goals be determined?
   * Who should be involved in managing eutrophication?
     – The government sector
     – The private sector

2. Assess the extent of information available about the waterbody
   * What information is necessary to assess the situation?
   * How can the necessary information be obtained?

3. Identify available options for management of eutrophication
   * Should one treat/manage the cause or the symptoms?
   * What management options are available?

4. Analyze all costs and expected benefits of alternative control strategies
   * What are the expected costs?
   * What resources are available to address the problem?
   * What are the costs and benefits of alternative management strategies?
   * What will happen if nothing is done about the problem?

5. Analyze adequacy of existing legislative and regulatory framework for implementing alternative eutrophication control programmes
   * Are existing institutions and regulations adequate?
   * How should new institutions or new regulations be developed?

6. Select eutrophication control strategy and distribute summary to affected parties prior to implementation
   * How can an appropriate eutrophication control programme be selected?
   * What sorts of information should be distributed, and to whom?
   * Should a public education programme be part of a plan?
   * Are there any other important considerations?

7. Provide periodic progress reports on the control programme to public and other interested parties

* How can the public be made better aware of a eutrophication control programme?
* What is the importance of public feedback?

Each of these steps is discussed in the following sections. In addition, the specific chapters in this book which provide further details on these topics are identified.

IDENTIFY EUTROPHICATION PROBLEMS AND ESTABLISH MANAGEMENT GOALS

Why is eutrophication a problem?

Scientists have accumulated considerable evidence linking accelerated lake and reservoir eutrophication to the excessive input of aquatic plant nutrients from point and non-point sources in the drainage basin (see Chapters 4 and 7). Consequently, nutrient loading concepts are frequently used in the assessment of nutrient control measures. Planners and managers often can use simple relationships or models to predict the potential impacts of management decisions on eutrophication-related water quality in lakes and reservoirs (see Chapter 6).

The main objective of traditional water pollution control efforts was to clean up raw wastewaters and gross industrial wastes, which are potential sources of pathogens and toxic materials. However, since treatment of such effluents is becoming more common, especially in industrialized nations, the environmental impacts of other types of pollution in the drainage basin have assumed greater importance in recent years. For example, pollution from non-point sources, such as urban and rural runoff, is now being seriously considered in the development of effective water pollution control programmes. Accordingly, there is a definite, continuing need to develop an integrated view of land, atmosphere and water interactions in the drainage basin, as they relate to the assessment and treatment of cultural eutrophication.

In industrialized countries, especially those with water surpluses, the development of an integrated view of land, atmosphere and water interactions in the drainage basin is necessary because of increasingly serious water quality problems, due both to point and non-point source pollutant inputs to surface and ground waters. In arid and semi-arid regions, socio-economic development can result in rapidly increasing water demands. In such cases, the availability of water can become an ultimate constraint to development. This is often due to the combined effect of large populations and climatically related water demands for increasing agricultural production. Furthermore, as population and development continue to increase, the uses to which water is being put are increasing. Yet, at the same time, the availability of suitable water for such uses is decreasing.

Eutrophication of lakes and reservoirs ranks as one of the most widely-spread water quality problems around the world. The effects of cultural eutrophication on a waterbody can render the water unsuitable for many uses, or else require that the water be treated (often expensive and time-consuming) prior to its use by humans. Eutrophication can also result in detrimental effects on the biological stability of a lake or reservoir ecosystem, affecting virtually all the biological populations and their interactions in the waterbody. Consequently, eutrophication of lakes and reservoirs can have significant negative ecological, health, social and economic impacts on man's use of a primary and finite resource (see also Chapters 1 and 4).

Establishment of eutrophication management goals

The designation of bad (unacceptable) versus good (acceptable) water quality in this book is based on the specific intended use or uses of the water resource. That is, water quality management goals for a lake or reservoir should be a function of the major purpose(s) for which the water is to be used.

Obviously, there are water quality conditions to be avoided because of their interference with water uses. Ideally, for example, a lake or reservoir used as a drinking water supply should have water quality as close to an oligotrophic state as possible, since this would insure that only a minimum amount of pre-treatment would be necessary to yield a water suitable for human consumption (see Table 11.1). For such a waterbody, the content of phytoplankton (and their metabolic products) in the water should be as low as possible to facilitate this goal. Further, if the water is taken from the bottom waters of a lake during the summer (usually the period of maximum algal growth), it should be free of interfering substances resulting from decomposition of dead algal cells. Eutrophic lakes and reservoirs can be used as a drinking water supply. However, extensive pretreatment would be necessary before the water is suitable for human consumption.

Some water uses may require no treatment at all, regardless of the existing water quality. Examples are fire-fighting purposes and the transport of commercial goods by ships. Further, in areas with extremely limited water resources, virtually all the water may be used for various purposes (with or without treatment), regardless of its quality.

Therefore, although humans can use waters exhibiting a range of water quality, there is a desirable or optimal water quality for virtually any type of water usage. Though it is not quantitative in nature, a summary of intended water uses and the optimal versus minimally acceptable trophic state for such uses is provided in Table 11.1. In addition, Tables 4.1 and 4.2 provide a summary of the values of several commonly measured water quality parameters corresponding to different trophic conditions. One primary exception is that a higher average, in-lake total phosphorus concentration of approximately 50 µg/l appears to be a more appropriate value for designating the eutrophic/mesotrophic boundary condition for tropical and sub-tropical waterbodies (see Table 4.6). Thus, it is possible to identify acceptable or optimal water quality for given water uses.

Given these factors, a prudent approach to setting eutrophication management goals is to determine the minimum water quality and trophic conditions acceptable for the primary use or uses of the lake or reservoir (see Table 11.1), and attempt to manage the waterbody so that these conditions are achieved. In a given situation, if the primary use or uses of a waterbody is hindered by the existing water quality, or else requires water quality or trophic conditions not being met in the waterbody, this signals the need for remedial or control programmes to achieve the necessary in-lake conditions.

Procedures for determining the trophic condition of a lake or reservoir are discussed in Chapters 4 and 8 and in Annex 1.

Who should be involved in addressing the problem?

The governmental role. It is recognized that a range of different forms of government, as well as economic conditions, exist around the world. Consequently it is difficult to provide general guidelines regarding the role of the government in environmental protection efforts that will cover all possible situations. However, virtually all nations also contain some type of civil service infrastructure which, if properly used, can be an effective instrument with which to address governmental concerns. Even so, as noted earlier, not all concerns identified in this chapter will receive the same degree of attention in all countries, in part because of differing governmental priorities and national perspectives.

Eutrophication management programmes are usually developed and implemented by a governmental entity. Consequently, all affected governmental agencies should be consulted. In this way, one can obtain the perspectives of the individual agencies involved, as well as bring their collective wisdom and experience to bear on the problem. Relevant agencies may include governmental units concerned with environmental quality, water quality, water supply, resource management, fisheries or aquaculture, power production, agriculture, commerce and/or public health. This inter-agency consultation is also a good planning strategy, because cooperation between governmental units, rather than confrontation, will concentrate more energy and resources on solutions to environmental problems.

The selection of effective eutrophication control measures depends on a number of scientific/engineering, sociological and political factors. Furthermore, lakes and reservoirs are complex aquatic environments. Consequently, eutrophication is a problem which the policymaker need not face alone. In the attempt to obtain an understanding of the eutrophication process, a multidisciplinary approach is highly desirable. Eutrophication policy and management decisions are usually best made in consultation with individuals in the following areas of expertise:

1. Municipal wastewater treatment engineer or consultant engineer — This expert can provide the planner with knowledge of the nutrient contributions of, and control strategies for, municipal wastewaters in the drainage basin.

2. Municipal chemist/consultant chemist — The chemist can supply information on nutrient concentrations in municipal wastewaters and industrial effluents, as well as other important point sources of nutrients.

3. Agriculturalist — The agricultural expert will have necessary knowledge of soils, land-use activities, feedlot and fertilizer practices, and other relevant farm operations, as well as methods for the control of soil erosion and associated nutrient runoff.

4. Hydrologist — Water movement and water balances play an important role in dictating the pathways of nutrients through the landscape. These factors can affect the nature and magnitude of the nutrient loads and concentrations reaching surface waters. The hydrologist can give guidance on these topics.

5. Limnologist — The limnologist can assess the impacts of excessive nutrient inputs on the aquatic environment with respect to plant growth, deteriorating water quality, fishery development, and general effects on the aquatic ecosystem. Relevant individuals include experts in the fields of algal physiology, fisheries, aquatic chemistry and water quality modeling.

6. Economist — Many eutrophication problems and control strategies require an economic evaluation as part of the assessment of alternative management strategies. The economist is essential for such endeavors.

7. Other professionals — The advice of legal, health and planning experts can be extremely valuable in the development of effective control programmes.

The public role. Where feasible, it can also be very helpful to seek the public's view regarding eutrophication problems and solutions. If the public's view is sought in a given situation, a readily usable forum for obtaining this viewpoint should be clearly identified. One example is the creation of a citizen's advisory committee. This type of committee can provide additional insight about the extent of a given eutrophication problem, and what the social and political consequences might be if the problem was left uncorrected. However, the policymaker often has to evaluate such public input subjectively (e.g. what is the 'value' of maintaining an aesthetically-pleasing lake versus its use as an industrial water source?). As noted earlier, the policymaker often must balance the interests of advocates of long-term benefits against those wishing more politically expedient solutions.

Interested citizens can be an asset in the development of effective eutrophication management programmes. As an example of the potential benefit of public input, actual water quality data from a lake may be scarce at the beginning of a control programme. In such cases, narrative descriptions of prior conditions, remembered by elder citizens and leaders, can be used as an initial reference point against which the potential effectiveness of a control programme can be assessed.

In the broad sense, such interactive communication can have at least two beneficial effects:

1. Knowledge gained through lifetime observations of a waterbody can be documented for use in developing management programmes; and
2. Persons encouraged to participate in the development of a programme are more likely to become advocates of the programme.

Knowledge gained in this manner by governmental personnel can be disseminated among the general population, preparing them for more informed future judgements and actions. Effective public participation requires that government officials be honest in their presentation of information and responsive to the views expressed to them. Nothing can be more damaging to public confidence in a new governmental initiative than a feeling by the public that the government did not listen to those participating in the process.

In some developing countries, financial constraints may limit the use of large structural solutions to eutrophication problems (e.g. municipal wastewater treatment plants). In such situations, the governmental entity may wish to make maximum use of community-based information and educational programmes on eutrophication control measures, especially those in which the public can most directly participate. In such cases, a communications specialist can be a valuable asset.

ASSESS THE EXTENT OF INFORMATION AVAILABLE ABOUT THE WATERBODY

Before a eutrophication monitoring or management programme is developed, one should attempt to determine the full scope of the problem. Previous studies and relevant case histories should be reviewed prior to developing a management programme. Likely sources of such information include drinking water and wastewater treatment agencies, universities and other types of research centers (including national, regional and local government laboratories), and the scientific and engineering literature dealing with aquatic ecosystems. While some control actions can be initiated in the absence of such knowledge, further refinement of management alternatives usually requires more knowledge.

If existing data are not sufficient to provide the necessary information for assessment or management purposes, it usually will be necessary to implement a drainage basin and/or in-lake monitoring programme. There are several reasons for collecting adequate monitoring data for any environmental assessment or management programme:

1. To establish past and present baseline conditions in order to confirm the problem, and to provide a reference against which progress can be assessed;
2. To identify significant information gaps; and
3. To develop a cost-effective monitoring programme.

An initial monitoring programme can be modest. However, the monitoring network should be designed to allow progressive expansion and revision, if necessary, to meet changing needs. Periodic progress reports to all concerned entities are also desirable, in order to facilitate communication and knowledge about the problem and to assess progress in meeting eutrophication management goals.

Data and documentation should be sufficient to support the undertaking of corrective measures. There probably will never be sufficient scientific understanding to convince every technical expert that a given management action is the ideal or timely one to be taken. New knowledge inevitably raises new questions. Nevertheless, if the policymaker delays implementation of any corrective actions until all questions are answered completely, the problem can become extremely difficult to correct.

Chapter 5 of this book discusses major factors that affect the extent of eutrophication of a lake or reservoir. Chapter 6 provides guidelines for the use of eutrophication models, while Chapter 7 provides guidelines for assessing the major aquatic plant nutrient sources and inputs in the drainage basin. In addition, Chapter 8 and Annex 1 identify necessary monitoring data and provide guidance for development of effective in-lake eutrophication monitoring programmes.

IDENTIFY AVAILABLE OPTIONS FOR MANAGEMENT OF EUTROPHICATION

Should one treat the causes or the symptoms?

There are several approaches towards assigning priorities to alternative eutrophication control programmes. The programmes can be directed either toward treating the basic causes or the symptoms (e.g. reducing aquatic plant nutrient inputs from the drainage basin versus periodic harvesting of excessive aquatic plant growths). In some cases, a combination of the two will be most useful. Alternatively, programmes can focus on treating primarily point sources or non-point sources of nutrients. Examples would be limiting 'pipeline' nutrient inputs from municipal wastewater treatment plants and controlling runoff from farms and urban areas, respectively. Further, the programme can be either structural or non-structural in form (e.g. building a municipal wastewater treatment plant versus changing agricultural fertilizer application practices). In a given case, the basic approach should be tied as closely as possible to the overall eutrophication management goals.

Where possible, it usually is most effective to attempt to treat the underlying and most readily controllable causes of eutrophication, rather than to attempt merely to alleviate the symptoms. In most cases, this treatment requires reduction or elimination of the excessive nutrient inputs that stimulate the excessive growths of aquatic plants in the first place. This approach will work to eliminate the basic problem, and usually is the most effective strategy over the long term.

The alternative strategy is to treat the specific symptoms of eutrophication. This is the logical and perhaps only option if the costs of treating the basic cause (excessive nutrient inputs) are too high, or if additional treatment is necessary in a given case. Other possible reasons for using this approach are the absence of an institutional framework for treating the cause or an inability to formulate and/or implement an effective management programme directed toward nutrient reductions. In such cases, several 'in-lake' treatment options (discussed in a following section) can offer temporary relief in varying degrees from the symptoms of eutrophication.

Consider full range of available control options

Reduction of nutrient inputs. The first control priority is usually to limit or reduce nutrient inputs to the waterbody from the sources in the drainage basin that contribute the largest quantities of the 'biologically available' forms of the nutrients (Rast & Lee, 1978; Lee *et al.*, 1980; Sonzogni *et al.*, 1982; also see Chapter 7). The control effort can be directed to both the point ('pipeline') and/or non-point (diffuse) nutrient sources in the drainage basin. For example, human and animal wastewaters contain large quantities of phosphorus and nitrogen, in chemical forms easily used by algae and other aquatic plants. Treatment to reduce the level of the nutrients in these wastewaters usually is a cost-effective approach to keep them from reaching surface waters (at least up to a certain advanced level of wastewater treatment).

Phosphorus and nitrogen are not the only nutrients needed by aquatic plants for growth. However, they are the most important nutrients from the management perspective, because their input to lake waters can be controlled significantly with existing technology (e.g. phosphorus removal from effluents at municipal wastewater treatment plants). Further, reduction of the quantities of phosphorus in phosphate-containing detergents can be an effective supplemental measure, especially in areas where the removal of phosphorus at municipal wastewater treatment plants is not practiced, or where there are a large number of septic tank waste disposal systems in a drainage basin.

Another method of reducing nutrient inputs to a waterbody is to divert municipal sewage wastewaters from the drainage basin of concern into a downstream basin. This latter method can be effective for the affected waterbody. However, it does not eliminate the basic problem; it merely shifts it to another waterbody which may or may not be more capable of handling it. There also are obvious social and political problems associated with this type of 'solution'.

A large number of nutrient control options directed to non-point sources in the drainage basin also exist in many cases. These various measures exhibit a wide range of costs and effectiveness.

In-lake control measures. Some treatment measures can be applied directly in a lake or reservoir in an attempt to alleviate the symptoms of eutrophication. They can also be used to augment other treatment methods, or to provide temporary relief from eutrophication symptoms while a long-term control strategy is being formulated or implemented.

Examples of in-lake methods include the harvesting of aquatic plants, the use of algicides, in-lake nutrient inactivation or neutralization, artificial oxygenation of bottom waters, dredging or covering of bottom sediments, increasing the water flushing or circulation rates, and 'biomanipulation'. Although such measures are usually less effective over the long term than external nutrient control programmes, they do offer an effective means of combating, at least temporarily, the negative impacts of eutrophication.

Valuable sources of relevant information regarding both point and non-point source eutrophication control measures include Monaghan (1977), Skimin *et al.* (1978), PLUARG (1978a), Johnson *et al.* (1978), Phosphorus Management Strategies Task Force (1980), E.B. Welch (1980), United States Environmental Protection Agency (1980; 1988), Bernhardt (1981; 1983) and Lester & Kirk (1986). In addition, Cooke *et al.* (1986) provide a comprehensive summary of in-lake treatment measures. These various reports discuss both the relative effectiveness and costs of the control measures considered.

The option of doing nothing. The environmental, social and economic consequences of doing nothing should also be considered prior to developing eutrophication management options, and even in deciding whether or not to implement such programmes in the first place. The consequences of doing nothing offer a basis for comparison with the potential impacts of initiating control programmes.

However, doing nothing is not usually a satisfactory solution to eutrophication problems. An inevitable consequence of human settlement of a drainage basin is a deterioration of water quality in the drainage basin over time, especially if these conditions continue to be ignored. An untreated waterbody exhibiting eutrophication symptoms sufficiently severe for control programmes to be considered are likely to become even worse over the long term. This deterioration will necessitate even more expensive control programmes at a later date, as well as the loss of an increasing number of water-use options. Further, while cultural eutrophication is largely reversible, a lake cannot be allowed to deteriorate indefinitely without diminishing the chances of a timely and successful rehabilitation.

In addition, it is pointed out that doing nothing when corrective actions are called for can also be a costly alternative over the long term because it can even create new costs. For example, long-term and presumably beneficial water uses may have to be curtailed or eliminated as a result of eutrophication-related water quality deterioration. The number of sources of good or acceptable water may also decrease over time. As noted earlier, the occurrence of parasitic diseases may be enhanced in some cases. Such impacts potentially could result in higher environmental, socioeconomic and/or political costs over the long term than initiation of appropriate remedial programmes at an earlier stage. Furthermore, some eutrophication prevention or control measures actually may result in monetary savings, or even profits, over time. An example is the use of nutrient-rich waters for irrigation or aquaculture purposes (see Chapter 10). Such possibilities can be an attractive part of an effective eutrophication control programme.

Chapter 9 in this book discusses the relative costs and effectiveness of a number of specific eutrophication control measures for both point and non-point nutrient sources in the drainage basin. Chapter 6 provides guidance in the selection and use of both simple and more detailed eutrophication models in assessing alternative eutrophication control options. Chapter 10 provides details on the positive use of the eutrophication process in relation to enhancement of fish production and other forms of aquaculture. Finally, Annex II provides a large number of case studies which highlight eutrophication control options previously used in different parts of the world.

ANALYZE ALL COSTS AND EXPECTED BENEFITS OF ALTERNATIVE MANAGEMENT STRATEGIES

Consider relative costs of control options

The costs of specific eutrophication control options vary substantially. The costs of phosphorus removal at municipal wastewater treatment plants, for example, will vary as a function of the treatment process used, the age of the plant, the number of people served, etc. Non-point source control measures also exhibit a wide range of costs and effectiveness. However, the long-term costs of some non-point source measures can be minimal (e.g. the use of sound land management practices). Furthermore, some non-point source control measures can actually result in monetary savings over the long term, even considering the initial costs of implementing the programmes. For example, for the North American Great Lakes Basin it was determined that, beyond a certain advanced degree of point source control, it was less expensive to implement some non-point source control measures than to implement further, more stringent point source control measures (PLUARG, 1978a; Johnson et al., 1978). In addition, some non-point source control measures can be relatively simple in concept. An example is the application of fertilizers in quantities which do not exceed the actual needs of the soil. Applying fertilizers in excess of soil needs can result in their transport to surface waters via agricultural runoff. Soil nutrient requirements can be readily determined by appropriate soil tests.

The costs of some eutrophication control measures (e.g. building and operating municipal wastewater treatment plants for nutrient removal) are sufficiently well-known by civil and sanitary engineers for reliable cost estimates to be obtained. Nevertheless, because of local conditions, eutrophication control alternatives still can show a wide range of costs. For example, the local costs of such factors as labor, energy, materials, etc., can vary in different regions and affect the ultimate costs of eutrophication control alternatives. Such factors will have to be considered in each situation.

It is beyond the scope of this chapter to provide a detailed discussion of the costs of all available eutrophication control options. However, Chapter 9 discusses the general costs and effectiveness of a large number of point and non-point source control measures.

Compare management goals, available resources and expected benefits

There is little practical value in designing or developing a substantial eutrophication control programme if the available resources or administrative structure are not adequate to carry out the programme. Consequently, the available resources should be identified and compared with the needs of the task to be undertaken. The resources would include such items as technical expertise, financial resources and manpower.

An initial assessment of resources can also help to identify what resources must be made available before an effective programme can be carried out. It must be recognized that effective eutrophication management programmes cost money. A well-formulated statute or regulation, for example, is of little value if it is not supported with an adequate administrative apparatus. The responsible agency must have sufficient funds to hire qualified staff, purchase necessary supplies, equipment and instruments. Responsible personnel must be able to travel within their country and examine similar problems, and to attend meetings (both inside and outside their native country) to learn of current approaches and practices from their fellow professionals elsewhere.

The initial control programme should not be so elaborate as to be unattainable. Rather, it should be designed and have sufficient scope that it has a realistic chance of achieving its management goals. A small, successful programme can be expanded as knowledge is gained and resources made available, building upon the success already achieved.

As suggested earlier, if the management goal is to alleviate the negative impacts of eutrophication, the most effective approach is usually to treat the most readily controllable cause of the problem – the input of excessive quantities of phosphorus and/or nitrogen from the drainage basin to the waterbody. The control programme should be directed toward the major sources of these nutrients in the drainage basin. These sources are primarily human and animal wastes (including municipal wastewater effluents and drainage from large animal feedlots). Non-point sources, especially runoff from urban and agricultural lands, also offer important nutrient control targets.

Changes in land usage patterns and/or land management practices offer a largely non-structural way of reducing nutrient loads associated with land runoff. However, the implementation of some non-point source methods may require a basic public education and/or a change in public attitudes regarding human use of the land. For example, a farmer located a long distance from a lake may not appreciate his role in causing or promoting nutrient runoff from his land to the lake. In such a case, the farmer understandably may resist the suggestion that a change in his fertilizer application practices or methods of plowing can help reduce the nutrient load to the lake by reducing the quantities of agricultural nutrient runoff. This is especially true if the suggested changes are more expensive or time-consuming than his current farming practices, or if no direct benefit to the farmer can be demonstrated. Obviously, if the suggested change can be shown to have a beneficial effect on farm operations, its usefulness is much easier to demonstrate. As suggested by PLUARG (1978a), such individual efforts may not seem impressive when viewed individually, but actually can be very significant on a cumulative scale.

It is reiterated that the enhanced biological productivity associated with eutrophication also can be a beneficial aspect of the eutrophication phenomenon in some situations. Examples are enhanced fish production and other forms of aquaculture for the production of needed food supplies. The use of harvested aquatic plants as a livestock food, or the use of nutrient-rich sewage sludge or lake bottom waters as a soil fertilizer, are also economically beneficial in some situations (see Chapter 10). Such possibilities can be positive components of environmentally acceptable, long-term eutrophication management programmes in a region or country.

Cost–benefit analysis

As noted earlier, the majority of the material in this book is concerned with the scientific and technical aspects of eutrophication. However, in many cases, 'non-scientific' concerns must also be taken into account when developing effective eutrophication control programmes. For maximum public acceptance of environmental protection programmes, the scientist or engineer cannot ignore economic and political realities, in favor of a solely technical approach. Likewise, the policymaker or manager cannot ignore environmental and engineering considerations. In fact, because of the tremendous growth in the size, scope and expenditures of central governments, the public reaction in some countries has been a reluctance to fund new environmental protection programmes without a thorough social and economic analysis of such programmes (in addition to environmental impact assessments).

One approach often used to assess the desirability or 'worthiness' of alternative management programmes is 'cost–benefit analysis'. Cost–benefit analysis has its basis in a branch of economic theory called 'welfare economics'. I.M.D. Little (1957) and Baumol & Oates (1975) provide a basic introduction to this topic.

In the broadest sense, cost–benefit analysis means a comparison of all the positive and negative elements of a decision, even if not all of the elements are measurable in strictly monetary terms. However, in practice, cost–benefit analysis usually means a comparison of the financial gains realized and the costs incurred for a particular programme or activity. If a value unit (e.g. dollars) is spent in a way that generates more wealth than is sacrificed, then the overall social welfare is increased. Thus, if the 'price' of a necessary or desirable activity (e.g. eutrophication control programme) does not exceed the expected benefits (e.g. enhanced water quality or water uses), it is usually considered desirable to proceed with the project. As a practical matter, a policymaker using cost–benefit analysis usually is asking whether or not the expected benefits of a eutrophication control programme are worth the investment of public funds.

It is obvious that, at the broadest scale, governments must choose between a wide array of potential projects and programmes, from national defense, to food subsidies, to environmental protection. Using solely economic criteria should stress 'efficiency' in assessing competing alternatives (i.e. the most efficient use of scarce resources is required to maximize their impact).

Unfortunately, a significant shortcoming of a strictly monetary-oriented approach is that it usually is done under the implicit assumption that a positive benefit:cost ratio alone is sufficient rationale to proceed with a given programme or activity. However, while the 'cheapest' solution to a eutrophication problem may be economically pleasing, it also may be environmentally short-sighted. This is because some elements of a eutrophication control programme may not be easily quantified, or else can be quantified only in an artificial or unrealistic manner. Examples of such elements are cultural values, the long-term sustainability of natural resources, political realities, societal and/or governmental structure and stability, and the national or regional distribution of wealth. It is because of such realities that the 'logical' solution to an environmental problem is not always the most socially-acceptable one.

Thus, using a strictly monetary-oriented cost-benefit analysis as the sole decision making tool may preclude realistic consideration of the long-term environmental, social and/or public health consequences of a given control programme. For example, if one cannot assign a realistic monetary value to the desirability of maintaining a particular fish species or the achievement of enhanced water quality, such factors may be ignored as a benefit when compared to the use of a water resource for industrial purposes or municipal waste assimilation. Non-scientific concerns also may require explicit consideration in the development of effective eutrophication control programmes.

Because of these types of problems, one can use cost–benefit analysis in a somewhat different manner for assessing management alternatives. Justification of eutrophication control programme expenditures can be based on an analysis of the expected benefits of alternative programmes, because different uses of monetary resources can be expected to yield different benefits. By comparing the expected benefits of alternative control programmes, one can attempt to select the most preferable option in a given situation. This can be done either by:

1. Comparing the benefit:cost ratios of alternative programmes and levels of expenditure;

2. Comparing the absolute values of the expected benefits of alternative projects, using a fixed level of monetary and other resources; or

3. Determining the minimum cost programme for achieving a specific goal or benefit.

The remaining discussion in this particular section assumes that the basic decision to proceed with development of a eutrophication control programme has been made.

Social concerns. As used here, social concerns are meant to cover the non-technical concerns related to development or implementation of an environmental protection or management programme. It is assumed that these concerns are measurable or definable in some way. Accordingly, four broad categories of social impact can be usefully distinguished, as follows:

1. Economic impacts – includes the 'efficiency' concept identified above, as well as such fiscal and social factors as employment rates, balance-of-trade, tax revenues, and the national and/or regional distribution of wealth. Consideration of the latter factor must include the observation that the expected benefits may not be equivalent, or may have different economic significance, for different income classes or occupational categories.

2. Demographic impacts – involves population distribution characteristics, such as shifts from urban to rural areas (whether self-sufficient or dependent on government agencies) or from one region to another. An example would be the potential impacts of construction of a centralized marine fish processing facility to replace local fish stocks lost due to advanced eutrophication. Changes in health parameters may also be an important consideration in some cases.

3. Environmental impacts – involves both natural (e.g. aesthetic appreciation of a pristine lake) and 'created' concerns (e.g. urbanization, crowding and increased noise as the result of a tourist influx to an aesthetically pleasing area). Real estate values sometimes reflect such concerns, especially within small areas.

4. Cultural impacts – involves the way that populations perceive and react to environmental changes. An example is a cultural or religious attitude regarding the sanctity of a pristine lake, as contrasted with a spartan work ethic supporting the notion that maximum use of a lake for economic productivity is critical.

A hypothetical example serves to illustrate how these concerns can affect the selection or implementation of a eutrophication control programme. It is assumed here that a lake or reservoir is exhibiting depleted oxygen levels in its bottom waters due to excessive nutrient runoff from agricultural lands. The oxygen depletion is having negative impacts on the recreational fishery of the waterbody. It is the task of the policymaker to determine whether or not a soil erosion (and associated nutrients) control programme based on minimum tillage farming methods should be implemented. The eutrophication control programme could possibly result in increased levels of desirable game fish species. The types of concerns the policymaker may have to consider in this example include the following:

1. A possible increased overall farm production, but also an increased

number of failing, small family farms due to new required capital investments. This could have the impact of promoting large-scale farming operations at the expense of small ones, and of changing prevailing farming practices;

2. A possible change in crop species, which could diminish the local production of a culturally significant commodity;

3. A possible increase in levels of toxic substances in fish, due to the transport to the waterbody of the increased quantities of herbicides used on the land surface. Increased herbicide use is often necessary with this method of farming;

4. A possible increased health risk from consumption of affected fish and ground water;

5. A possible increased recreational fishery, with larger and more plentiful game fish, but a decreased commercial fishery; and

6. A possible increased tourist influx (with attendant economic benefits), but accompanied by increased traffic noise, congestion, etc.

In analyzing the importance of such often competing factors in development of an effective eutrophication control programme, some mechanisms for prioritizing, 'weighting' and/or integrating them into the decision-making process is desirable.

One way to accomplish this weighting, particularly when several control options are being considered, is to develop a simple matrix which considers the relative 'social impact' of each control option. Since multiple control options may be available in a given situation, one can rank (even if subjectively) the major criteria of concern for each of the control programmes or options being considered. For example, if one assumes a ranking scale of -5 to $+5$, with zero being 'no effect' and 5 being a large impact in a positive or negative direction, a social impact ranking matrix of alternative control programmes can be developed. In this example, Control programme A refers to phosphorus removal at municipal wastewater treatment plants, programme B refers to in-lake nutrient inactivation and programme C refers to urban non-point source control measures (see Chapter 9 for discussion of alternative eutrophication control options).

The resulting ranking matrix can be presented as follows:

	criterion 1 (e.g. health effects)	criterion 2 (e.g. desirable fish)	criterion 3 (e.g. aesthetic quality)
Control programme A	$+1$	$+2$	-5
Control programme B	-3	$+5$	-2
Control programme C	-1	$+2$	$+4$

The ranking numbers for each control programme can be generated quantitatively or qualitatively. For example, using the occurrence of conjunctivitis (an eye infection associated with swimming in degraded waters), a ranking scale for health effects could be developed as follows:

	ranking
up to 5% increase in conjunctivitis	−1
up to 10% increase...	−2
up to 25% increase...	−3
up to 50% increase...	−4
greater than 50% increase...	−5

The reverse would be true for a decrease in the occurrence of conjunctivitis.

If all criteria used in the ranking matrix above were of equal social value, then the overall rankings of the alternative control programmes would be as follows:

Programmes A: +1 +2 −5 = −2   Rank: 3
Programmes B: −3 +5 −2 = 0   Rank: 2
Programmes C: −1 +2 +4 = +5   Rank: 1

In many situations, however, some factors may be more important than others. In such cases, the policymaker can assign a value to the ranking criteria which signifies its relative social importance, thereby integrating this value into the ranking process. The relative importance could be assigned directly by the policymaker or manager based on personal experience or knowledge of the particular situation, or can be obtained by other means (e.g. public opinion polls, referendum votes, a canvass of opinion from scientific experts, etc.).

In this example, if health effects were the primary concern, and the policymaker had determined that the relative social importance of the three factors considered above were 7, 2 and 1 (out of a possible 10), respectively, then the overall rankings of the alternative control options would change as follows:

Control programme A: 7(+1) + 2(+2) + 1(−5) = +6   Rank: 1
Control programme B: 7(−3) + 2(+5) + 1(−2) = −13   Rank: 3
Control programme C: 7(−1) + 2(+2) + 1(+4) = +1   Rank: 2

Clearly, the relative weights (however assigned), as well as the specific ranking criteria used in this simple analysis, can significantly affect the eutrophication control programme ultimately selected. The monetary costs of a control programme also can be included as one (but not the only) ranking criterion. The primary difficulty with such an analysis is in establishing the relative weights of the individual criteria, even if the necessary data can be obtained. This process often ends up as a political one.

As a practical matter, the policymaker already implicitly performs simple analyses of this type in choosing between management alternatives. The point made here is that it is usually beneficial to perform such an analysis explicitly (whatever form it may take), especially if socioeconomic or political factors are of significant concern in selecting between alternative eutrophication control programmes. In this way, individuals whose input to the selection process is strictly technical can see why some of their recommendations may have to be subordinated to non-technical ones. Such an approach would also educate the scientist and engineer as to the additional factors which the policymaker and/or manager must consider, in addition to the strictly technical ones.

An example of a simplified approach used to determine the minimum cost nutrient control strategy for the North American Great Lakes Basin is provided by PLUARG (1978a) and Johnson *et al.* (1978).

The reader is referred to other reports for more details regarding the important topic of cost–benefit analysis in relation to development of effective pollution control programmes. Useful information sources include OECD (1974), D.W. Henderson (1974), Krutilla & Fisher (1975), Sinden & Worrell (1979), Pineau *et al.* (1985), Conn (1985) and Thornburn (1986).

Chapter 9 of this book outlines available control methodologies for correcting or preventing eutrophication of lakes and reservoirs, while Chapter 10 discusses the eutrophication phenomenon in relation to enhanced production of fish and other forms of aquaculture. Chapter 6 discusses the use of both simple and complex eutrophication models for assessing alternative eutrophication control options. Finally, Annex 2 provides references to many case studies which illustrate the effectiveness of various eutrophication control programmes in different parts of the world.

ANALYZE ADEQUACY OF EXISTING LEGISLATIVE AND REGULATORY FRAMEWORK FOR IMPLEMENTING EUTROPHICATION CONTROL PROGRAMMES

Institutional concerns

As noted earlier, a range of different forms of government, national priorities, customs and socio-economic conditions exist around the world. Consequently, guidelines for addressing institutional and regulatory concerns in one country may not be appropriate for other countries. Thus, the following concerns will have different priorities in different countries.

The legislative and regulatory frameworks for addressing eutrophication should be examined as a necessary component of an effective eutrophication control programme. There is little point in developing a complex eutrophication monitoring network, for example, if the legislative or regulatory framework for implementing or enforcing the eutrophication control programmes does not exist. Conversely, a well-formulated statute is of little value if the necessary monitoring or pollution alert network for determining compliance with the statute is inadequate.

At the central government level, it is usually most efficient to assign environmental programmes to a single agency structured to manage multiple environmental concerns (e.g. air, water and land resources) than to create a separate governmental unit to deal with each problem as it arises. Furthermore, the responsible agency or institution for carrying out such programmes should be identified clearly. As noted earlier, the public especially may have concerns or suggestions about various aspects of eutrophication, but be frustrated by the lack of a clearly identified and readily accessible forum for expressing such concerns to the policymaker or administrator.

An effective eutrophication control programme also may contain elements, or be involved with problems, which overlap political boundaries and/or governmental agency concerns. If there is an existing agency with which a new control programme is compatible, that agency is the logical one to carry out the new function. However, care must be taken to prevent a new programme from being assigned to an existing governmental unit having a conflicting purpose or goal. For example, an agency responsible for promoting commercial fisheries (thereby interested in enhancing the overall productivity of a waterbody) is not necessarily the best agency to be given the task of protecting a lake as a drinking water supply (for which increased algal production is not desirable). The optimal water quality for these two uses is markedly different. If a central agency is charged with governmental programmes which have conflicting purposes, much effort can be expended in resolving the conflict rather than addressing the problem, or else one purpose will advance at the expense of the other.

If it is necessary to create a new programme for the management or control of eutrophication, a definite term for the existence of the programme should be enacted. Approximately five to ten years would be a reasonable period of time. At the end of this time interval, the programme would either be terminated or re-enacted. Even though this provision creates some risk that a beneficial agency or programme might be terminated, this limited term provides an incentive for governmental accomplishment, mandates timely attention by the policymaker and/or the public, and provides an opportunity for necessary updating and refinement of the programme.

Regulatory concerns

Lengthy, complex, detailed regulations should be discouraged, if not prohibited, because of the potential difficulty both in understanding and administering them. The proliferation of agency-oriented regulations that typically burden many environmental programmes certainly should be avoided. Many environmental regulatory statutes adopted in the United States in the 1970's, for example, were initiated without full appreciation of the magnitude of the duties ultimately required of the relevant governmental regulatory body, usually the Environmental Protection Agency. As a result, the required duties often were more than the available manpower and money could support or accomplish within the prescribed deadlines. If a new statute cannot be implemented because of unrealistic components, not only is achievement of the worthwhile objective delayed, but respect for the timely compliance with laws is also generally eroded. Over the long term, an orderly progression in statutory complexity and development, from one legislative session (or its equivalent) to another, is preferable to the confusion that can result from attempting to accomplish too much, too soon, with too little information, manpower and money.

The administering agency should record its significant decisions, especially those that are controversial or strongly contested. These recorded decisions provide a mechanism for guiding both the administering agency and the affected parties in the future. In addition, eutrophication may be the only concern addressed in a given statute. However, in drafting such statutes, consideration also should be given to relating eutrophication to other environmental problems (and their solutions) in the future.

Initial regulatory statutes should contain provisions for securing sufficient data and knowledge useful for periodically re-evaluating a eutrophication control programme. In the United States, for example, many environmental improvement programmes began as attempts to assist local governmental units (e.g. States, regions, counties or cities) in understanding and addressing their specific needs. Such assistance as training grants to develop technical staffs, money for personnel, assistance to develop monitoring networks, support for students to secure advanced degrees in colleges and universities, and research grants to study the issues and problems, contribute to future refinements in the regulatory scheme. It also furthers our general scientific understanding of the complexities of the eutrophication process.

The degree to which local and regional resources can be focused on solving environmental problems depends in part on the institutional framework, and in part on the strengths and interests in a particular country. The size of the country also affects the choices to be made. As an example, the United States has learned that more effective programme administration within its large geographic area can be achieved in a cooperative hierarchial arrangement, with its relevant federal, State, and local governmental entities.

SELECT EUTROPHICATION CONTROL STRATEGY AND DISSEMINATE SUMMARY TO AFFECTED PARTIES PRIOR TO IMPLEMENTATION

Select and summarize desired control strategy

A specific eutrophication control programme should be selected only after thorough consideration of the types of concerns discussed above. Obviously, such factors as the magnitude of the eutrophication problem, desired uses of the waterbody, and the available control options, and their relative costs and effectiveness, will be major considerations in developing effective control programmes.

A basic methodology for selecting a eutrophication control programme in a given situation is discussed in Chapter 11, and the reader is referred to that chapter for further details.

Once a specific control programme is selected, it is useful and desirable to develop a detailed working plan of the chosen programme, in order that regulators, implementors and all other interested individuals/agencies will have adequate documentation of the tasks to be undertaken, and the goals and objectives to be met. Such an approach will usually work to foster cooperation, rather than confrontation, between involved governmental units and between governmental agencies and the public. As a minimum, the working plan should identify the specific goals of the control programme and the obligations of the involved governmental agencies.

A brief overview of the control programme can also be prepared for all interested parties, both inside and outside of government. This overview should contain a clear statement of the goals of the control programme, and should be disseminated widely prior to implementation of the programme. In countries where such groups exist, members of some type of citizen's advisory committee can be a valuable link in disseminating such documentation.

To foster a greater understanding of the complexities of the eutrophication process, and the public's role in both causing and mitigating its negative impacts, a clearly-articulated education programme can be valuable. Such a programme can be administered by a governmental unit, a concerned community group, or a component of the public education system. This education programme should include a periodic evaluation of the general effectiveness of the implemented control programme (based on post-treatment monitoring data), and an interactive communication between governmental officials and the public. It can also be the basis for development of periodic progress reports to all interested parties.

Other practical considerations

As a practical matter, a lake or reservoir usually does not respond instantaneously to a eutrophication control programme, especially those based on reducing the external nutrient input. Rather, there usually is a time interval ('lag period') between the implementation of the control programme and the observable results in the waterbody. The lag period represents the time necessary for a waterbody to flush itself, or otherwise neutralize the effects of its internal store of nutrients, following implementation of a control programme based on reducing the external nutrient supply to the waterbody. In contrast, in-lake methods, such as harvesting aquatic plant growths, may show smaller or no lag periods, since this latter approach directly addresses the symptoms of eutrophication, rather than the underlying cause. However, as noted earlier, the symptoms often reappear within a short period.

Efforts should be made to inform the public that such lag periods are not unexpected. Otherwise, lack of an immediate response to a control programme may be prematurely, and erroneously, interpreted to mean that a control programme has failed. Methods for calculating the expected duration of the lag period are discussed in Chapter 11.

Although control programmes have been successful in many cases in addressing the negative impacts of eutrophication, there is always some remaining element of uncertainty regarding the ultimate success of any individual control programme. In most cases, however, this usually is not sufficient reason to delay development and implementation of needed eutrophication control programmes.

If a control programme does not produce the desired results, the only reasonable alternatives are to consider further, usually more stringent, control measures, or to be satisfied with the results obtained with the original programme. Fortunately, even when a control programme is unsuccessful in achieving all the desired goals, such programmes usually still work to the positive ecological benefit of the waterbody.

An additional factor to consider in selecting a eutrophication control programme is the expected duration of its effectiveness. Programmes designed to be effective over the long term are usually preferable (although often more costly) to programmes effective only over the short term.

PROVIDE PERIODIC PROGRESS REPORTS ON THE CONTROL PROGRAMME TO THE PUBLIC AND OTHER INTERESTED PARTIES

The role of public awareness

Where it is feasible, public participation in developing an effective eutrophication control programme can be important, particularly with regard to lakes or reservoirs used extensively for recreational purposes. Many individuals may have experienced eutrophication-related problems in such waterbodies in the past, or else may have been exposed to media coverage of such problems. The result can be a 'collective memory' of poor water quality conditions in certain waterbodies, which can lead to a certain degree of public curiosity about lake/reservoir management programmes. Greater public awareness of water-related issues can usually be developed by making details of new eutrophication control programmes, and expected improvements in water quality, available to the public. Such communication efforts can also provide governmental feedback to the public in the form of answers to public questions regarding a given lake or reservoir.

The type and extent of information, and the format used, will probably vary considerably with the target audience. Appropriate media for public information purposes include the press, television and radio, and popular scientific publications. In view of the non-technical background of the lay audience, general information is often most informative (e.g. a new municipal wastewater treatment plant is being built to reduce nutrient levels in Lake X; these nutrient reductions, in turn, should lead to the elimination of algal blooms and related water quality degradation in the lake). Appropriately illustrated information can be very useful in such public communications, and the use of specific technical jargon should be kept to a minimum. A more detailed technical discussion is appropriate for an audience of scientific and/or engineering peers. Water users such as agriculturists or industrialists would probably require information at an intermediate level between these two extremes.

The value of public feedback

The management of water resources is often done at the local level, with little recognition or appreciation given to the long-term needs of a region or country. Furthermore, costs are frequently the only criterion used in developing and/or choosing between management options. Consequently, where feasible, public awareness and feedback can be an important component of effective eutrophication control programmes. If the public can be persuaded of the severity of a eutrophication problem (and its environmental, health and/or economic consequences if left untreated), the public can appreciate more easily the need for eutrophication control programmes. The result can be the development of a proprietary interest by the public in the work involved, and can even make the public more amendable to the associated expenses. This is especially true if the public's experiences with past pollution control programmes have been positive (i.e. if control programmes have been successful in the past). Thus, public awareness and feedback can be an important part of eutrophication control.

The reader is referred to the interesting experiences of the PLUARG (1978b, 1978c) public participation panels for details of the ways in which the public can work together with the government in developing effective environmental management strategies.

CHAPTER 4

CHARACTERISTICS OF EUTROPHICATION

THE CONCEPT OF EUTROPHICATION AS RELATED TO TROPHIC STATUS

Lakes and reservoirs can be broadly classified as oligotrophic (Greek for 'little food') or eutrophic (Greek for 'well fed'). These terms were originally used to describe the soil fertility of peat bogs in northern Germany (Weber, 1907). Subsequently, Thienemann (1918) and Naumann (1919) applied these terms to lakes. A third descriptive term, mesotrophic, is generally used to describe waterbodies in a transition state between oligotrophic and eutrophic. An interesting discussion of the historic evolution of these terms is provided by Rodhe (1969).

Although these trophic descriptions have no absolute meaning, they are generally used today either to denote the nutrient 'status' of a waterbody, or else to describe the effects of the nutrients on the general water quality and/or trophic conditions of a waterbody. As discussed further in a later section, a lake or reservoir eutrophication control programme can be based on achieving certain desired trophic conditions (and characteristic water quality) in the waterbody. Consequently, attempts have been made to relate these descriptive trophic terms to specific 'boundary' values for certain water quality parameters. One example, based on the OECD (1982) International Cooperative Programme on Monitoring of Inland Waters, provides specific boundary values of total phosphorus, chlorophyll *a* and Secchi depth for these trophic conditions in a range of temperate zone lakes (Table 4.1).

There are limitations, however, to using 'fixed' values to delineate lake and reservoir trophic conditions. Some overlap is inevitable, i.e., some waterbodies can be classified in one trophic condition based on one parameter, and in another trophic condition based on a second parameter. OECD attempted to overcome this limitation, using a statistical evaluation of its data base. The resultant 'open boundary' classification scheme is

Table 4.1 OECD boundary values for fixed trophic classification system (modified from OECD, 1982)

Trophic Category	TP	mean Chl	maximum Chl	mean Secchi	minimum Secchi
Ultra-oligotrophic	<4.0	<1.0	<2.5	>12.0	>6.0
Oligotrophic	<10.0	<2.5	<8.0	>6.0	>3.0
Mesotrophic	10-35	2.5-8	8-25	6-3	3-1.5
Eutrophic	35-100	8-25	25-75	3-1.5	1.5-0.7
Hypertrophic	>100	>25	>75	<1.5	<0.7

Explanation of terms:
TP   = mean annual in-lake total phosphorus concentration (ug/l);
mean Chl   = mean annual chlorophyll *a* concentration in surface waters (ug/l);
maximum Chl   = peak annual chlorophyll *a* concentration in surface waters (ug/l);
mean Secchi   = mean annual Secchi depth transparency (m);
minimum Secchi = minimum annual Secchi depth transparency (m).

illustrated in Table 4.2. In this latter system, a waterbody can be considered correctly classified if no more than one of the parameters in Table 4.2 deviates from its geometric mean value by ± 2 standard deviations (OECD, 1982; see Chapter 8 for discussion of geometric mean values). In the absence of absolute trophic standards, the overlap in the range of values in Table 4.2 attests to the still subjective nature of trophic classification schemes (also see Annex 1).

It is clear, therefore, that the terms 'oligotrophic' and 'eutrophic' can have different meanings in different limnological situations. Individuals accustomed to oligotrophic lakes or reservoirs (characterized by generally good water quality for most uses) will probably be more sensitive to the more degraded conditions of eutrophic lakes than individuals on areas normally characterized by more productive waterbodies. Taylor *et al.* (1980) have summarized commonly used trophic classification criteria for temperate lakes from a large number of literature sources.

As noted in Chapter 1, eutrophication is usually considered undesirable, since the effects of eutrophication can interfere significantly with man's use of the water resource for many purposes (e.g. drinking water supply, recreational use, irrigation, etc.). Consequently, eutrophic waters have more constraints on their general usage than do oligotrophic waters. However, it should also be noted that the increased productivity at all trophic levels inherent in the eutrophication process can be a positive feature under some conditions (see Chapter 10).

SYMPTOMS OF EUTROPHICATION

General observations

Although further studies of tropical water systems are still needed, the available evidence suggests that the characteristics of oligotrophic and

Table 4.2 OECD boundary values for open trophic classification system (annual mean values)* (modified from OECD, 1982)

Parameter		Oligo-trophic	Meso-trophic	Eutrophic	Hyper-trophic
Total	$\bar{x}$	8.0	26.7	84.4	
phosphorus	$\bar{x} \pm 1$ SD	4.85–13.3	14.5–49	48–189	
(μg P/l)	$\bar{x} \pm 2$ SD	2.9–22.1	7.9–90.8	16.8–424	
	Range	3.0–17.7	10.9–95.6	16.2–386	750–1200
	n	21	19 (21)	71 (72)	2
Total	$\bar{x}$	661	753	1875	
nitrogen	$\bar{x} \pm 1$ SD	371–1180	485–1170	861–4081	
(μg N/l)	$\bar{x} \pm 2$ SD	208–2103	313–1816	395–8913	
	Range	307–1630	361–1387	393–6100	
	n	11	8	37 (38)	
Chlorophyll a	$\bar{x}$	1.7	4.7	14.3	
(μg/l)	$\bar{x} \pm 1$ SD	0.8–3.4	3.0–7.4	6.7–31	
	$\bar{x} \pm 2$ SD	0.4–7.1	1.9–11.6	3.1–66	
	Range	0.3–4.5	3.0–11	2.7–78	100–150
	n	22	16 (17)	70 (72)	2
Chlorophyll a	$\bar{x}$	4.2	16.1	42.6	
peak value	$\bar{x} \pm 1$ SD	2.6–7.6	8.9–29	16.9–107	
(μg/l)	$\bar{x} \pm 2$ SD	1.5–13	4.9–52.5	6.7–270	
	Range	1.3–10.6	4.9–49.5	9.5–275	
	n	16	12	46	
Secchi	$\bar{x}$	9.9	4.2	2.45	
depth	$\bar{x} \pm 1$ SD	5.9–16.5	2.4–7.4	1.5–4.0	
(m)	$\bar{x} \pm 2$ SD	3.6–27.5	1.4–13	0.9–6.7	
	Range	5.4–28.3	1.5–8.1	0.8–7.0	0.4–0.5
	n	13	20	70 (72)	

*the geometric means (after being transformed to base 10 logarithms) were calculated after removing values which were greater than, or less than, two times the standard deviation obtained (where applicable) in the first calculation.
x = geometric mean
SD = standard deviation
( ) = the value in brackets refers to the number of variables (n) used in the first calculation.

eutrophic waterbodies are basically similar in temperate and tropical/subtropical regions. The differences that do exist between these two limnological settings are related primarily to differences in magnitude and/or timing, rather than substance (J.A. Thornton, 1987). Preliminary evidence suggests that the basic characteristics of oligotrophic and eutrophic waters also apply to sub-arctic waterbodies (McCoy, 1983; V.H. Smith et al., 1984; personal communication, Paul Woods, U.S. Geological Survey, 1985).

In the following discussions, the general characteristics of oligotrophic and eutrophic waters, as observed in temperate zone lakes and reservoirs, are highlighted. This is followed by a brief consideration of differences between (natural) lakes and (man-made) reservoirs, as they relate to the eutrophication process. Finally, there is a discussion of the differing limnological significance of various trophic parameters in tropical/subtropical waterbodies, as contrasted with temperate zone systems.

As shown in Table 4.3, oligotrophic lakes and reservoirs are usually characterized by low nutrient concentrations in the water column, a diverse plant and animal community, a low level of primary productivity and biomass, and good overall water quality for most uses. By contrast, eutrophic waterbodies have a high level of productivity and biomass at all trophic levels, frequent occurrences of algal blooms, anoxic bottom waters (hypolimnion) during periods of thermal stratification, often fewer types of plant and animal species, enhanced growth of littoral zone aquatic plants, and poor water quality for many uses. The overall trophic response of lakes and reservoirs to increased eutrophication is highlighted in Table 4.4.

It is important to recognize that cultural eutrophication is a specialized aspect of the productivity of lakes and reservoirs (OECD, 1982). The undesirable qualities associated with eutrophic waterbodies (from the viewpoint of human use of the water) are not due directly to enhanced levels of nutrient or in-lake productivity, but rather to the resultant impacts of these factors on the overall water quality and water use. As noted earlier, these impacts can necessitate treatment of the water, often expensive and time-consuming, before it can be used by man for a multitude of purposes. Excessive nitrogen levels in water can also be directly hazardous to human health, as well as being an aquatic plant nutrient. Some water uses sensitive to the effects of eutrophication are drinking water supply, industrial uses, stock watering, irrigation and recreation (see Chapter 11 and Annex 1).

The notable exception to this generally negative view of eutrophication is where the increased productivity is desired for fish production or other types of aquaculture, as discussed in Chapter 10. In general, as nutrient inputs to a waterbody increase, the overall fish production also increases. Thus, eutrophic lakes and reservoirs can be extremely valuable as fisheries. In some societies, for example, the artificial stimulation of fish pond productivity has reached a high degree of sophistication, and is an integral part of annual food production. Although overall fish production is generally much higher in eutrophic waterbodies than in oligotrophic waterbodies, the former generally contain more 'coarse' fish species than in oligotrophic waters. Whether or not this is a net 'gain' in regard to a given waterbody depends on local customs, and preferences regarding the relative worth and use of the available fish species. Additional details concerning the eutrophication process, and its effects on water quality and trophic conditions in temperate zone lakes and reservoirs, are given by Sawyer (1966), Fruh *et al.* (1966), Stewart & Rohlich (1967), Vollenweider (1968), National Academy of Sciences (1969), Lee (1971), Landner (1976), E.B. Welch (1980) and OECD (1982).

Table 4.3 General characteristics of oligotrophic and eutrophic lakes and reservoirs in the temperate zone (compiled from P.S. Welch, 1952; Sawyer, 1966; Fruh et al., 1966; Lee, 1971; Golterman, 1975; OECD, 1982)

Parameter	Type of waterbody	
	Oligotrophic	Eutrophic
I. *Biological*:		
Aquatic plant and animal production	low	high
Number of plant and animal species	many	many; can be substantially reduced in hypertrophic waters
General levels of biomass in waterbody	low	high
Occurrence of algal blooms	rare	frequent
Relative quantity of green and blue-green algae	low	high
Vertical extent of algal distribution	into hypolimnion (bottom waters) in thermally stratified waterbodies	usually only in surface waters
Aquatic plant growth in shallow shoreline area (littoral zone)	can be sparse or abundant; if present, usually consists of submerged and emergent vegetation	often abundant; usually an increase in the presence of filamentous algae and a decrease in macrophytes
Daily migration of algae	extensive	limited

Some characteristic algal groups	Green algae:
	  Desmids
	  Staurastrum
	
	Diatoms:
	  Tabellaria
	  Cyclotella
	
	Golden-brown algae:
	  Dinobryon
		Blue-green algae:
		  Anabaena
		  Aphanizomenon
		  Microcystis
		  Oscillatoria
		
		Diatoms:
		  Melosira
		  Fragilaria
		  Stephanodiscus
		  Asterionella

Some characteristic zooplankton groups	*Bosmina obtusirostris* *B. coregoni* *Diaptomus gracilis*	*Bosmina longirostris* *Daphnia culcullata*
Characteristic bottom animals	*Tanytarsus*	Chironomids
Characteristic fish types	deep-dwelling, cold water fishes (salmon, trout, cisco)	surface-dwelling, warm water fishes (pike, perch, bass)

(continued)

Table 4.3 *continued*

Parameter	Type of waterbody	
	Oligotrophic	Eutrophic
II. *Chemical*:		
Oxygen content of bottom waters (hypolimnion)	high throughout year	can be low or absent during thermal stratification period
Total salt content of water (specific conductance)	usually low	sometimes very high
III. *Physical*:		
Mean depth of waterbody	often deep	often shallow
Volume of hypolimnion	often large	can be small or large
Temperature of hypolimnion waters	usually cold	cold water usually minimal, except in deep eutrophic waterbodies
IV. *Water usage*:		
Water quality for most domestic and industrial uses	good	often poor
Impairment of multi-purpose use	normally little impairment	often considerable impairment

Differences between lakes and reservoirs

As indicated in Chapter 1, this document is meant to be useful for lakes and reservoirs. As used here, the term 'lake' denotes a natural waterbody, while the term 'reservoir' denotes a man-made waterbody (even though many reservoirs are called lakes). Reservoirs are usually formed by the construction of a dam across a river or stream, resulting in the impoundment of the water behind the dam structure.

There are some differences between these two types of waterbodies. K.W. Thornton *et al.* (1982) have compared a number of variables common to natural lake and reservoir systems (Table 4.5). The comparison involved 309 natural lakes and 107 reservoirs, located primarily in the temperate region of the United States. As shown in Table 4.5, reservoirs generally have larger drainage basins, surface areas and drainage basin to surface area ratios than do natural lakes. The mean and maximum depths are usually greater in reservoirs than in lakes. Reservoirs also have greater areal waters loads and shorter hydraulic residence times. The larger drainage basin to surface area ratio and greater water load usually means that the areal nutrient loads to reservoirs are larger than those to lakes. However,

Table 4.4 Trophic criteria and their responses to increased eutrophication[1] (modified from Brezonik, 1969; Taylor et al., 1980)

Physical	Chemical	Biological[2]
Transparency (D) (e.g. Secchi depth)	Nutrient concentrations (I) (e.g. spring maximum)	Algal bloom frequency (I)
Suspended solids (I)	Chlorophyll a (I)	Algal species diversity (D)
	Electrical Conductance (I)	Phytoplankton biomass (I)
	Dissolved Solids (I)	Littoral vegetation (I)[3]
	Hypolimnetic oxygen deficit (I)	Zooplankton (I)
	Epilimnetic oxygen supersaturation (I)	Fish (I)[4]
		Bottom fauna (I)[5]
		Bottom fauna diversity (D)
		Primary production (I)

[1](I) signifies the value of the parameter generally increases with the degree of eutrophication; (D) signifies the value generally decreases with the degree of eutrophication.
[2]The biological criteria have important qualitative (e.g. species) changes as well as quantitative (e.g. biomass) changes, as the degree of eutrophication increases.
[3]Aquatic plants in the shallow, nearshore area may decrease in the presence of a high density of phytoplankton.
[4]Fish may be decreased in numbers and species in bottom waters (hypolimnion) beyond a certain degree of eutrophication, as a result of hypolimnetic oxygen depletion (personal communication, O. Ravera, EURATOM, 1984).
[5]Bottom fauna may be decreased in numbers and species in high concentrations of H_2S, CH_4 or CO_2, or low concentrations of O_2 in hypolimnetic waters (personal communication, M. Straškraba, Czechoslovak Academy of Science, 1985).

Table 4.5 Comparison of the geometric mean values of selected parameters of lake and reservoir systems (modified from K.W. Thornton et al., 1982)

Parameter	309 natural Lakes	107 Reservoirs	Probability mean values are equal
Drainage area (km²)	222.0	3,228.0	<0.0001
Surface area (km²)	5.6	34.5	<0.0001
Drainage:Surface area ratio	33.0	93.0	<0.0001
Mean depth (m)	4.5	6.9	<0.0001
Maximum depth (m)	10.7	19.8	<0.0001
Shoreline development ratio	2.9	9.0	<0.001
	$(n = 34)$[a]	$(n = 179)$[b]	
Areal water load (m/yr)	6.5	19.0	<0.0001
Hydraulic residence time (yr)	0.74	0.37	<0.0001
Total phosphorus (µg/l)	54.0	39.0	<0.02
Chlorophyll a (µg/l)	14.0	8.9	<0.0001
Areal phosphorus load (g P/m².yr)	0.87	1.7	<0.0001
Areal nitrogen load (g N/m².yr)	18.0	28.0	<0.0001

[a]From Hutchinson (1957)
[b]From Leidy & Jenkins (1977)

mainly because of the shorter hydraulic residence times and greater mean depths, the average phosphorus and chlorophyll concentrations in reservoirs are usually lower than those in lakes (K.W. Thornton *et al.*, 1982), even though Kimmel & Groeger (1984) have reported that phytoplankton primary productivity of reservoirs appeared to be greater than for natural lakes.

Reservoirs are generally constructed to store or control the flow of rivers and streams, primarily in regions where either excesses or shortages of water occur. The former results in flooding, whereas the latter requires the seasonal or long-term storage of water (Kennedy *et al.*, 1985). Consequently, reservoirs are geologically younger than natural lakes. In the United States, for example, many reservoirs are less than 50 years old (K.W. Thornton, 1984). The world-wide distribution pattern for lakes shows a bimodal pattern. The largest percentage lies between approximately 35°–55° latitude in both the northern and southern hemispheres, and between approximately 15° North to 20° South latitude around the equator (K.W. Thornton, 1984; Kennedy *et al.*, 1985). In contrast, reservoirs do not exhibit this bimodal distribution. Kennedy *et al.* (1985) showed that, in the United States, most U.S. Army Corps of Engineers reservoirs are located in the middle latitudes, which is also the region containing the minimum number of natural lakes. It is not clear whether or not this reservoir distribution pattern observed in the United States applies globally.

Natural lakes usually occupy natural depressions in the local topography. Consequently, they are located in the center of relatively symmetrical and contiguous drainage basins, and have tributary inputs entering the lake at several points around the lake. By contrast, reservoirs are usually constructed at the downstream boundary of a drainage basin. Thus, reservoir drainage basins are often narrow and elongated, with only a small portion of the basin being contiguous with the reservoir. In many (if not most) cases, a majority of the annual water, sediment and nutrient inputs enter the reservoir through a single, large tributary, located at an upstream point some distance from the dam.

Because of these characteristics, there often is a spatial gradient in sediment and nutrient concentration patterns along the body of a reservoir, especially in long, narrow, dendritic reservoirs. This is accompanied by a spatial gradient in biological productivity and water quality in the reservoir. It is noted that this feature may not always occur in all reservoirs (e.g. see Wells & Gordon, 1982).

A schematic view of the spatial patterns often seen in reservoirs is provided in Figure 4.1. This figure illustrates the difference in the characteristics of the river/stream input end of the reservoir and the deeper, more 'lake-like' downstream end of the reservoir and how these differences can be reflected in a spatial heterogeneity in water quality (K.W. Thornton *et al.*, 1982; Kimmel & Groeger, 1984).

Characteristics of eutrophication

• Narrow, channelized basin	• Broader, deeper basin	• Broad, deep, lake-like basin
• Relatively high flow	• Reduced flow	• Little flow
• High suspended solids; low light availability at depth	• Reduced, suspended solids; light availability at depth	• Relatively clear; light more available at depth
• Nutrient supply by advection; relatively high nutrient levels	• Advective nutrient supply reduced	• Nutrient supply by interval recycling; relatively low nutrient levels
• Light-limited primary productivity	• Primary productivity relatively high	• Nutrient-limited primary productivity
• Cell losses primarily by sedimentation	• Cell losses by sedimentation and grazing	• Cell losses primarily by grazing
• Organic matter supply primarily allochthonous	• Intermediate	• Organic matter supply primarily autochthonous
• More eutrophic	• Intermediate	• More oligotrophic

Figure 4.1 Longitudinal zonation in environmental factors controlling the trophic condition of reservoir systems (modified from Kimmel & Groeger, 1984)

There is an additional factor which can affect the water quality in various regions of a reservoir. Natural lakes discharge surface waters by uncontrolled overflow. By contrast, water withdrawals from reservoirs are usually done via submerged gates in the dam, often located at various depths in the water column. Thus, selective withdrawal of water from specific depths in a reservoir is possible.

The net effect of these features is that reservoirs can exhibit physical, chemical and biological gradients not normally observed in natural lakes. These gradients can differ in both time and space in given situations. Consequently, sampling strategies developed specifically for reservoirs exhibiting water quality gradients are discussed in Chapter 8. The differing nature of lake and reservoir systems can be important with respect to eutrophication modeling efforts (see Chapter 6), for calculating annual nutrient loads (see Chapter 7) and for developing effective in-lake sampling programmes (see Chapter 8). These differences should be taken into account when engaging in these activities, as discussed in the cited chapters.

Because of these differing features, therefore, a basic question is whether or not a eutrophication management approach, useful for both natural lake and reservoir systems can be developed. However, in terms of developing effective eutrophication control programmes, these differences appear to be much less significant. In fact, the main impact of these differences appears to be that some of the control options available for natural lakes may not work effectively in reservoir systems. This latter aspect is discussed further in Chapter 9. The reader is referred to the reports of Kennedy et al. (1985), K.W. Thornton (1984), K.W. Thornton et al. (1981; 1982), Kimmel & Groeger (1984), Walker (1981; 1982; 1985), Clasen & Bernhardt (1980) and OECD (1982) for further discussion of the differences between lakes and reservoirs, and how these differences should be considered in relation to effective eutrophication control.

Differences between temperate and tropical lake systems

As noted by J.A. Thornton (1987), not all the symptoms of eutrophication normally observed in temperate zone lakes and reservoirs necessarily occur in tropical/sub-tropical systems. It is not that the eutrophication process is influenced by different factors in tropical systems, but rather that the eutrophication 'symptoms' in these systems may not be indicative of the same water quality and/or trophic conditions as in temperate zone waterbodies. Although there is a surprisingly large literature addressing lake and reservoir limnology in tropical/sub-tropical settings, this literature is relatively dispersed in many (often hard to obtain) sources. A recent work, which brings together many studies of tropical/sub-tropical lakes and reservoirs, is the review of J.A. Thornton (1987). Based on his review, Thornton has reached several important conclusions regarding the effective control of eutrophication in tropical water systems.

Compared to the temperate zone, tropical lakes and reservoirs are characterized by highly seasonal (predominantly summer) rainfall and a more limited annual temperature cycle of 10 °C or less. The mean annual temperature (approximately 25 °C) is higher than in temperate zones (approximately 10–15 °C). There is no annual freeze/thaw cycle in tropical/sub-tropical regions, so that the growing season generally extends throughout the year. Stable temperature gradients are also produced, although they are generally less pronounced than in temperate lakes. Oxygen depletion in the hypolimnion can occur in tropical lake/reservoir systems regardless of their trophic status. Consequently, hypolimnetic oxygen depletion has little meaning as a trophic state indicator (Mitchell & Marshall, 1974; Walmsley & Toerien, 1977, as cited in J.A. Thornton, 1987). Because of the more favorable, year-round growing season, tropical lake and reservoir productivity is usually higher than that of temperate zone waterbodies. Phytoplankton blooms can occur at any time of the year, therefore, without following an annual cycle.

The 20–30 μg P/l phosphorus concentration often used as the mesotrophic–eutrophic boundary value in temperate zone lakes and reservoirs may be low when applied to tropical systems (Table 4.6). A concentration of 50–60 µg P/l was suggested as a more realistic mesotrophic–eutrophic boundary value for Lake McIlwaine in Zimbabwe (J.A. Thornton & Nduku, 1982, as cited in J.A. Thornton, 1987). Similar phosphorus boundary values were reported for Australian waterbodies (Williams & Wan, 1972; Wood, 1975, both as cited in J.A. Thornton, 1987). By contrast, lower nitrogen concentrations between 20–100 µg N/l are suggested as mesotrophic–eutrophic boundary ranges for Australian and African lake systems (several authors, as cited in J.A. Thornton, 1987), as compared to the higher range for the temperate zone.

Tropical systems often develop extremely low N:P ratios, thereby favoring the dominance of blue-green, N_2-fixing algae. Bioassay studies of nutrient limitation in African lakes (see following discussion of limiting nutrient concept) suggest initial phosphorus-limitation under natural conditions, shifting to nitrogen-limitation as nutrient inputs increase further, which is similar to temperate zone waterbodies. It is generally believed that nitrogen is the primary nutrient which limits the maximum algal biomass levels in tropical/subtropical systems (e.g. see Table 4.6; Henry et al., 1985). The available evidence, however, produces a less clear picture of specific nutrient limitation in such systems. For example, Henry & Tundisi (1982) identified nitrogen as the primary limiting nutrient in Lobo Reservoir in Brazil, based on algal bioassay tests. Bioassay tests also suggest nitrogen limitation for Lake Jacaretinga in Brazil (Zaret et al., 1981; Henry et al., 1985) and Lake

Table 4.6 Comparison of temperate and tropical boundary values for several trophic state indicators (modified from J.A. Thornton, 1985; 1987)

Trophic indicator	Mean boundary value between mesotrophic and eutrophic conditions		References
	Temperate lakes	Tropical lakes	
Mean primary productivity (g C/m^2.day)	1.0	2–3	Robarts (1982)
Chlorophyll a (µg/l)	10–15	10–15	Walmsley and Butty (1980)[1]
Total phosphorus (µg P/l)	30	50–60	J.A. Thornton and Nduku (1982)[1]
Total nitrogen (µg N/l)	50–100	20–100	Wood (1975)[1]
Nutrient limitation often by	phosphorus	nitrogen	Toerien et al. (1975)[1]
Dominant algal types	diatoms	blue-green algae	J.A. Thornton (1985)
Photosynthetic efficiency	<1%	>2–3%	Wetzel (1975)[1]

[1] As cited in J. A. Thornton (1987)

Valencia in Venezuela (Lewis, 1983). Similar conclusions were reached by Viner (1975) for a number of tropical rivers and lakes in Uganda. Vincent et al. (1984) suggested a persistent cellular deficiency of nitrogen for phytoplankton of Lake Titicaca in Peru-Bolivia, based on five types of algal physiology bioassays. Lake Chapala in Mexico also appears to exhibit nitrogen limitation (personal communication, J. Limón Macias, University of Guadalajara, Mexico).

In other studies of tropical/sub-tropical lakes and reservoirs, however, phosphorus was shown to be a primary growth limiting nutrient. For example, Moss (1969) reported that either nitrate or nitrate plus phosphorus was the algal growth-limiting nutrient in nine central African waterbodies, based on algal bioassay studies. Lindmark (1979) showed that phosphorus was the algal growth-limiting nutrient in Lake Paranoa in Brasilia, Brazil. Lindmark also suggested that nutrients play a major role in controlling the seasonal patterns of algal growth, since the growing season is essentially continuous in tropical lake systems, and since temperature and light are usually above levels which would limit algal growth (see Chapter 5 for discussion of temperature and light). Melack et al. (1982) used experimental fertilization of natural algal populations taken from an equatorial African soda lake, and showed phosphorus limitation of phytoplankton biomass. They concluded that phytoplankton in tropical waters appeared to be equally amenable to the types of nutrient limitation seen in temperate zone lakes. J.A. Thornton & Nduku (1982) reported phosphorus limitation for Lake McIlwaine, a man-made impoundment in Zimbabwe, Africa. Zutshi & Wanganeo (1984) reported phosphorus limitation in sub-tropical Manasbal Lake in India, based on algal bioassay studies.

Light limitation of phytoplankton growth (see Chapter 5) also occurs in tropical/sub-tropical waterbodies, probably to a greater extent than in the temperate zone. Overall, tropical/sub-tropical lakes and reservoirs appear to be able to tolerate higher phosphorus loads than temperate zone waterbodies, while the boundary nitrogen levels appear to be lower in tropical settings than in temperate settings.

Enhanced algal growth has the same trophic significance in both temperate and tropical settings. Specific algal species, however, may not be useful as eutrophication indicators in tropical/sub-tropical waterbodies because of the normally wider range of habitats available in these systems, compared to the temperate zone. As an example, enhanced macrophyte growth is not always indicative of culturally eutrophic lakes and reservoirs in tropical areas (Gaudet & Denny, 1981). Nevertheless, aquatic macrophytes (see Chapters 5 and 10) are more often a problem in tropical/sub-tropical waterbodies than in temperate zone waterbodies.

The effects of altering water residence times (i.e. flushing rates) on in-lake phosphorus concentrations and trophic status (see Chapters 5 and 7) appears to be similar in temperate and tropical waterbodies. Furthermore, until recently, a lack of seasonality of climatic conditions has been thought to be an important feature of tropical limnology. It appears, however, that distinct seasonality also has been noted for most tropical lakes and reservoir systems, negating this assumption (J.A. Thornton, 1987; Davies & Walmsley, 1985). The effect of the rainy season on annual runoff patterns is significant in tropical settings, since the period of maximum algal production in tropical/sub-tropical lakes and reservoirs often takes place one or two months after the rainy season. In spite of the observed differences in the two water systems, present evidence supports the belief that control of eutrophication in tropical/sub-tropical lakes and reservoirs can be considered in essentially the same manner as in the temperate zones (also see Hill & Rai, 1982). Available evidence suggests the same is also true for sub-arctic lake systems (V.H. Smith *et al.*, 1984).

As noted previously, the Pan American Center for Sanitary Engineering and Environmental Science (CEPIS), whose headquarters are in Lima, Peru, is coordinating an international study of the eutrophication of warm-water lakes. The objective of the CEPIS study is to develop simplified methodologies for evaluating eutrophication in warm-water tropical lakes. CEPIS defined warm water lakes as lakes with a minimum temperature of 10°C and an average annual temperature greater than 15°C. It is anticipated that this study will provide a basis for the development of a quantitative framework (including models) for assessment of eutrophication in tropical regions similar to that developed in the OECD (1982) eutrophication study on temperate zone lakes and reservoirs (Salas, 1983). The CEPIS study, when completed, should provide more specific guidance about the assessment and control of eutrophication in tropical/sub-tropical lakes and reservoirs.

Another noteworthy effort is the report of Hart & Allanson (1984), which discusses limnological criteria for management of water quality in the southern hemisphere. This latter work, based on a four-day workshop with participants from developed and developing countries in the southern hemisphere, discusses a number of environmental concerns related to lakes and reservoirs in this area of the globe. Specific topics discussed at the workshop included eutrophication, catchment management, stream regulation, turbidity and fisheries. The proceedings of the workshop (Hart & Allanson, 1984) also discuss the important management implications of these topics, as they apply in the southern hemisphere.

THE LIMITING NUTRIENT CONCEPT

As noted in Chapter 3, because aquatic plant nutrients are essential for the growth of algae in a waterbody, one of the most effective long-term strategies for controlling cultural eutrophication of lakes and reservoirs is to reduce the quantity of the nutrients entering the waterbody. Assuming that algal growth is not controlled by a non-nutrient factor (e.g. inadequate light, suboptimal temperature), reducing or 'limiting' the input of the nutrients should also limit the resulting algal biomass in the waterbody. Numerous experiments in the laboratory and in the field have demonstrated the major role of phosphorus, nitrogen and, in some cases, silicon, in influencing the dynamics of algal populations, including algal concentrations and species composition. In recent years, whole-lake experiments in Canada have further substantiated this causative role of nutrients (Schindler & Fee, 1974; Schindler, 1977). Therefore, limiting the input of these 'macro-nutrients' represents a primary eutrophication control option. This section focuses on the basis for this approach, the limiting nutrient concept. In addition, techniques for identifying the limiting nutrient in a given waterbody are also discussed. This latter aspect is especially important, because it will help identify which nutrient should be the primary focus of a given eutrophication control programme.

The limiting nutrient concept was probably first advanced by Justus Liebig in 1840, who found that the yield of terrestrial crops was often limited by the materials needed by the crop in minute quantities. All organisms need essential nutrients and other materials for their growth, and these needs can be very species-specific. Liebig pointed out that the ultimate yield of a crop was limited by the essential nutrient which was most scarce in the environment, relative to the specific needs of the crop (Odum, 1971; E.B. Welch, 1980). The concept has since been applied to phytoplankton growth in lakes and reservoirs. Although subjected to some misuse because of over-simplification (Wetzel, 1975), this concept still highlights a fundamental fact useful for the control of eutrophication, namely, that both the absolute and relative quantities of essential nutrients in a lake or reservoir are primary factors regulating the algal biomass in the waterbody. The concept is supported by the successful formulation of several empirical, quantitative relationships between nutrients and algal biomass, as well as other indicators of lake trophic state (see Chapter 6). Prominent examples include the phosphorus–chlorophyll relationships of Sakamoto (1966), Dillon & Rigler (1974), Vollenweider (1976a), J.R. Jones and Bachmann (1976), Rast & Lee (1978), Larsen & Mercier (1976), Ryding (1980), and OECD (1982).

These studies clearly support the belief that macro-nutrients are a major variable controlling algal biomass in lakes and reservoirs. It is also important, however, to recognize that other variables can markedly influence the algal biomass in a waterbody under some conditions. For example, in some cases, trace elements and organic factors (e.g., sodium, potassium, trace metals and vitamins) can be important variables influencing the quantity and types of phytoplankton found in lakes and reservoirs (Provasoli, 1969). As noted above, light availability and water temperature also have important roles in controlling phytoplankton growth under some conditions (e.g. see Stråskraba, 1980; Tilman et al., 1982).

Nevertheless, available evidence indicates that the phytoplankton biomass in a waterbody (e.g. chlorophyll concentration) appears to be proportional to its nutrient load, at least up to a certain point. Beyond that point, some factor (or factors) other than nutrients can exert primary control over in-lake algal levels. This latter condition is illustrated by the 'plateau' region of the characteristic curvilinear growth curve of phytoplankton. The growth curve (i.e. algal response to nutrients) generally shows a linear increase with increasing nutrient concentrations, up to a certain nutrient level. Beyond this point, the phytoplankton exhibit a relatively constant plateau value, showing no further increases in algal growth, regardless of further nutrient increases. Lee *et al.* (1980), Cowen & Lee (1976) and Sridharan & Lee (1977) discuss how one can assess the approximate position of a waterbody on this curvilinear algal growth response curve, as an aid to designing optimal eutrophication control strategies.

The limiting nutrient concept has its basis in the photosynthesis reaction. Although the exact stoichiometry of the reaction is still debated, the chemical needs of phytoplankton photosynthesis can be presented conceptually (Stumm & Morgan, 1970) as follows:

$$CO_2 + NO_3^- + PO_4^{2-} + H_2O + H^+ \,(+ \text{ trace elements; sunlight})$$
$$\underset{respiration}{\overset{photosynthesis}{\rightleftharpoons}} \text{algal protoplasm} + O_2 \qquad (4.1)$$

A discussion of the specific energetics of this reaction, including differences specific to other photosynthetic organisms, is provided by Golterman (1975).

As mentioned earlier, cultural eutrophication can be viewed as a specialized aspect of the overall productivity of a lake (OECD, 1982). In a practical sense, therefore, if one can control the latter, one should also be able to control the former. Equation 4.1 shows that several variables are involved in the photosynthesis reaction. Each variable theoretically could be subjected to control efforts. However, since our knowledge regarding algal dynamics and physiology is lacking in many areas (e.g. see Morris, 1980; Platt, 1981) and, based on the practical experience in lake/reservoir restoration efforts over the past several decades, it is clear that control or limitation of the external nutrient supply (especially phosphorus and nitrogen) remains one of the most effective long-term measures for attempting to control or alleviate the impacts of eutrophication (Golterman, 1975; Schindler, 1977; OECD, 1982; Rast & Kerekes, 1982).

Some assumptions in using the limiting nutrient concept

It is emphasized that the limiting nutrient concept applies to both the absolute and relative quantities of nutrients in a waterbody. In using this concept, it is assumed that the ratio at which the nutrients are taken up and used by algae reflect the relative composition of these elements in their cellular material (exceptions are discussed below). Fleming (1940) and Redfield (1934) (both as cited in Redfield et al., 1963) examined the organic matter in sea water samples to determine the cellular content of carbon, nitrogen and phosphorus in phytoplankton and zooplankton. The average atomic ratio of these three elements in the plankton samples was found to be 106 to 16 to 1 (i.e. an atomic ratio of 106C:16N:1P). This ratio has become a widely cited standard reference value for assessing the limiting nutrient in waterbodies.

It is mentioned here that carbon normally becomes an algal growth-limiting nutrient only when the water is saturated with both phosphorus and nitrogen, when light availability and temperature are high, and when the transport of carbon dioxide from the atmosphere to the water column is slow. Although this situation can occur in well-fertilized ponds and sewage lagoons, it normally is not seen in lakes and reservoirs (Fuhs et al., 1972; Maloney et al., 1972; James & Lee, 1974; Schindler, 1977; Rast & Lee, 1978). Even in those unusual situations where it does occur, Shapiro (1973) and James & Lee (1974) have shown that carbon limitation usually only affects the types, rather than quantity, of algae occurring in a waterbody. Consequently, further discussion in this section focuses on the role of phosphorus and nitrogen as primary algal growth-limiting nutrients.

Odum (1971) points out several precautions in the practical use of the limiting nutrient concept. First, strictly speaking, the concept applies only under steady-state conditions. Under such conditions, 'the essential material available in amounts most closely approaching the critical minimum needed will tend to be the (growth) limiting one.' The limiting nutrient concept is far less useful in 'transient state' situations (e.g. waterbodies which receive substantial, but intermittent, inputs of phosphorus or nitrogen).

Another consideration is that a waterbody normally contains a number of different algal species, each with relatively specific nutrient requirements. Different algal species can assimilate nutrients in different quantities and at different rates (Tilman et al., 1982). How well one species can compete with another for available nutrients depends on such factors as nutrient uptake and assimilation rates. This is the basis of the concept of 'resource competition' by algal populations (Hutchinson, 1961; Kilham & Kilham, 1978; Tilman et al., 1982). Tilman et al. (1982) provide support for this concept with the observation that diatoms appear to be superior competitors for phosphorus, but inferior competitors for nitrogen and silica. Thus, diatoms grow better under low phosphorus levels than other algal species. Lang & Brown (1981) reported that some blue-green algae were more efficient in phosphorus uptake under phosphorus-limited conditions than some green algae. R.E.H. Smith & Kalff (1982) indicated that, under nutrient-deficient conditions, smaller algal cells are usually more efficient at nutrient uptake than large cells. Tilman et al., (1982) even suggested that one should not speak of 'phosphorus limited' or 'nitrogen limited' lakes. Rather, one must recognize that 'individual algae – not lakes – are limited by a particular nutrient'. An interesting discussion of limiting nutrients in relation to the resource competition and phytoplankton ecology is provided by Tilman et al. (1982).

The relative proportions of aquatic plant nutrients in a lake or reservoir can also change seasonally and year-to-year. Thus, one nutrient may limit algal biomass levels at one time, and another nutrient may be limiting at another time. Practical use of the limiting nutrient concept, however, assumes that a single nutrient required by algae will be the limiting factor, at least during the period of water quality concern. In fact, this latter conclusion has been basically substantiated by Droop (1973) and Rhee (1978), who showed in their experiments that algal growth was regulated by the single nutrient in shortest supply, relative to the proportional needs of the algae (as contrasted to a multiple nutrient effect).

As a practical observation, phosphorus is often singled out for primary attention in eutrophication control strategies. This is because it is relatively easily removed from municipal and other wastewaters by standardized, relatively inexpensive water treatment methods, compared to other nutrients. Furthermore, Golterman (1975) suggested that it is not important whether or not phosphorus is the limiting nutrient in a given situation. The fact is that 'it is the only essential element that can easily be made to limit algal growth'. This 'controllability' factor is of major importance in development of an effective eutrophication control strategy (see Chapter 11).

It also is necessary to distinguish here between the phytoplankton standing crop in a waterbody and the rate at which it is produced. The major algal problems associated with eutrophication are caused by the density of the algal accumulation in a waterbody at a given time. The algal biomass represents the quantity of algae per volume or area of the waterbody, and is usually expressed in units of mass (e.g. μg chlorophyll/l). Thus, it is a static measure of eutrophication symptoms. By contrast, the rate of algal production (primary productivity) is a dynamic process and defines the rate at which algal biomass is produced (e.g. mg C fixed/m^2.day). Although algal biomass in a waterbody can be directly proportional to algal primary productivity, this is often not the case. For example, a waterbody can exhibit a high rate of primary productivity, but a small algal biomass, because of such factors as zooplankton grazing of the phytoplankton, a fast flushing rate (which can remove algal cells from the waterbody before they can accumulate to nuisance levels) and sedimentation (which can remove algal cells from the water column). It is noted that an increase in the rate of primary production in a waterbody may provide an earlier indication of potential algal problems than elevated algal biomass levels (Rast & Kerekes, 1982).

Nevertheless, although the level of productivity can influence how long excessive algal growths will last, as well as the rate of algal succession in a waterbody (Lund, 1969), productivity *per se* is not the major problem associated with cultural eutrophication. Rather, it is the excessive algal (or macrophyte) biomass which interferes with man's use of the water resource. It is for this reason that primary attention in this book is directed toward the control of algal biomass levels in lakes and reservoirs.

The use of nitrogen and phosphorus concentrations and ratios to assess nutrient limitation

The use of the limiting nutrient concept assumes that the growth of algae in a waterbody is proportional to the quantity of nutrients in the waterbody (assuming appropriate temperature and light conditions already exist). It is also assumed that the nutrient content of a waterbody is a function of the nutrient load to the waterbody. Finally, it is assumed that phytoplankton take up and use nutrients from the water column in the above-noted atomic ratio of 106C:16N:1P. Thus, one should be able to limit the maximum algal biomass in a lake or reservoir by controlling the quantity and/or ratio at which these nutrients are supplied to the waterbody. One can measure the concentrations of these essential nutrients in the water column, and calculate the C:N:P ratio of the waterbody. In addition to examining the absolute concentrations of these nutrients in the waterbody, comparing the calculated N:P ratio to the 16N:1P reference value, it usually is possible to determine which of these nutrients is inadequate in a waterbody. The identified nutrient (or nutrients) is likely to be the factor that will limit the maximum algal biomass that can grow in the waterbody.

Attention normally should be given to the 'biologically available' forms of phosphorus and nitrogen in a waterbody during the growth season, normally the period of maximum algal biomass levels. This is also usually the period of maximum use (and maximum degradation) of the water resource for secretion, water supply, fishing, etc. Measurement of the nutrient concentrations in the water column during this period will allow one to determine which nutrient is present in excessive quantities ('left over') after the 16N:1P atomic ratio needed by phytoplankton is satisfied. The excess nutrient represents the difference between the supply of the nutrient to the waterbody and the amount needed by the algae for growth and reproduction, and is not likely to be the primary factor limiting the maximum algal biomass in the waterbody (assuming minimal light and temperature conditions are met). Utilization of this approach to help identify the potential limiting nutrient requires knowledge of both the in-lake concentrations and ratios of nitrogen and phosphorus.

Nitrogen and phosphorus concentrations. To use in-lake nutrient concentrations to help identify the limiting nutrient, water samples should be restricted to the epilimnion, the water layer in which most algal photosynthesis normally occurs in a waterbody. In fact, except for highly transparent waterbodies in which light penetration extends into the hypolimnion (see Chapter 8), water samples can be collected from approximately the 0.5–1 m depth. Samples should be collected during the algal growing season, normally the period of maximum biomass levels and water usage. Ideally, measurements also should be made over the annual cycle for maximum information on nutrient limitation dynamics.

Algal blooms are relatively short-term events. Therefore, the quantities of the 'biologically available' forms of phosphorus and nitrogen, rather than total nitrogen and phosphorus, are of primary interest. Although there is debate as to which chemically defined nutrient fractions comprise the biologically available nutrients, these fractions can be approximated by the dissolved reactive phosphorus concentration (expressed as P), and the sum of the ammonia and nitrate (and nitrite, if available) concentrations (expressed as N).

Initially, one should examine the measured concentrations of these nutrients in the waterbody, in order to determine if one or both of them has been depleted to algal growth-limiting levels. As rough boundary values for lakes and reservoirs, practical experience suggests that concentrations of biologically available phosphorus of less than 5 μg P/l indicate potential phosphorus limitation, while concentrations of biologically available nitrogen less than 20 μg N/l suggest nitrogen limitation. If both nutrients are present in concentrations less than these boundary values, both nutrients may be limiting. Conversely, concentrations of both nutrients above these values suggest neither nutrient is the primary algal growth-limiting factor.

Nitrogen to phosphorus ratios. If the absolute concentrations of the biologically available phosphorus and nitrogen in the waterbody have not decreased to growth-limiting levels, the in-lake ratios can give an indication of which nutrient may become the limiting nutrient. The calculated atomic ratio of the nutrients measured in the water column can be compared to the previously cited reference value of 16N:1P. Assuming the algal populations use the nutrients in this ideal ratio, any deviation from the ratio can be used to determine the potential limiting nutrient in the waterbody.

At a practical level, it is easier to use the mass ratios of the biologically available forms of nitrogen and phosphorus, rather than the atomic ratios, to calculate the limiting nutrient. If one measures the in-lake nutrient concentrations in similar units (e.g. mg/l), the 16N:1P atomic ratio reference value corresponds to a mass ratio of 7.2N:1P. Therefore, if the ratio of the measured concentrations is less than about 7N:1P, nitrogen is the potential limiting nutrient; if the ratio is greater than 7, phosphorus is the potential limiting nutrient. If the ratio is approximately 7, then both nutrients or some other factor (e.g. light or temperature) may be limiting.

There are several precautions in using N:P ratios to assess the limiting nutrient. As a practical matter, one should not be overly strict in applying the 7N;1P value as the exact limiting ratio. The cellular content of nitrogen and phosphorus in algae can vary, at least under laboratory conditions. Thus, the ratio at which the algae use these nutrients may also vary to some degree. Chiaudani & Vighi (1974), for example, have used an N:P mass ratio of 5 or less to define potential nitrogen limitation, and a ratio of 10 or more for potential phosphorus limitation. A ratio between 5–10 suggests that either (or neither) nutrient could be limiting. Forsberg & Ryding (1980) used a mass ratio of dissolved nitrogen:phosphate phosphorus of 5:1 or less to denote nitrogen limitation, and a ratio of 12:1 or greater to denote phosphorus limitation, in Swedish waterbodies. A mass ratio between 5-12 indicated that either (or both) nutrient could be limiting. They also reported total phosphorus:total nitrogen ratios of <10 as a boundary value for potential nitrogen limitation and >17 as a boundary value for potential phosphorus limitation; values between these boundaries suggest either or both nutrients could be limiting. In laboratory studies, Rhee (1978) used an N:P mass ratio of 13:1 to assess nitrate limitation in *Scenedesmus* in chemostat experiments.

As an additional consideration, in mixed algal populations, N_2-fixing blue-green algae can be limited by phosphorus, while non-N_2-fixing algae can be limited by nitrogen, simultaneously in the same lake. In addition, Torrey & Lee (1976) have reported that N_2-fixing algae will not always fix atmospheric nitrogen, even in waterbodies with low concentrations of available nitrogen. Luxury uptake of phosphorus by algae, under conditions of phosphorus excess in a waterbody (Gerloff & Skoog, 1954) can also confuse the situation.

Nevertheless, although the N:P ratio may vary from species to species, wide ranges in algal N:P ratios are not usually found in natural waters. Overall, a mass ratio of the biologically available nutrients of about 7-8N:1P appears to be a reasonable approximate boundary for defining the potential limiting nutrient, recognizing that mass ratios between approximately 5-10N:1P have been used as rough boundary values by others (Rast & Lee, 1978). Schindler & Fee (1974), for example, used a mass ratio of 8N:1P to define the limiting nutrient in their whole-lake experiments in the Experimental Lakes Area of Canada. Fuhs *et al.* (1972) and Rast & Lee (1978) provide further discussion of the ranges of nutrient limitation found with various types of algae.

Primary attention in assessing the limiting nutrient should be given to the period in which excessive algal biomass levels can significantly affect human use of the water resource. This usually is the growing season, the period of maximum algal biomass. It can be misleading to use an N:P ratio measured during other times of the year to determine the potential limiting nutrient during the growing season, since the in-lake nutrient concentrations can change significantly over the annual cycle. As reported by Rast & Lee (1978), an N:P mass ratio based on the annual cycle can give a different picture of potential limitation than one based solely on growing season conditions. Furthermore, it must be recognized that the N:P ratio will only indicate potential nutrient limitation, since this approach assumes that algae will continue to utilize nutrients in the above-cited reference ratio, and that the supply of nutrients to the waterbody will not change significantly over the annual cycle.

Because of these factors, one always should use both the absolute concentrations and the ratios of the biologically available forms of the nutrient in assessing the potential limiting nutrient. The N:P ratio should never be used alone to assess nutrient limitation in a waterbody. This is because an in-lake ratio of these two elements can always be calculated, even if both are present in excessive quantities. Obviously, an N:P ratio calculated for a waterbody containing excessive (i.e. non-limiting) quantities of both phosphorus and nitrogen will have little meaning in terms of identifying appropriate nutrient control targets.

The N:P ratio in a waterbody can be useful as a diagnostic tool for assessing the types of algae likely to exist under different nutrient conditions. V.H. Smith (1983), for example, related nutrient ratios to the concept of resource competition as a major factor affecting phytoplankton community structure. He reported that low total N: total P ratios appear to favor blue-green algal dominance in natural lakes in the temperate zone. Blue-green algae were rare in such lakes when the total N:total P mass ratio was greater than about 29:1. The relative fraction of blue-green algae increased proportionately, as the ratio decreased below this boundary value. Barica *et al.* (1980) added nitrogen to enclosures in small prairie lakes and ponds to manipulate the structure of algal populations. Prior to nitrogen addition, the algal populations were dominated by the N_2-fixing blue-green algae, *Aphanizomenon flos-aquae*. Following nitrogen addition, the *Aphanizomenon* biomass was significantly reduced and the development of green algae was stimulated. Such findings suggest that nutrient competition can ultimately affect both algal biomass and species composition. Thus, the enhancement or elimination of specific types of algae in a waterbody should be considered when assessing eutrophication control alternatives. Otherwise, unexpected results may occur in some situations (see Chapter 11).

Uhlmann (1982) presented a simple, graphical approach for using the N:P ratio to assess nutrient limitation. The deviation of the reference atomic ratio of 106C:16N:1P from the ratio of the measured concentrations can be used to calculate a 'dimensionless' concentration of dissolved nitrogen or phosphorus (N^* and P^*, respectively, in Figure 4.2). The reference atomic ratio of 106C:16N:1P corresponds to a value of 1 in Figure 4.2 (i.e. $C^* = 1$,

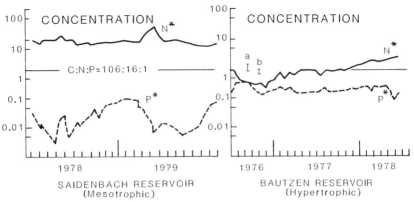

Figure 4.2 Dimensionless concentration of inorganic nitrogen (N*) and dissolved inorganic phosphorus (P*) in the upper water layer of two reservoirs of different trophic state (see text for description of terms; modified from Uhlmann, 1982)

N* = 1 and P* = 1). A nitrogen or phosphorus value > 1 signifies that the nutrient is present in excess quantities, while a value < 1 signifies that the nutrient is deficient. Thus, a ten-fold excess in dissolved nitrogen would give an N* value of 10 (i.e. C* = 1, N* = 10 and P* = 1). This corresponds to an atomic ratio of 106C:160N:1P instead of the reference value of 106C:16N:1P. Conversely, if the phosphorus level were deficient by a factor of ten relative to the other nutrients, this would give a P* value of 0.1 (i.e. C* = 1, N* = 1 and P* = 0.1), which corresponds to an atomic ratio of 106C:16N:0.1P).

Using this approach, a graphical representation of the seasonal changes in nitrogen and phosphorus can be a simple way to attempt to identify the algal growth-limiting nutrient. In Figure 4.2, the mesotrophic Saidenbach Reservoir has an N* value of about 10 throughout the year, indicating that nitrogen levels exceed algal needs by one order of magnitude. The reservoir exhibits a shortage of dissolved reactive phosphorus, with a P* value ranging between 0.1–0.01. Thus, the phosphorus levels are one to two orders of magnitude below algal needs. By contrast, in the hypertrophic Bautzen Reservoir, the phosphorus deficiency is much less severe. In fact, the dissolved nitrogen in the reservoir appears to be at growth-limiting levels at times (e.g. when N* < 1.0)

The use of algal bioassays to assess nutrient limitation

If one cannot identify the limiting nutrient based on the in-lake nutrient concentrations and ratios, or if one simply wishes to substantiate conclusions based on in-lake nutrient measurements, algal bioassays can also be used to assess the limiting nutrient. Algal bioassays basically consist of adding known quantities of nutrients (and/or trace elements) to standard algal cultures or natural algal populations under optimal conditions of light and temperature. If one of the nutrients is limiting maximum algal biomass levels, its addition to the algal culture would normally result in an increase in algal biomass levels, proportional to the quantity of added nutrients. By using standard laboratory algal cultures in combination with natural lake or reservoir waters, one also can determine whether or not algal growth-inhibiting substances (e.g., toxic chemicals) are present in a waterbody.

A fairly standard algal bioassay involves taking water samples from the surface waters during the period when algal blooms are affecting water use. Particles (including algal cells) are normally removed from the test waters by filtration. Uniform quantities of standard test algae (e.g. *Selenastrum*) and a series of known quantities ('spikes') of biologically available phosphorus, nitrogen and other suspected limiting chemical factors are then added to the test water samples. Untreated algal cultures serve as control cultures. A series of standard solutions is also prepared, identical to the spiked test cultures, except that distilled laboratory water is used in place of the pretreated lake/reservoir water. All the cultures are then incubated under optimal light and temperature conditions for several weeks. The algal biomass in the cultures is monitored until a growth plateau is attained.

The algal cultures are then examined to see how the nutrient additions, either singly or in combination, have affected the algal biomass levels. For example, if the addition of phosphorus, or phosphorus plus nitrogen, produced a larger algal biomass in the treated samples than in the untreated, control cultures, and the nitrogen addition resulted in little or no growth response, then phosphorus is likely to be the limiting nutrient. The opposite would be true if nitrogen, or nitrogen plus phosphorus, produced a growth response, while phosphorus addition alone did not. Little or no growth response to any nutrient addition would suggest that neither nutrient was the growth-limiting factor. Furthermore, if the algal growth response in the test water cultures was less than in the standard cultures (which contain the same quantities of nutrients and algae as the test cultures, but use distilled water in place of the test water), one must consider the possibility that the waterbody from which the test waters were taken may contain substances (e.g. chemicals) toxic to algae.

As a practical matter, one should only use small quantities of nutrients in the nutrient spikes. The use of high nutrient level spikes could artificially induce nitrogen or phosphorus limitation in the algal test cultures. Further, one should recognize that the optimal light and temperature conditions in laboratory cultures may mask actual light or temperature limitation in the natural environment.

A disadvantage of algal bioassays, however, is that they do not consider nutrient recycling in the waterbody (which may be significant in some cases). Thus, the results obtained refer to maximum attainable algal biomass, not to total algal growth (Golterman, 1975). Further, Golterman noted that algal bioassays show what will happen in the short term as one adds nutrients to algal cultures. Bioassays may give erroneous results, however, in regard to the longer term, especially under field conditions. Lean & Pick (1981), for example, provide evidence that the ^{14}C bioassay used to determine limiting nutrients in natural waters might provide misleading conclusions. Their preliminary studies showed that, when phosphorus-deficient algal populations were exposed to higher levels of phosphorus, photosynthesis was temporarily reduced, rather than enhanced. Contrary to previous assumptions, this produced a temporary depression of the uptake rate of ^{14}C by algae, rather than a rapid increase. Healey (1973) made somewhat similar observations in his earlier work on algal physiology. This decreased uptake rate could be interpreted erroneously to mean that the test algae exhibited no response to the phosphorus additions, thereby leading to inaccurate conclusions regarding nutrient limitation.

Nevertheless, when properly used and interpreted, algal bioassays can provide useful information about the nutrient status of lakes and reservoirs. Analytical details regarding algal bioassays can be found in several references, including Lund et al. (1971), United States Environmental Protection Agency (1971; 1974), Powers et al. (1972), Vollenweider (1974), Maloney & Miller (1975), Sridharan & Lee (1977), Miller et al. (1978) and Wetzel & Likens (1979).

The use of physiological indicators to assess nutrient limitation

If in-lake nutrient concentrations/ratios, and algal bioassays, are inadequate to define clearly the limiting nutrient, or if the results of such tests are contradictory or misleading, the physiological and morphological properties of algal cells also can provide useful information regarding algal nutrient limitation. However, obtaining this type of information usually requires more sophisticated resources than the simpler approaches discussed above. Nevertheless, where such techniques are feasible, they often provide more detailed knowledge of the physiology of algal populations in lakes and reservoirs than the simpler approaches.

Healey (1973) earlier provided an extensive review of physiological indicators of nutrient deficiency for a number of algal species. He reported that general effects of nutrient deficiency on algal populations included a decrease in algal cell content of chlorophyll, proteins and nucleic acids, as well as decreased photosynthesis and carbon assimilation rates. At the same time, there was an increase in cellular carbohydrate content, an increased cellular uptake rate of the deficient nutrient, and an increased ability of the algae to use organic forms of the deficient nutrient. There was also an increase in cell size (in some species) and in alkaline phosphatase activity for phosphorus-deficient algae. Fuhs et al. (1972) reported similar findings for phosphorus-limited diatoms in Lake Canadarago, New York. Healey

& Hendzel (1975, 1980) provide a summary of the approximate boundary values for various physiological indicators (Figure 4.3).

In a study of small prairie lakes, Healey & Hendzel (1976) also found that the cellular content of RNA, chlorophyll, phosphorus and nitrogen, and the protein: carbohydrate ratio increased during the early part of the algal growth phase. However, these variables decreased later in the growth phase as phosphorus deficiency continued.

To cite other examples, Tilman et al. (1976) showed that the number of cells/colony in steady-state populations of *Asterionella formosa* varied according to the nutrient status. With phosphorus limitation, the number of cells/colony decreased linearly in proportion to growth rate, to less than two cells/colony at very low growth rates. The inverse situation occurred with silicate limitation, with exponential increase to > 20 cells/colony at low growth rates. Rhee (1978) grew *Scenedesmus sp.* in phosphorus-deficient waters and reported that the cellular amino acid and RNA content remained constant, regardless of the N:P ratio, although excess nitrogen accumulated in the protein. The cellular carbon content was higher under phosphorus-limited conditions than under nitrogen-limited conditions, even though the assimilation rate was constant in both cases. R.E.H. Smith & Kalff (1981), using continuous flow cultures, also showed that phosphatase activity (an indication of phosphorus limitation) varied inversely with growth rate under phosphorus limitation.

In another study, Chiaudani & Vighi (1982) presented a multi-step approach for identifying the limiting nutrient in eutrophic Adriatic coastal waters. They assessed the utility of nutrient concentrations and ratios, algal bioassays, the assimilation coefficient, the ^{14}C-dark uptake: photosynthetic ^{14}C-uptake ratio, the ammonium ion enhancement ratio, the chlorophyll content per unit of algal biomass, and the alkaline phosphatase activity. Their study showed that the ammonium ion enhancement ratio and the alkaline phosphatase activity were the best indicators of nitrogen and phosphorus deficiency, respectively. While the other parameters were useful for assessing general nutrient deficiency, they were not as useful for identifying the specific deficient nutrient.

Zevenboom et al. (1982) presented a contrary view of the usefulness of some physiological indicators, at least for hypertrophic waterbodies. Using chemostat cultures of the blue-green algae, *Oscillatoria agardhii*, they concluded that cellular nutrient and pigment contents were not good physiological indicators of nutrient limitation in their hypertrophic systems, unless one also had a prior knowledge of growth rates and light conditions. In addition, neither the extracellular or intracellular N:P ratios appeared to be good indicators of nutrient limitation in their test system. However, they also reported that the uptake capacity value of their test algae for the deficient nutrient appeared to be a useful indicator of the algal growth-rate limiting factor. Under hypertrophic African conditions, Robarts & Zohary

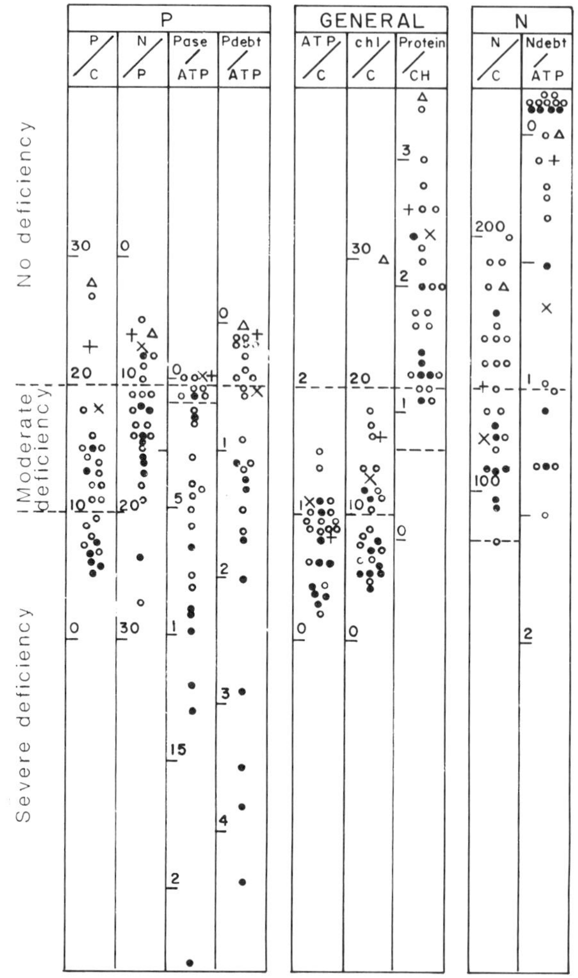

Explanation of terms: P and N headings = phosphorus and nitrogen deficiency, respectively; General = indicators which cannot be ascribed with certainty to either N- or P-deficiency. The boundary values between severe, moderate and no deficiency are based on measurements on algal cultures ● = average of 4–5 summer epilimnetic samplings in each of 9 ELA lakes; o = average of 2 mid-summer samplings in each of 16 small prairie lakes (Erickson district); X = average of early, mid and late summer samplings at 6 stations distributed throughout Southern Indian Lake; + = average of early, mid and late summer sampling of Notigi Reservoir (east basin); △ = average of 4 mid-summer samplings from the south basin of Lake Winnipeg. Composition ratios are μg particulate, P, N, ATP or chlorophyll a (chl)/mg particulate C; μg particulate N/μg particulate P; μg protein/μg carbohydrate (CH). Nutrient debts are μmoles ammonium or phosphate removed in 24-hour darkness/μg particulate ATP. Alkaline phosphatase activity (P_{ase}) is μmole o-methylfluorescein phosphate hydrolyzed/h.μg particulate ATP. The group of data points at the top of the N debt column indicates a net release of ammonium. No attempt is made to show the distribution of the actual values.

Figure 4.3 Mean summer epilimnetic values for various physiological indicators of nitrogen and phosphorus deficiency in several central Canadian lakes (modified from Healey & Hendzel, 1980)

(1984) reported that the colony size of *Microcystis aeruginosa* was a function of mixing, with small colonies being produced in a waterbody under well-mixed conditions. However, whether or not these various findings, based on hypertrophic conditions, also apply to less productive lakes and reservoirs is not clear.

Overall, these various studies suggest that many algal species appear to exhibit similar physiological responses to nutrient deficiency. Consequently, the physiological and morphological characteristics of algae, and their responses to nutrient-limited conditions, if properly interpreted, can provide useful information for assessing nutrient limitation in specific cases. It is also clear, however, that knowledge regarding nutrient limitation is far from complete. Thus, a degree of caution is always prudent before assuming a given methodology will be reliable in all cases.

CHAPTER 5

FACTORS AND PROCESSES AFFECTING THE DEGREE OF EUTROPHICATION

FACTORS RELATED TO THE DRAINAGE BASIN

Natural factors

Natural factors which can affect the degree of eutrophication in lakes and reservoirs relate mainly to the drainage basin and include location, climate, hydrology, geology and the physiography and geochemistry of the drainage basin (Figure 5.1). Each of these factors can significantly influence the input of nutrients to a waterbody and, hence, its overall biological productivity. A variety of cultural factors also may be simultaneously affecting the productivity of a waterbody, as discussed in the following sections.

Climate. Climate can influence lake productivity by affecting the annual energy and water input to the waterbody, the hydrology of the catchment area and flushing rate of the waterbody, and the nutrient and sediment transport to the waterbody (Barnes & Mann, 1980). The type and patterns of climate can differ as a function of location (latitude, longitude) and altitude. In turn, the climate can affect the annual water temperature, length of growing season, direction and velocity of winds, the quantity of precipitation and the thermal structure of a waterbody. The availability of solar energy is a very important factor controlling phytoplankton production during the growing season.

Lake productivity (phytoplankton photosynthesis) and both lake latitude and altitude appear to show a strong inverse relationship. Thus, the closer a waterbody is to the Equator, generally the greater its annual productivity (although during the growing season, even sub-arctic waterbodies can be highly productive). Of 93 physical variables examined in the International Biological Programme (IBP), Brylinsky & Mann (1973) showed that the latitude of a waterbody (a parameter which integrates the effects of day

Control of eutrophication

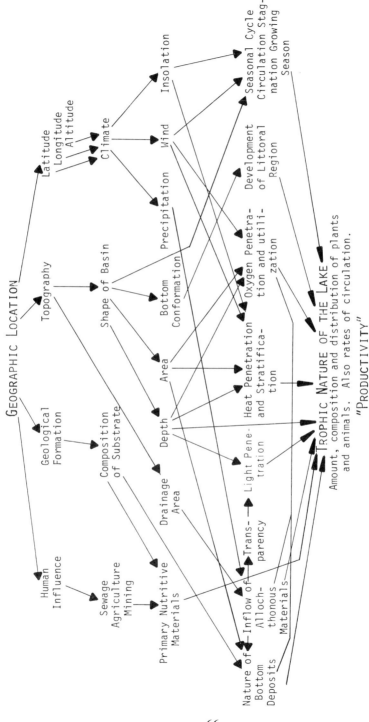

Figure 5.1 Interrelationships of factors affecting the metabolism of lakes and reservoirs (from Rawson, 1939, as modified by Stewart & Rohlich, 1967)

length, angle of sunlight, length of growing season, and air and water temperature) was the variable most strongly correlated with phytoplankton productivity data from lakes studied on a global scale.

In equatorial regions, lakes can receive a relatively constant amount of sunlight energy throughout the year, while the amount of solar radiation reaching the lake's surface in temperate regions usually changes markedly with the seasons. By contrast, in polar regions, energy inputs can drop to zero for part of the year. Thus, possible reasons for the high correlation between phytoplankton photosynthesis and latitude include the availability of light energy, and perhaps the greater rate at which nutrients are recycled at higher water temperatures.

Straškraba (1980) has examined the effects of variables related to the latitude of a waterbody on phytoplankton productivity and concluded that the effects can be substantial. In addition to the latitude-related variables which can directly influence phytoplankton productivity (e.g. solar radiation and temperature), the latitude of a waterbody can also affect the thermal stratification and mixing conditions in lakes and reservoirs. Further, Straškraba reported that the higher intensity light levels in tropical waterbodies can inhibit phytoplankton productivity in the surface waters. Overall, the net effect is that relatively less light may be available for phytoplankton productivity in deep lakes in the equatorial region. However, the potential impact of other variables (e.g. water transparency) obscure this relationship to phytoplankton productivity in these waters.

This climate-productivity relationship also extends to higher trophic levels, via food chain relationships, as measured by commercial harvest data. Solar input provides light energy for primary production, as well as heat energy, both of which can accelerate the rate of in-lake metabolic processes. For example, analogous to the relation for phytoplankton production, commercial shrimp yields also were inversely correlated with latitude (Turner, 1977). Further, for a number of intensively-fished lakes located between 62° N and 15° S latitude, climate accounted for 74% of the observed variability in the maximum sustained fish yield (Schlesinger & Regier, 1982). Such factors can also extend the growing season and increase fish productivity through more rapid growth and/or decreased generation times (Ryder, 1982). Fish yield and other forms of aquaculture are discussed further in Chapter 10.

Hydrology. Climate can also affect hydrology in drainage basins which have experienced human development. Generally, for a given drainage basin, the greater the rainfall and/or snowfall, the greater the quantity of water and nutrients transported to a waterbody over the annual cycle. However, the transport of sediments and associated nutrients may not continue to increase indefinitely with increasing rainfall. Although there is an increasing amount of water available with increasing rainfall to erode soil and transport it to a waterbody, the density of vegetation also increases with increasing rainfall, thereby acting to prevent erosion. For example, in areas receiving less than 120 mm annual rainfall, the rainfall is usually not sufficient to seriously erode and transport soil, and vegetation is usually absent. About 300 mm annual precipitation is sufficient to remove soil, but not enough to support a continuous, dense vegetation cover. Consequently, sediment erosion and transport is usually maximal in this range. Annual precipitation above 750 mm usually results in sufficient vegetation to reduce erosion (Satterlund, 1972). In forested and agricultural areas, for example, annual erosion rates were about 0.01 and 0.05 mm, respectively, while annual erosion rates of 10-20 mm have been reported for the northcentral plains of the United States (Sly, 1978).

In addition to the quantity of rainfall over the annual cycle, the timing of the rainfall is an important factor related to nutrient inputs to lakes and reservoirs from soil erosion. The greatest input of erosional material to waterbodies normally coincides with the temporal rainfall pattern. In temperate zones, this is usually the rainy period in the spring. In tropical/subtropical regions, the highest levels of biological productivity are usually seen in lakes and reservoirs two to three months after the rainy season (and maximum annual input of erosional materials). However, one also must consider that certain land usage factors can enhance the erosion process. For example, there can also be a large input of erosional material to a waterbody in the autumn in agricultural regions characterized by plowing activities during this period. These activities can leave the soils exposed to rainfall action. In such cases, there can be significant inputs of sediment and erosion-associated materials to lakes and reservoirs during the autumn and winter.

Geology and physiography of catchment area. A lake and its catchment area are a basic ecosystem unit (see Figures 7.5 and 7.6), since the terrestrial and aquatic portions of the watershed are intimately linked by movement of materials from land to water (Likens & Bormann, 1974). Consequently, the chemical composition of lake waters is also greatly influenced by the geological composition, size and topography of the drainage basin. Since nearly all the ions in a lake originate in its surrounding drainage basin, the mineral content of freshwaters can vary within a wide range, due to differences in the regional geochemical and climatic conditions.

As mentioned in the previous chapter, the belief that the magnitude of the external phosphorus and nitrogen supply is a primary factor in the differences seen in lake and reservoir productivity has received widespread acceptance in recent years (Schindler, 1971a, b; Schindler & Fee, 1974; Rast & Lee, 1978; Ryding, 1980; V.H. Smith & Shapiro, 1981; OECD, 1982). In watersheds exhibiting little or no cultural impact, phosphorus is usually supplied to a waterbody by direct atmospheric precipitation and by weathering of the watershed geology and subsequent runoff to the waterbody. Phosphate (in the form of apatite) occurs in igneous rocks in the range of 0.07-0.13%; volcanic rocks have intermediate concentrations; and sedimentary rocks are generally highest in phosphates (Golterman, 1973). In the absence of human influences, the nutrients supplied to a waterbody from precipitation, and from soils and vegetation in a given basin are often directly proportional to the total catchment area, and inversely proportional to the volume of the lake (Schindler, 1971a). Furthermore, for undisturbed watersheds, with similar ratios of catchment area to lake volume, and with other factors being equal, differences in the nutrient regime are usually related to the soil fertility in the watershed.

Several studies of regional limnology demonstrate clearly the relationship between the characteristics of the catchment area and the mineral composition of drainage streams and lake waters. Examples are the studies of Deevey (1940), Moyle (1956), Rawson (1960), J.R. Jones (1977), PLUARG (1978a), and J.R. Jones & Bachmann (1978). Important causative factors include the composition of the underlying rock structure and the type of soil in the watershed. For example, in northeastern Minnesota (USA), the surface geology is dominated by igneous and metamorphic rock, relatively resistant to dissolution and erosion processes. As a result, streams draining this region usually exhibit inorganic nitrogen concentrations $< 200\,\mu g/l$ and total phosphorus concentrations $< 15\,\mu g/l$ (although the biological activity of soils can also affect the nitrogen characteristics of waters, as discussed below). Since the lakes in this region receives an abundance of water and usually have active outlets, the lakes have low concentrations of dissolved salts, phosphorus and nitrogen, and are usually oligotrophic in character. By contrast, in southwestern Minnesota, the catchment basins are composed of mineral-rich glacial drift materials. Evaporation exceeds precipitation in this area and the lakes often have intermittent outlets that function only during wet years, thereby allowing the accumulation of salts in the lakes. As a result, streams draining this region generally have total inorganic nitrogen concentrations $> 1500\,\mu g/l$ and total phosphorus concentrations $> 100\,\mu g/l$. Likewise, the lake waters are rich in dissolved salts and nutrients, and show an enhanced productivity. Lakes and streams located between these areas tend to have waters of intermediate quality (Omernick, 1977).

J. R. Jones & Bachmann (1978) showed that waterbodies located in the young glacial soils of northern Iowa (USA) exhibited specific conductance values (a measure of the content of dissolved ions) around $400\,\mu S/cm$, while waterbodies in the older soils of southern Iowa averaged about $230\,\mu S/cm$. Specific conductance values may range from $< 50\,\mu S/cm$ in dilute waters in tropical regions to $> 5000\,\mu S/cm$ in waterbodies in arid tropical settings, e.g. East African Rift lakes (Talling & Talling, 1965).

Deevey (1940) showed that lakes located in soluble sedimentary and igneous rock basins in Connecticut (USA) generally exhibited elevated phytoplankton levels, while lakes located in metamorphic formations showed little productivity. The phosphorus loads in Canadian streams draining igneous watersheds of volcanic origin averaged about 15 times that of streams draining igneous watersheds of plutonic geological origin (Dillon & Kirchner, 1975).

The natural loss of nutrients from a watershed can also be affected by the presence of an 'upstream' waterbody in the watershed. This is because lakes and reservoirs usually retain a portion of their annual nutrient inputs in their bottom sediments, thereby acting as nutrient sinks over the annual cycle. Consequently, a lake or reservoir which receives a large proportion of its water input from an upstream waterbody in the same drainage basin usually has a smaller nutrient input than might be expected on the basis of geology, vegetation and climate (Dillon & Kirchner, 1975).

The physiography or morphology of the drainage basin can affect the nutrient input in several ways. For example, the nutrient inputs to lakes and reservoirs will usually be greater in watersheds with steeper slopes than in those with shallow slopes (Dugdale & Dugdale, 1961; Mackenthun et al., 1964; PLUARG, 1978a). Furthermore, the morphology can affect the hydrodynamics in a waterbody by affecting the wind patterns in the basin. A waterbody in a relatively flat catchment area with little tree cover will usually experience more wind-induced mixing of the water column than one with hills or significant cover near the waterbody (Brezonik, 1969; PLUARG, 1978a). Furthermore, the nutrient load to a waterbody is a function of the product of the inflow nutrient concentration multiplied by the inflow water volume. Therefore, an increased lake/reservoir inflow volume will result in an increased nutrient load over time even if the influent nutrient concentration remains unchanged. Thus, if both are located on the same river, a smaller lake or reservoir will receive a larger nutrient load over the same period of time than a larger waterbody, for the same water inflow.

In specific reference to nitrogen cycles, a gradual increase in the nitrate concentration in rainfall can produce increased nitrate concentrations in streams, lakes and reservoirs over time. The nitrogen input to a waterbody can also be affected to a significant degree by biological (primarily microbial and plant) activity in the soil. For example, biological nitrification and denitrification can affect the nitrogen flux in a watershed. Further, nitrate levels are negatively related to the percentage of marshland area within the watershed. Marshlands are areas of anaerobic decomposition, resulting in denitrification and the loss of nitrogen in the form of nitrogen gas (J.R. Jones et al., 1976). The soil horizon which contains humus is an important reservoir for nitrogen for a watershed. A freezing and thawing of the upper soil can result in increased nitrate release from the humus (Likens et al., 1977).

It is mentioned that there has been a gradual increase over time in the nitrate content of surface and ground water in many places around the world. This increase, especially in Europe, is related primarily to an increased usage of mineral nitrogen fertilizers (e.g. see Prochazkova, 1975).

Anthropogenic factors

Point sources. Point sources in the drainage basin (e.g. municipal wastewater treatment plants) can be the major source of the aquatic plant nutrient load to a waterbody. Lake Zurich, in the foothills of the European Alps, provides an example of the effects of sewage effluent on the trophic condition of a waterbody. This lake has two distinct basins, separated by a narrow passage. The deeper basin, originally oligotrophic in character, became strongly eutrophic for a time because of the input of urban sewage effluents from communities along the lake shoreline (this situation has since been alleviated by advanced wastewater treatment for phosphorus removal from sewage waters). In contrast, the shallow basin receives no major urban sewage effluents and, as a result, has retained its oligotrophic character (Hasler, 1947).

Lake Washington (USA) is an example of a lake which experienced severe phytoplankton problems resulting from sewage inputs. Following complete diversion of the sewage from the lake, it has exhibited a significant recovery (Figure 5.2) from the symptoms of eutrophication (Edmondson, 1970, 1972). It has been suggested that a sewage treatment plant effluent phosphorus restriction of 1 mg P/l would reduce concentrations of this nutrient in the central and eastern basins of Lake Erie and Lake Ontario from >20 $\mu g/l$ to about 10 $\mu g/l$ (Chapra & Robertson, 1977; Vallentyne & Thomas, 1978).

Land usage and other non-point source factors. While geologic and physiographic factors establish a regional potential for affecting lake water quality in most cases, human alterations and disturbances of a watershed can result in greater nutrient export to a receiving waterbody than can natural factors. For example, differences in land use patterns and fertilizer usage in the drainage basin can result in significant differences in the chemical composition of runoff waters from these areas, as shown in a binational, land use-water quality study conducted in the North American Great Lakes Basin by the United States and Canada (PLUARG, 1978a; also see Chapter 7). Although a wide range in nutrient unit area loads were found in this study, a higher nutrient load was correlated with increased urban and agricultural land usage, in contrast to the nutrient load from relatively undisturbed forested or wooded areas.

In one example for the Ozark Plateau region of Missouri (USA), the nutrient concentration in tributaries draining the watershed was positively correlated with an increasing intensity of land usage in the watershed

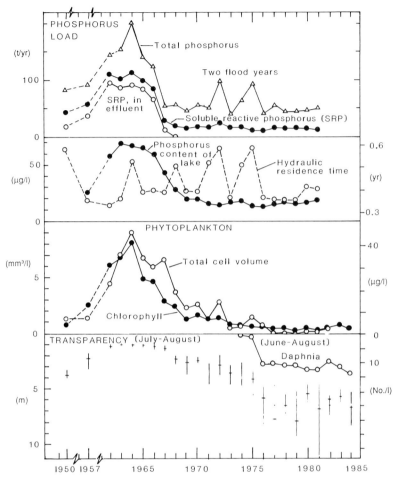

Figure 5.2 The effects of sewage inflow and its subsequent diversion from Lake Washington (modified from Edmondson, 1985)

(Table 5.1). Strong correlations were found between the percentage of pasture area in a watershed and the concentrations of most parameters in drainage streams. Such studies show that, at least in temperate regions, ion concentrations are generally lowest in streams draining forested watersheds, intermediate in streams draining pastured watersheds and greatest in streams draining urban areas (Smart *et al.*, 1985). The effects of such non-point source nutrients derived from land usage (as well as municipal point sources) on the water quality of Lake McIlwaine, Zimbabwe, provide examples of these factors as they occur in tropical Africa (see J.A. Thornton, 1982).

Factors other than land usage also can affect the nutrient and other

Table 5.1 Mean nutrient concentrations in streams draining watershed of differing land usage in the Missouri Ozarks, U.S.A. (From Smart et al., 1985)

	Land use			
Variable	Urban ($n=3$)	Pasture ($n=3$)	Forest & pasture ($n=9$)	Forest ($n=3$)
Total phosphorus (μg P/l)	106	45.9	30.6	20.3
Total nitrogen (mg N/l)	11.5	3.4	1.5	0.9
Nitrate nitrogen (mg N/l)	1.8	1.1	0.25	0.01

n = number of samples

material contributions to receiving waters, especially from non-point sources. These factors include the efficiency of chemical cycling and hydrologic processes, as well as anthropogenic activities, in the drainage basin (see Chapter 7).

As an example of such factors, low chemical concentrations in streams draining forested areas are due in part to an efficient cycling of ions among components of the forest ecosystem (Likens et al., 1977). In undisturbed forest ecosystems, most mineralized nitrogen is recycled to vegetation. Removal of this vegetation can disrupt this cycling process, resulting in an accelerated loss of ions from the drainage basin to a lake in the basin. Destructive disturbance of the watershed can increase the soil temperature and moisture availability, and accelerate nitrogen mineralization at a time when the uptake of nitrogen by vegetation is greatly reduced. This enhanced nitrogen mineralization can then be lost to streams and ground waters (Vitousek et al., 1979). Further, the combined effects of removing forest vegetation and losses from agricultural activities can increase the chemical content of pasture streams (Timmons et al., 1970; J.R. Jones et al., 1976; Sharpley & Syers, 1981). In urban streams, man-induced nutrient inputs, impervious surfaces and a decreased contact time between the runoff water and the soil in urban areas account for the usually elevated chemical concentrations (Sartor et al., 1974).

Available data suggest that the loss of ions in a watershed is usually proportional to the degree of disturbance to the chemical cycling processes in the watershed. Thus, all other factors being equal, lakes located in watersheds with substantial amounts of agricultural or urban land usage are usually more eutrophic than lakes located in forested watersheds. For example, the export of total phosphorus from both igneous and sedimentary watersheds in Ontario, Canada, was significantly lower in forested watersheds than in watersheds containing both pasture and forest (Dillon &

Kirchner, 1975). Further, about 25% of the annual phosphorus input to Lake Erie is due to runoff from intensive agricultural areas–an amount equal to 1.5 times the annual phosphorus input from weathering processes in the basin (Chapra & Robertson, 1977; PLUARG, 1978a).

In the previously-cited PLUARG (1978a) study on land use pollution in the North American Great Lakes Basin, the fundamental factors affecting nutrient and sediment loads from non-point sources (via land runoff) included land form, land use intensity and materials usage. Land form characteristics refer to soil texture, soil chemistry, type of soils, drainage density and slope. Overall, the soil texture was found to be the single most important factor affecting nutrient inputs to the Great Lakes. Greatest surface runoff and associated nutrient and sediment inputs to the Great Lakes was correlated with fine-grained (clay) soils. Calcareous soils often produce high phosphorus concentrations in land runoff. Land with a high degree of channel development, or steep slopes, also exhibited increased water and nutrient runoff in the PLUARG study. Furthermore, intense land cultivation exposes soils to erosional forces, incuding runoff losses. Excessive fertilization was a major source of nutrients in land runoff from agricultural areas (PLUARG, 1978a). These various factors are discussed further in Chapter 7 in relation to estimation of nutrient loads.

FACTORS RELATED TO THE WATERBODY

Although the basic causes of cultural eutrophication are nearly always external to a waterbody, the characteristics of the waterbody itself can significantly modify (both positively and negatively) the effects of these basic causative factors. External nutrient loads or in-lake nutrient concentrations alone do not exclusively control the overall productivity of a waterbody. Other factors can affect the productivity indirectly, by affecting the distribution, availability and utilization of the nutrient inputs (Brezonik, 1969). The effect of a reduced nutrient input from the catchment area on the nutrient availability and algal productivity of a waterbody depends to a large degree on the physical and biotic structure of the waterbody. Thus, even with the same annual nutrient load, the responses of similar lakes can still be quite different. The reasons for such differing responses include internal nutrient cycling (particularly food-web structure and sediment regeneration) and specific lake basin properties such as morphology and in-lake hydrodynamics.

Morphology of the lake basin

The mean depth of both the whole waterbody and the hypolimnion can substantially affect the impacts of an increased nutrient load to a lake or reservoir. The oxygen content in the hypolimnion during periods of thermal stratification, and the related processes of water quality deterioration and nutrient regeneration in this water layer, depend to a large degree on the mean depth (z_H) of the hypolimnion. For the same quantity of phytoplankton biomass produced in the euphotic zone, the oxygen consumption per unit volume of the hypolimnetic water column (due in part to microbial decomposition of the algal biomass) usually will be much greater in waterbodies with small hypolimnetic mean depths than in those with large hypolimnetic mean depths (Figure 5.3).

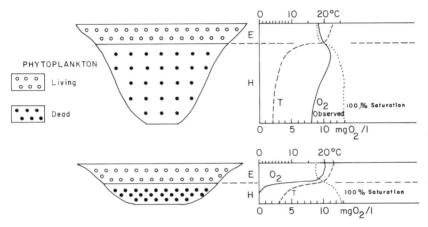

Explanation of terms: E = epilimnion; H = hypolimnion; T = temperature; dotted line = dissolved oxygen (O_2) concentration corresponding to 100 percent saturation at temperature T. It it assumed that the phytoplankton die and sink into the hypolimnion, where they undergo bacterial decomposition.

Figure 5.3 Vertical distribution of dissolved oxygen concentration at end of summer stratification period in lakes with different hypolimnetic volumes (after Thienemann, 1918, as modified by Uhlmann, 1979)

In many cases, the hypolimnetic oxygen consumption due to microbial decomposition (R_H; expressed as g O_2/m^3.day) can be approximated as follows:

$$R_H = L_H/z_H \tag{5.1}$$

where: L_H = areal surface loading of hypolimnion (g/m².day) with easily degradable organic material (e.g. decaying phytoplankton, expressed as oxygen equivalent COD); and

z_H = mean depth of hypolimnion (m).

If one considers L_H to be a constant, Equation 5.1 describes an inverse hyperbolic relationship, with R_H as the dependent variable and z_H as the independent variable.

The following example illustrates the depth-dependent relationship expressed in Equation 5.1:

Given an L_H value of 0.3 g/m².day (a realistic value for a eutrophic lake), calculate the hypolimnetic oxygen consumption, R_H, for waterbodies with hypolimnetic mean depths of 3 m and 9 m, respectively. Determine also if the hypolimnetic dissolved oxygen concentration at the end of a 122 day stagnation period will be at least 4 mg/l (often cited as the minimum concentration necessary to maintain fish life in the hypolimnion), assuming an initial hypolimnetic oxygen concentration of 11 mg/l at the beginning of the stagnation period:

(1) For the 122 day stagnation period, the maximum daily oxygen consumption rate to maintain a minimum oxygen concentration of 4 mg/l (assuming a constant consumption rate) is calculated as follows:

$$\frac{11 \text{ mg/l} - 4 \text{ mg/l}}{122 \text{ days}} = 0.05 \text{ mg/l.day}$$

(2) For a hypolimnetic mean depth of 3 m:
$R_H = L_H/z_H = (0.3 \text{ g/m}^2.\text{day})/(3 \text{ m}) = 0.1 \text{ g/m}^3.\text{day}$

For a hypolimnetic mean depth of 9 m:
$R_H = L_H/z_H = (0.3 \text{ g/m}^2.\text{day})/(9 \text{ m}) = 0.03 \text{ g/m}^3.\text{day}$.

These calculations show that the oxygen consumption rate becomes smaller per unit volume of water as the mean depth of the hypolimnion increases; it tends to become very small for z_H values of 10 m or greater. However, in meromictic or oligomictic lakes, in which the hypolimnion is not subjected to sufficient atmospheric aeration during the overturn periods, the hypolimnetic oxygen concentration at the end of the stagnation period can be very low or even zero, even if R_H is also very low.

Thus, there will usually be more oxygen in the hypolimnion of a deeper lake than in shallow ones of equal productivity. As pointed out earlier by Thienemann (1918) for Baltic lakes, a waterbody mean depth of greater than about 18 m appears to be a prerequisite for oligotrophic conditions (i.e. for an orthograde oxygen curve; see upper part of Figure 5.3). A mean depth of 15 m or greater is assumed to be characteristic of 'morphometrically' oligotrophic lakes (Technical Standard, 1982).

If one considers the hypolimnetic volumetric oxygen depletion rate in relation to the optical gradient, rather than to the thermal gradient, a more accurate calculation of R_H is usually possible.

As indicated earlier, the longitudinal shape of a waterbody can also affect its water quality. For example, with the creation of a reservoir, the impoundment of meandering rivers and their floodplains can often produce a long, narrow, dendritic waterbody which receives a majority of its water and nutrient inflows from a single tributary. As a result, such reservoirs often exhibit longitudinal gradients in water quality and trophic conditions (J.R. Jones & Novak, 1981; K.W. Thornton et al., 1982). Appropriate sampling techniques for such waterbodies are discussed in Chapter 8.

In-lake nutrient sources

With regard to controlling or reversing lake or reservoir eutrophication, the role of bottom sediments on the trophic status of the waterbody can be of major concern. The role of the sediments on the dynamics of phosphorus cycling between the sediments and water column is of particular interest in eutrophic waterbodies, especially shallow, homothermal ones. In oligotrophic lakes, a substantial part of the imported nutrients is retained in the sediments (i.e., the sediments act as a nutrient 'sink').

A net deposition of phosphorus into the bottom sediments also occurs annually in many eutrophic lakes and reservoirs. However, in overloaded lakes, in which the bottom sediments have become heavily enriched with phosphorus over time (Figure 5.4), the release of phosphorus from the sediment can exceed the flow into the sediments for periods of time in the summer, especially if the hypolimnion becomes anoxic in thermally-stratified waterbodies (e.g., see Ryding, 1981b). Furthermore, in lakes in which the external phosphorus load from domestic or industrial sewage has been drastically decreased in a short period of time, a net outflow of phosphorus from the waterbody throughout the entire annual cycle has been observed for a period of time (Ryding & Forsberg, 1977). Both of these situations will negate, at least temporarily, the effectiveness of eutrophication control measures based on reduction of the external phosphorus load.

The release of phosphorus from the bottom sediments back into the water column is a complicated process, involving the interaction of a number of physical, chemical and biological mechanisms. These mechanisms

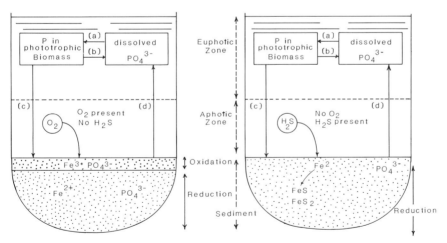

Figure 5.4 Mobilization of dissolved orthophosphate from bottom sediments of a soft-water lake as result of oxygen depletion and formation of hydrogen sulfide at mud–water interface (illustrating destruction of barrier layer of iron (III) hydroxide and adsorbed phosphate; after Ohle & Hayes, from Uhlmann, 1979)

include the mineralization of organic matter, desorption, dissolution of salts, ligand exchange equilibria, etc. (Figure 5.5).

A number of recent reports have focused on the main factors believed to regulate the release process (e.g., see Golterman *et al.*, 1983). Such factors as the redox conditions, the nitrate concentration, mineralization, gas bubble formation, bioturbation, effects of phytoplankton and macrophytes, different sediment characteristics, high pH values, diffusion and wind turbulence have been suggested as influencing phosphorus release rates from the sediments. However, recent experiences (Ryding, 1985), based on whole-lake experiments in a variety of lakes, emphasize that the dynamics and magnitude of sediment phosphorus release in shallow, eutrophic lakes are affected primarily by such physical factors as water temperature (controlling the seasonal variations) and water renewal (regulating the year-to-year differences). These factors, in turn, induce chemical and microbiological processes which regulate the exchange of substances between the sediments and the water column. Furthermore, wind can affect both seasonal and annual variations in sediment phosphorus release patterns.

In the initial state of lake eutrophication the remobilization of phosphorus is often retarded by the sorption and chemical bonding of phosphate to the bottom sediments, as indicated in the left side of Figure 5.4. However, the phosphorus release from sediments can become significant once the sediments are saturated with phosphate. The corresponding delay (δ_3 in Figure 5.6) can cover a period of many years, and may cause the public to conclude prematurely that a eutrophication control effort was unsuccessful (Ryding, 1981b). A long duration of δ_3 is expected for lakes with slow

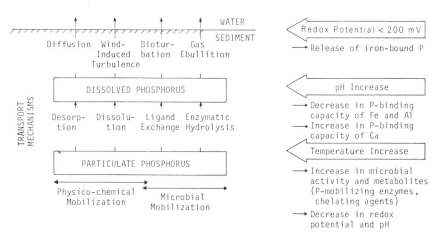

Figure 5.5 Diagrammatic representation of important processes and environmental factors affecting the release of phosphate from lake sediments (modified from Boström *et al.*, 1982)

Factors and processes affecting the degree of eutrophication

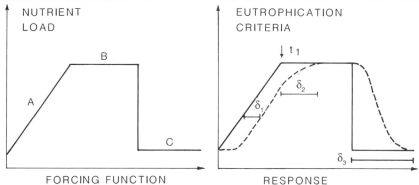

Explanation of terms: A = period of increasing phosphorus load; B = period of constant, high phosphorus input; C = period following step decrease in phosphorus load; δ_1, δ_2, δ_3 = response time of waterbody to A, B and C, respectively; see text for discussion.

Figure 5.6 Diagrammatic representation of the delay in the response of a lake to altered external phosphorus loads (from Uhlmann & Hrbáček, 1976)

flushing rates, and it may easily exceed the theoretical water residence time. In such cases, additional in-lake eutrophication control measures may be needed (see Chapter 9).

In contrast to sediment phosphorus, practical experience suggests that nitrogen does not normally exhibit a continuing remobilization from sediment once the external and internal phosphorus load is brought under control.

Flushing rate

The extent to which nutrients accumulate in a waterbody also depends greatly on both rainfall and flushing rate. For example, in closed lake basins in arid regions, dissolved phosphorus concentrations can be as high as levels in waste treatment lagoons, even with negligible cultural eutrophication in the drainage basin. In the following section, the inflow of water to a lake is assumed to be primarily overland flow from the drainage basin. However, if ground water inflow also is significant, this latter source should be quantified as accurately as possible.

If the inflow volume (Q) of a lake or reservoir is very high compared to its basin volume (V), phytoplankton can be flushed out of the waterbody before they can grow to nuisance levels. This can occur even in waterbodies containing high nutrient concentrations. This situation can be expressed in a general way as follows:

$$dx/dt = (u \cdot x) - (D \cdot x) \quad \text{with } u > D \tag{5.2}$$

where: x = biomass of the phytoplankton species under consideration;
t = time;
u = specific daily growth rate of phytoplankton; and
D = flushing or dilution rate.

Further, the flushing rate (D; 1/yr) can be defined generally as:

$$D = Q/V = 1/t_w \qquad (5.3)$$

where: t_w = theoretical water mean residence time (yr)
(= hydraulic residence time);
Q = discharge (m³/yr); and
V = lake volume (m³).

If the growth rate is less than the flushing rate (i.e., $u < D$), then $dx/dt < 0$ and, consequently, the phytoplankton will be flushed from the lake prior to its reaching nuisance levels.

Experience suggests that a hydraulic residence time (t_w) greater than about three days is a prerequisite for excessive phytoplankton growths. This is important to remember for the proper design both of sedimentation basins, where phytoplankton growth is undesirable, and of pre-reservoirs (also called pre-impoundments or cascade reservoirs in some countries) of drinking water reservoirs, where such growth is promoted for the control of non-point phosphorus sources (Benndorf & Pütz, 1985; also Chapter 9). The calculation of accurate phosphorus loads for lakes and reservoirs with rapid flushing rates is discussed in Chapter 7.

Light intensity

The mixing depth (z_m) in a lake is often taken to mean the thickness of the epilimnion. The mean light intensity (I) in the mixed layer can be calculated as follows (Stefan *et al.*, 1976):

$$I = [I_o/(\varepsilon \cdot z_m)] [1 - e^{-\varepsilon \cdot z_m}] \qquad (5.4)$$

where: I_o = surface light intensity;
ε = mean light attenuation coefficient (1/m); and
z_m = mixing depth (m).

As shown in Equation 5.4, the mean light intensity in a waterbody (I) will be low if I_o is low, or if the value of $\varepsilon \cdot z_m$ (equivalent to the extinction depth, d) is high. This condition can exist in a highly turbid or colored waterbody, or in a waterbody with a large mixing depth (z_m). This suggests that one can attempt to reduce phytoplankton growth by artificially mixing a waterbody (Oskam, 1978; see also Chapter 9). However, at a mean depth ($\bar{z}$) of less than about 10 m, the mixing depth (z_m) may actually become too small to reduce phytoplankton growths. In such a case, the growth of phytoplankton can actually be increased by artificial mixing.

Biological controls

One can also attempt to control phytoplankton biomass with the use of higher trophic level organisms, notably zooplankton. An extended version of Equation 5.2, which shows the significance of zooplankton grazing, is as follows (Uhlmann, 1971):

$$dx/dt = (u \cdot x) - (D + B + G) \cdot x \qquad (5.5)$$
$$\text{growth} \qquad \text{losses}$$

where: B = sedimentation coefficient (l/day); and
G = grazing coefficient (l/day).

The significance of zooplankton grazing to control phytoplankton biomass is highlighted in waste treatment lagoons, which do not normally contain fish. In such lagoons, zooplankton grazing of phytoplankton can reduce the phytoplankton biomass to very low levels, even if the lagoon contains very high concentrations of dissolved phosphorus and nitrogen (Uhlmann, 1958). Species of the 'water-flea' (*Daphnia*) are particularly effective in controlling phytoplankton densities, although rotifers and copepods (calanoids) can also be significant in some cases. In the same manner, one can control the small fish which selectively feed on filtering zooplankton (which help maintain a low phytoplankton density) in lakes and reservoirs by enhancing the populations of predatory fish (Hrbáček, 1969). This is an example of so-called 'biomanipulation' (Shapiro et al., 1975; Andersson et al., 1978; Benndorf et al., 1981). To be effective, however, this technique requires detailed knowledge of the food-web structure of a lake or reservoir, including the size distribution and productivity of the different fish species under consideration. Such information is still lacking for most waterbodies.

Macrophyte growths

Normally, submerged macrophytes and bottom algae cannot compete successfully with phytoplankton in lakes and reservoirs containing very high dissolved phosphorus and nitrogen concentrations. This is due to the shading effect of high algal densities and/or high turbidity on bottom plants. As a result, submerged macrophytes are not usually present in waterbodies exhibiting dense phytoplankton growths. One must be careful, therefore, that control measures designed to reduce phytoplankton biomass do not inadvertently produce light conditions favorable for the excessive growth of phytobenthos.

As an example, shallow lakes rich in nutrients can exhibit two different steady states of the food-web structure as a result of this phenomenon (Figure 5.7). The uncertainty associated with this 'bistability' has to be considered in lake management efforts. An increase in water transparency (due to phosphorus control, food-web manipulation or other control measures) can significantly improve the quality of water for drinking or industrial water supply. However, such measures often are not useful for lakes used for recreation or bathing if they result in the phytoplankton

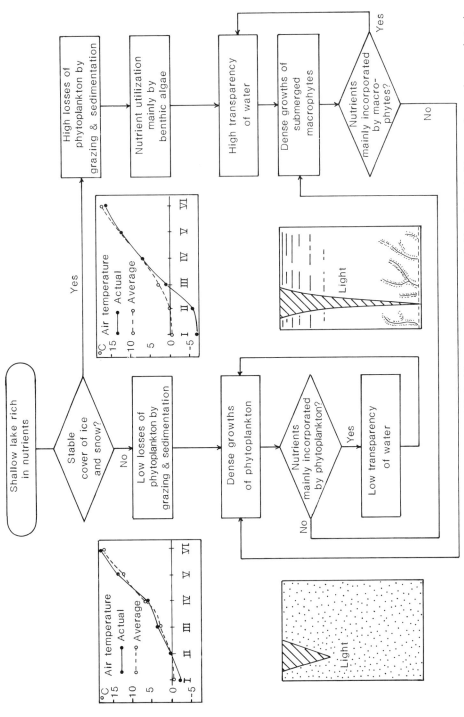

Figure 5.7 Schematic representation of possible causes of 'bistability' with respect to dominance of either phytoplankton or submersed macrophytes in a shallow, highly eutrophic lake (modified from Klapper, as cited by Uhlmann, 1980)

biomass being replaced by dense growths of cattails or other associated macrophytes or filamentous algae (Breck *et al.*, 1979).

Because rooted macrophytes depend more on the nutrient content of the bottom sediment than of the water column (Carignan & Kalff, 1980), allowances must be made for the fact that, in contrast to phytoplankton, macrophytes are relatively insensitive to control by nutrient flushing. Nevertheless, control of submersed macrophytes by herbivorous fish (mainly the grass carp, *Ctenophyoryngodon idellus*) has been successful in the temperate zone of Eurasia (Barthelmes, 1981).

It is also mentioned, however, that in eutrophic waterbodies, effective control programmes may have the effect of replacing macrophyte growths with higher phytoplankton concentrations. Thus, one must consider the overall impacts of controlling macrophytes versus controlling phytoplankton in developing effective eutrophication control programmes.

In lakes with very clear waters, sunlight can penetrate down into the hypolimnion. In such situations, both photosynthetic oxygen and 'sealing' of the mud/water interface by algae or other phototrophs (e.g. *Chara*) can be a buffer against the regeneration of dissolved nutrients in the hypolimnion.

Available models for assessing macrophyte-infested waterbodies are presented in Chapter 6, while procedures for reducing macrophytes, or for using them for food or commercial products, are discussed further in Chapter 10.

Cage fish farming

Cage fish farming involves the partitioning of areas in a lake or reservoir, usually with float-supported nets. These areas constitute protected cages or 'corrals', which are used for the production of fish (often salmonids or other high quality fish). At the beginning of the growing season, the cages are stocked with small, young fish and nutrient-rich food pellets are added at frequent intervals. The fish grow rapidly in the protected, caged areas and are usually harvested at the end of the growing season.

Cage fish farming is expanding in many countries, especially in Europe. Since the addition of the necessary nutrients for the fish (as well as the fish excretion products) can result in the accelerated eutrophication of a waterbody, the environmental impacts of cage fish farming have been studied in Sweden (see Enell, 1984). The quantity of nutrients added to a waterbody from cage fish farming depends on the density of the fish populations in the caged areas. The Swedish study indicated that, for each tonne of fish produced, the resultant nutrient input to the waterbody was 85–90 kg phosphorus and 12–13 kg nitrogen.

In addition, the nitrogen addition to the waterbody from the cage fish farming operations was about 85% in the dissolved form, mainly urea and ammonia from fish excretions. The remaining 15% was in particulate form, mainly from the loss of fish food pellets. In contrast, only 15–20% of the phosphorus input to the waterbody from the cage fish farm operations was in a dissolved form. The remainder of the phosphorus input was in a particulate form, which settled to the bottom sediments. In the Swedish study, however, about 5–10% of the sedimented phosphorus was regenerated to the water column, as a result of anoxic chemical conditions and biological processes in the bottom waters. Ackerfors and Enell (1989) provide details of more recent experiences in the field of cage fish farming.

Beveridge (1984) has recently examined the environmental impacts of the cage culture of fish in tropical regions.

The reader is referred to Chapter 10 for further discussion of the eutrophication process as it relates to enhancement of fish production and other forms of aquaculture.

CHAPTER 6

THE USE OF MODELS

GENERAL TYPES OF MODELS

In the simplest sense, a model is an approximation of a real world system. The object of experimenting with an approximation is to gain insight into the real world system, without having to duplicate it completely. The model is often represented by a series of mathematical expressions which attempt to describe the phenomena occurring within the system. The significant parameters that can affect the system, and the factors which influence them, are usually included within the mathematical expressions in some manner. Thus, the accuracy of a model reflects the state of knowledge regarding the system being modeled.

The mathematical description of an environmental system is usually the product of the individual describing the system. Thus, the type of mathematical model used in a given situation is dictated largely by the nature of the problem being assessed, and the type and complexity of the answers being sought. Biswas (1976) has offered general guidelines for developing models (including eutrophication models), which include the following:

1. Develop the simplest possible model for the system being modeled. The model simplicity should be dependent on the information being sought. As a general rule, water quality managers appear to prefer simple models that can be understood, even if the models are based on qualitative structures, broad assumptions and limited data. This is in preference to complex models which may be more scientifically rigorous, but which also may be more difficult to comprehend;

2. Do not develop generalized all-purpose models. Such models are often unwieldy, expensive to use and have large data requirements;

3. Try to involve the expected user of the model in its development. User involvement will assure familiarity with the working principles of the

model and increase its potential for acceptance and use as a management tool;

4. Model development and data collection should proceed in parallel;
5. Provide adequate documentation, so that others can use the model without personal instruction from the developer of the model;
6. Continually update the model as more data becomes available and/or as understanding of the system being modeled increases;
7. The model developer should attempt to become acquainted with the environment for which the model was intended, and should have a good working relationship with the managers and decision-makers who will be using the model.

Models used for understanding and controlling eutrophication can be classified as:

1. watershed (nutrient load) models;
2. waterbody models; and
3. management models.

The broad objective of watershed models in eutrophication control efforts is to provide estimates of the nutrient loads reaching a lake or reservoir. This is usually done by identifying and quantifying the nutrient contribution from sub-regions and sub-watersheds (the origin of the majority of the nutrient load; see Chapter 7). Such knowledge is a prerequisite for development of a cost-effective eutrophication control strategy based on nutrient reduction. Watershed models can be static or simulation models (i.e. focusing either on the average conditions or the actual temporal and spatial dynamics of the system being modeled, respectively). In many cases, the available or attainable data are insufficient for the application of simulation models. The use of static models is necessary in such cases. Fortunately, although not capable of simultaneously considering as many variables as simulation models, static models have been found to be both useful and accurate for many eutrophication assessment/prediction purposes. As an example, static models are often used to estimate the annual nutrient (phosphorus) load, a major determinant of both water quality and trophic state in lakes and reservoirs (see Chapter 7).

Both static and simulation watershed models can be used for descriptive and planning purposes. Descriptive models can provide a better understanding of a test system, while planning models can be used to assess alternative nutrient control options in the watershed (e.g. the efficiency of phosphorus removal at municipal wastewater treatment plants; see Chapter 9).

Eutrophication waterbody models range from simple empirical models to more detailed ecological models. Virtually all of them are based, at least in part, on nutrient mass balance considerations. These models often are used to estimate the water quality or biological 'response' of a lake or reservoir to nutrient load reductions. As with watershed models, waterbody models can be either static or simulation in nature. The same concerns expressed above with regard to simple versus complex watershed models also apply to waterbody models.

As an additional component, management models are often used to determine the 'optimal' (often defined as the most 'cost-effective') eutrophication control strategy for a given waterbody–watershed system. As discussed in Chapter 9, management models can incorporate information about necessary phosphorus load reductions to achieve a desired water quality or trophic condition, and allow the manager or decisionmaker to select between alternative control strategies. In general, management models are 'optimization' models, used to select the optimal combinations of all the variables considered. Optimization models often include a planning-type watershed model, a waterbody model, cost estimates of various nutrient control/removal measures, and associated constraints.

As a practical matter, only watershed and waterbody models will be discussed in this chapter. Because of the practical focus of this book, discussion will focus primarily on simpler models, which are usually the initial assessment tools in eutrophication control efforts. For further information on the topic of management models, however, the reader is referred to the reports of Biswas (1976), Somlyódy (1983) and Somlyódy & van Straten (1986).

WATERSHED MODELS

As mentioned above, watershed or drainage basin models can provide information on the nutrient sources to a waterbody, including the total nutrient input to be expected and the effects of the individual parameters controlling or regulating the nutrient input. Watershed models are particularly useful for estimating non-point source nutrient inputs. These models allow one to obtain estimates of source magnitudes and the effectiveness of control measures. They vary greatly in their complexity, data requirements and outputs. At present, the most practical models are based on relatively simple approaches.

The selection of an appropriate watershed model depends largely on the methods used to describe waterbody eutrophication. It also depends on the nature of the problem being analyzed, the types of answers sought, the extent of available data and resources, and the skills of the modeler. Since many of the simpler lake eutrophication models (see Section 6.3) are based on steady-state total phosphorus loadings, a watershed model which predicts the average annual total phosphorus inputs from point and non-point sources in the drainage basin is often sufficient. In contrast, more complex (and perhaps more realistic) models can be used to attempt to distinguish between biologically available and unavailable nutrients, and also to consider time variations in nutrient loadings. The latter types of watershed models should also be capable of predicting seasonal variations in dissolved and solid-phase nitrogen and phosphorus inputs to a waterbody.

Non-point source nutrients are contained in runoff and baseflow or ground water flow (see Chapter 7). The former is a rapid drainage response to precipitation or snowmelt events, usually reflected by increases in stream discharges during and shortly after an event. Baseflow is the drainage water which primarily moves through saturated soil zones, and its magnitude is influenced by the water table elevations in the drainage basin. Thus, during wet months, stream baseflows can be high, reflecting extensive ground water movement to channels. The nutrient content of runoff is a function of processes occurring on the soil surface, while the nutrient levels in baseflow are influenced by the nutrient status of the soil profile. Baseflow in arid climates, however, may consist solely of point source, treated municipal wastewater effluents.

As discussed in more detail in Chapter 7, non-point nutrient sources can be broadly classified as rural or urban. Rural sources include runoff and baseflow from both agricultural and forested areas. The magnitude of the nutrient contribution depends primarily on the character of the soil and the vegetative cover conditions. By contrast, urban non-point source pollution is largely a runoff phenomenon, strongly influenced by the extent of impervious surfaces in the drainage area. Consequently, urban non-point nutrient inputs to a waterbody are relatively more variable and sensitive to individual storm events than are rural sources. A reliable watershed model must be able to reflect these various nutrient sources and inputs to a lake or reservoir.

Empirical watershed models

The simplest models for estimating the nutrient load to a waterbody are of the general forms:

$$A = a_o + a_1 X_1 + a_2 X_2 + \ldots + a_m X_m \tag{6.1}$$

or

$$B = b_o + b_1 Y_1 + b_2 Y_2 + \ldots + b_m Y_m \tag{6.2}$$

where: A = average annual nutrient load to a waterbody (kg/yr);
X_1 = area (ha) of watershed with land use 1, 2, 3..., etc.;
B = average nutrient concentration in streamflow (mg/l);
Y_1 = fraction of watershed occupied by land use 1, 2, 3...etc; and
a_o, a_1, b_o, b_1, etc. = coefficients.

Equations 6.1 and 6.2 are used to predict the annual nutrient load. Specifically, Equation 6.1 provides an estimate of the annual nutrient load based on the watershed characteristics. An estimate of the annual nutrient load can be obtained by multiplying the annual average stream nutrient concentration (B) by the annual stream discharge. Thus, the units of a_j and b_j are kg/ha and mg/l, respectively.

Equations 6.1 and 6.2 are empirical, since the values of the coefficients are inferred from water quality sampling data. In the simplest application (i.e., $a_o = 0$), X_j are areas of different major land uses, and the a_j are known as 'nutrient export coefficients' (Rast & Lee, 1983) or 'unit loads' (Novotny & Chesters, 1981). Further discussion of unit area loads, and tables of representative unit area loads for various land use activities, are provided in Chapter 7. Comparable coefficients are also available for the concentration relationships in Equation 6.2 (Omernik, 1977).

Ideally, such coefficients are based on actual water quality sampling in small watersheds with a single predominant land use. The coefficients should be based on long-term sampling (including storm events) in a number of sub-watersheds with single land uses within the same geographic region as the waterbody of interest. However, since this approach often requires considerable effort, it is frequently not done in practice. The best compromise, therefore, is probably the identification of small, single land use catchments within the watershed of interest, and measurement of the total nutrient flux from each catchment for several years. This flux can then be divided by the number of years of monitoring and the catchment land area to produce nutrient export coefficients (unit area loads).

Since nutrient export coefficients are usually based only on broad land use categories, they may not be useful for identifying sub-classes of nutrient sources (e.g. a particular crop or soil type) or for evaluating alternative nutrient control options. However, Equations 6.1 and 6.2 can be derived from regression techniques, with the dependent variables representing a variety of watershed characteristics. Coote *et al.* (1979), for example, derived regression equations for dissolved and total nitrogen and phosphorus, based on two years of water quality data from ten Canadian agricultural watersheds. They measured twenty-four watershed parameters, including detailed land uses (row crops, corn, vegetables, etc.), nutrient sources (fertilizer and manure applications, soil extractable phosphorus, etc.) and physical characteristics (soil erosion and clay content, watershed slopes, etc.). These regression equations can be used to estimate the effects of such control alternatives as fertilizer reductions and erosion prevention on the average annual nutrient loads to waterbodies.

Empirical models, including nutrient export coefficients, are a popular and easily-used means of estimating annual nutrient inputs to lakes and reservoirs. However, because such models are usually based on sets of water quality measurements reflecting specific weather conditions, land use and management patterns in the monitored catchments, etc. their extrapolation to other watersheds may not be justified. Further, the desired long-term sampling programme for deriving the most accurate coefficients usually requires extensive monitoring efforts. Consequently, such empirical models may not provide as much detailed quantitative information on specific nutrient sources and/or control programmes as desired.

Nevertheless, empirical models involving nutrient export coefficients do allow one to make estimates of both runoff and baseflow nutrient loadings to a waterbody and, if properly used, can provide useful management information on nutrient inputs and source identification. Loading estimates derived in this manner are normally used with simple eutrophication models which require only average annual phosphorus input values for a waterbody. Nutrient export coefficients and their use in estimating the nutrient input to a waterbody are discussed in more detail in Chapter 7.

Simulation watershed models

A finer resolution of the nutrient loading to a waterbody usually requires models which simulate the physical and biochemical processes which can affect the nutrient sources. Conceptually, such models describe mathematically the water fluxes and associated nutrient flux from the land surface and soil profile. The complexity of specific watershed simulation models depends on the time interval of interest, and the extent to which important biochemical processes are considered in the model. Simple simulation models usually consider detailed (e.g. daily) time frames, and simulate the physical processes of water and sediment movements and chemical washoff. More complex simulation models consider even finer time intervals and more detailed biochemical phenomena. The following sections provide generalized examples of simple watershed simulation models.

Simple simulation models. These simple models usually consider distinct nutrient sources within a watershed, often distinguishing between the solid and dissolved nutrient sources. The general relationships are:

$$D_{kt} = d_{kt} Q_{kt} TD_{kt} \tag{6.3}$$

and

$$S_{kt} = s_{kt} X_k TS_{kt} \tag{6.4}$$

where: D_{kt} = dissolved nutrient loss (kg) in runoff or baseflow from area k during time period t;
S_{kt} = solid-phase nutrient loss (kg) in runoff or baseflow from area k during time period t;
Q_{kt} = runoff or baseflow (m^3);

d_{kt} = dissolved nutrient concentration in runoff or baseflow (kg/m^3);
s_{kt} = solid-phase nutrient concentration in sediment (kg/t);
X_{kt} = sediment loss (t);
TD_{kt} = fraction of dissolved nutrients which reaches a waterbody from area k; and
TS_{kt} = fraction of solid-phase nutrients which reaches a waterbody from area k.

Thus, the total nutrient load to a waterbody during time period t is calculated as the sum of the nutrient losses from each source in the watershed.

Equations 6.3 and 6.4 are completely general and can be used for both urban and rural areas. Their use requires methods for estimating nutrient concentrations, runoff and sediment losses. Further, such methods should be appropriate for the waterbody's location. In rural areas, for example, runoff estimates (Q_{kt}) should be based on methods generally used to predict rural runoff in the region. In the United States, a suitable method often used is the Soil Conservation Service Curve Number Equation (Ogrosky & Mockus, 1964). In other parts of the world, however, other predictive equations can be developed and may be more appropriate in a given situation.

Simple simulation models for estimating rural nutrient loads. Applications of Equations 6.3 and 6.4 to estimate nutrient loads from rural areas in the northeastern United States are described by Haith & Tubbs (1981) and Haith (1982). Most of the procedures are suitable for other regions as well. For example, source areas (k) are categorized by a distinct cover and soil type. Dissolved concentrations (d_{kt}) are obtained from flow-weighted average concentrations in runoff from field plots. These concentrations can vary with both land cover and time of year. Solid-phase concentrations (s_{kt}) can be determined from sediment samples from drainage channels in the rural area. Alternatively, the nutrient contents of soil samples can be multiplied by enrichment ratios. Solid-phase concentrations are assumed not to vary with time.

Transport factors (TD_{kt}, TS_{kt}) reflect a complex chain of physical and biochemical processes which can affect nutrient movement from a watershed area to a waterbody. These complexities have not yet been accurately described in any watershed model. Hence, it is often assumed that $TD_{kt} = 1$ and TS_{kt} is equivalent to the watershed's sediment delivery ratio (Haith & Tubbs, 1981).

Sediment loss (X_{kt}) from rural source areas is given by erosion or soil loss equations, such as the Universal Soil Loss Equation used in the United States (Wischmeier & Smith, 1978). In principle, this equation can be applied to regions outside the United States (Hudson, 1971; Haith et al., 1984).

However, local erosion predictions should still be used whenever possible.

As noted earlier, rural runoff estimates can also be based on the Soil Conservation Service Curve Number Equation. Descriptions of this equation are given in a number of references (Ogrosky & Mockus, 1964; Kibler, 1982; Novotny & Chesters, 1981; Wanielista, 1979; McCuen, 1982). However, application of the equation outside the United States requires that one classify watershed soils into one of the four Soil Conservation Service hydrologic groups based on minimum infiltration rates (McCuen, 1982; Schwab et al., 1981; Musgrave & Holtan, 1984).

Overall, Equations 6.3 and 6.4 can be used to identify and quantify the major sources of nutrients in rural runoff, as well as to evaluate non-point source management options based on reductions in runoff (Q_{kt}) or erosion (X_{kt}).

With regard to baseflow nutrient sources, baseflows are ground water inflows from the saturated soil zone. The associated nutrients are primarily in the dissolved form. Often, the rural baseflow load can be higher than the runoff load. If it is assumed that ground water nutrient concentrations are relatively stable, the dissolved nutrient loads can be calculated as:

$$D_k = d_k Q_{kt} \tag{6.5}$$

where: d_k = dissolved nutrient concentration in ground water flow from area k (kg/m³); and
Q_{kt} = baseflow volume (m³).

Concentrations (d_k) may be based on water quality samples taken during low flow periods in watershed tributaries free of point sources. However, the ground water flux (Q_{kt}) is often difficult to determine, since it depends on the distribution of saturated soil zones in the watersheds. In humid areas, the annual baseflow may be estimated by subtracting runoff and evapotranspiration from the total precipitation. There do not appear to be comparably simple means available for calculating the annual baseflows for shorter time intervals.

Simple simulation models for estimating urban nutrient loads. Saturated ground water flow in urban areas is often collected in sanitary or storm sewers and is reflected in measurements of point source discharges. Watershed modeling in urban areas, therefore, is usually limited to runoff sources. Although different urban runoff models are available (Huber & Heaney, 1980), most urban nutrient loading estimates are based on the following version of Equation 6.4:

$$S_{kt} = s_k X_{kt} \tag{6.6}$$

where: X_{kt} = runoff of solids (t) from land use k during time period t;
s_k = nutrient concentration (kg/t) in X_{kt}; and
S_{kt} = total nutrient load (kg) during time period t.

Although Equation 6.6 applies, in principle, only to solid-phase nutrient losses, it is often used to estimate total nutrient loads in urban runoff. In cases where urban drainage is transported via self-scouring channels or sewers, the transport factor (TS_{kt}) is assumed to have a value of 1.0. In developing countries, however, runoff may be carried along dirt channels in all but the largest urban centers, or along natural water courses. The accumulation of debris in the former case shows that such dirt channels may not be self-scouring.

Concentrations of nutrients in urban solids can be obtained from samples of dirt and debris taken from streets and other impervious surfaces, or from water quality samples from storm drainage. In the latter case, s_k is calculated as the chemical load divided by the solids load. Examples are given in Kibler (1982), Wanielista (1979) and Loucks et al. (1981).

The washoff of solids from urban surfaces is based on a mass balance model of solids accumulation, and removal by cleaning and washoff, as follows:

$$Y_{k,t+1} = Y_{kt} + y_k - X_{kt} - z_{kt}$$
(6.7)

where: Y_{kt} = solids on land use k at beginning of time period t;
y_k = solids accumulation (t) per time period t; and
z_{kt} = solids removal (t) by street cleaning during time period t.

Measurements of y_k are based on water quality samples of urban runoff drainage for large runoff events, for which it is assumed all accumulated solids have washed off. Typical accumulation rates are given by Kibler (1982), Wanielista (1979) and Novotny & Chesters (1981). The removal of solids by street cleaning is given by $f_t Y_{kt}$, where f_t is the cleaning efficiency during the period of street cleaning. If there is no street cleaning activity, then $f_t = 0$, and the removal term z_{kt} will also be zero.

The most common form of the washoff function (X_{kt}) is:

$$X_{kt} = Y_{kt}(1 - e^{aq_{kt}})$$
(6.8)

where: q_{kt} = runoff (cm) from land use k during time period t; and
a = washoff rate constant.

Thus, calculation of washoff requires an associated runoff model. Calibration may be used to estimate the appropriate value of a. In many applications in the United States, it is assumed that 90% of the accumulated solids will be removed by 1.27 cm (0.5 in) of runoff (i.e. $a = 1.8\,\text{cm}^{-1}$).

The simplest urban runoff model is:

$$q_{kt} = C_k(R_t + M_t)$$
(6.9)

where: R_t = rainfall (cm) during time period t;
M_t = snowmelt (cm) during time period t; and
C_k = runoff coefficient for land use k.

One can subtract depression storage from Equation 6.9 (Loucks *et al.*, 1981). The Soil Conservation Service Curve Number Equation can also be adapted to predict urban runoff (Kibler, 1982; McCuen, 1982; Wanielista, 1979).

A practical example of a watershed model for estimating the total nutrient load for lakes and reservoirs is provided in Chapter 7.

WATERBODY MODELS

Empirical waterbody models

Many simple lake and reservoir eutrophication models have been developed to address either phosphorus loadings or concentrations (e.g. Sakomoto, 1966; Vollenweider, 1968, 1975, 1976a; Dillon & Rigler, 1975; J.R. Jones & Bachmann, 1976; Larsen & Mercier, 1976; Rast & Lee, 1978; Ryding, 1980; Fricker, 1980a; Clasen & Bernhardt, 1980; Reckhow & Simpson, 1980; OECD, 1982). Over the past 15 years, such models have proven to be very useful for lake eutrophication assessment, planning and management. Their use and popularity results in large measure from the simplicity of the mathematics involved and the relatively limited data needs for application of the model. OECD (1982) provides a good example of the development and use of simple, empirical phosphorus models to assess eutrophication in a wide range of lakes and reservoirs.

As a general observation, the selection of the most appropriate model for a given waterbody should begin with careful consideration of specific eutrophication control objectives. For lake management issues that are best studied with predictions of long-term, lake-wide conditions, a simple model usually is best. These models (see above references) are largely empirical, and are frequently designed to predict the total phosphorus concentration in a lake or reservoir as a function of the annual phosphorus loading. Other empirical models, primarily developed as extensions of the phosphorus model, are available to predict such water quality parameters as chlorophyll *a* concentrations (Dillon & Rigler, 1974; Vollenweider, 1975, 1976a; V.H. Smith & Shapiro, 1981; OECD, 1982), Secchi depth transparency (Carlson, 1977; Rast & Lee, 1978; OECD, 1982), and dissolved oxygen concentrations (Reckhow, 1978; Rast & Lee, 1978; Ryding, 1980; Vollenweider & Janus, 1981).

One advantage of such simple models is that they can be subjected rather easily to uncertainty analysis (see Reckhow, 1979), which is an important attribute of this type of model. Simply stated, uncertainty provides a measure of the value of a model prediction. Without this estimate of uncertainty, there may be little basis by which to weigh the information developed by the modeler against other items or concerns which may be relevant to the initiation of eutrophication control measures.

As a practical matter, for simple empirical models, the prediction error for in-lake phosphorus concentrations can be ±30 percent or more. The prediction error for variables which are a function of in-lake phosphorus concentrations (e.g. chlorophyll *a* concentrations) may even be larger in a given situation. Nevertheless, even with this range of potential prediction error, many applications of these simple phosphorus models are possible, based on minor modifications that can greatly reduce the prediction error. Rast *et al.* (1983) assessed the predictive capability of simple load-response models of the type developed in the OECD (1982) international eutrophication study, by comparing the measured responses of a number of temperate lakes (which had undergone phosphorus load reductions) with the predicted responses of the models. They found that all the predicted mean total phosphorus and chlorophyll *a* concentrations were within a factor of ±2 of the measured values in most of the waterbodies, and within a factor of ±3 in all the lakes.

Examples of the use of some of these simple models, and the assessment of potential prediction error, are provided below. As discussed in Chapters 4 and 11, available evidence suggests these approaches are largely appropriate for temperate, tropical and sub-arctic waterbodies, although the models may require some calibration or adjustment when used in non-temperate settings.

Simple phosphorus models based on multiple lake phosphorus loading characteristics. Assessment of the trophic state or eutrophication potential of several lakes in a region can be done with any simple empirical model that has been verified satisfactorily for the waterbodies in the region. However, a more accurate assessment can usually be made if a model is developed which directly uses the lakes of the region itself as a statistical data base. This approach was used, for example, in the previously identified international eutrophication study conducted by OECD (1982).

A prominent example of a simple phosphorus model derived from multiple lake data is that developed by Vollenweider (1975). As a refinement to his earlier approach (see Vollenweider, 1968), this updated model uses information on areal water loading ($\bar{z}/t_w$) and areal phosphorus loading to assess the likely trophic state for a lake or reservoir, based on the responses of a large number of similar waterbodies to such inputs. All the waterbodies in a region can be plotted on the graph for easy visual comparisons (Figure 6.1).

Reckhow (1979) has subsequently shown how uncertainty analysis can be applied to phosphorus loading plots, to yield quantitative estimates of the probabilities of a waterbody having a specific trophic condition. OECD (1982) also has provided a simple approach for classifying lakes and reservoirs which illustrates the probabilistic nature of trophic classification. This classification scheme (see Table 4.2; also see Figures A.2–A.5 in Annex 1.) is based on several easily measured water quality parameters.

Control of eutrophication

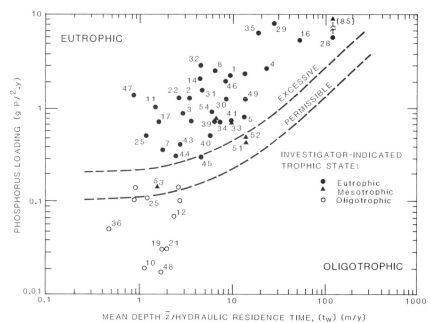

Figure 6.1 US OECD data applied to modified Vollenweider phosphorus loading diagram (from Rast & Lee, 1978)

In a refinement to his earlier model, Vollenweider (1976a) subsequently developed another phosphorus loading model (see Figure 6.2), based on the simple empirical relationship:

$$TP = (L_p/q_s)(1 + \sqrt{t_w}) \quad (6.10)$$

where: TP = average annual in-lake total phosphorus concentration (mg/l);
L_p = annual areal phosphorus loading (mg/m².yr);
q_s = annual areal water loading (m/yr) (= $\bar{z}/t_w$);
t_w = hydraulic residence time (yr); and
$\bar{z}$ = mean depth (m).

The standard error of this model is 0.193 (base 10 log TP). Chapra & Reckhow (1979) have subsequently developed plots for error analysis and trophic state classification probabilities associated with this type of model.

An updated version of this model is based on the results of the OECD international eutrophication study (OECD, 1982), as follows:

$$[P]_\lambda = 1.55\{[P]_j/(1 + \sqrt{t_w})\}^{0.82} \quad (6.11)$$

where: $[P]_\lambda$ = average annual in-lake total phosphorus concentration (μg/l); and
$[P]_j$ = average annual inflow total phosphorus concentration (μg/l) (= $L_p q_s$; see equation 6.10).

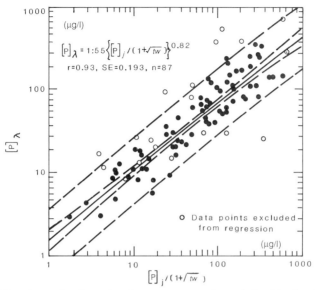

Figure 6.2 Phosphorus loading phosphorus concentration relationship (modified from OECD, 1982)

With this refined model (Figure 6.2), the correlation coefficient (r) is 0.93 and the standard error (SE) is 0.193.

It is noted that J.A. Thornton (1979, 1980; J.A. Thornton & Walmsley, 1982) and V.H. Smith et al. (1984) have reported that these types of simple phosphorus models appear to be appropriate for tropical/sub-tropical and sub-arctic lake and reservoir systems, respectively (see also Chapter 4).

Though usually based on less extensive data bases, variations of the simple phosphorus mass balance models illustrated above have been developed by Dillon & Rigler (1974), Larsen & Mercier (1976), J.R. Jones & Bachmann (1976) and Reckhow & Simpson 1980).

Baker et al. (1985) has even assessed the predictive capability of these types of simple phosphorus models in a large number of Florida lakes in the sub-tropical southern United States. Florida is characterized as having phosphate-rich soils. Therefore, many Florida lakes are nitrogen-limited, rather than phosphorus-limited (see Chapter 4 for discussion of the limiting nutrient concept). Since the simple phosphorus models identified above were developed primarily for phosphorus-limited lakes, their use in Florida lakes would violate one of the basic assumptions of the models. Baker et al. (1985) showed their sub-tropical study lakes in Florida had less chlorophyll a per unit phosphorus than temperate lakes. They also suggested boundary total phosphorus concentrations for mesotrophic and eutrophic waterbodies of 25 µg/l and 50 µg/l, respectively. This is consistent with the suggestions of J.A. Thornton (1980; 1985) regarding sub-tropical lakes and reservoirs (see Table 4.6). Even with these differences, Baker et al. (1985) were able to use statistical procedures similar to those used with the simple phosphorus models identified above, and develop simple nitrogen loading models which were good predictors of trophic status in nitrogen-limited Florida lakes. They also suggested that if one is faced with the task of predicting the average chlorophyll a concentration in a lake of unknown nutrient limitation, both a phosphorus-based and a nitrogen-based model should be used. Based on the limiting nutrient concept (see Chapter 4), the correct predictor should be the one that produces the lower chlorophyll a estimate.

Canfield et al., (1983; 1984) has also studied the applicability of simple models for predicting chlorophyll a concentrations in 233 Florida lakes. Their analyses also show that the models must be modified somewhat for application to nitrogen-limited lakes. Statistical analysis of their data base showed that the best models for the lakes were: (1). log chlorophyll a (μg/l) = $-2.49 + 0.269$ log total phosphorus (μg/l) + 1.06 log total nitrogen (μg/l); and (2). log chlorophyll a (μg/l) = $-2.49 + 1.06$ log (total nitrogen/total phosphorus) + 1.33 log total phosphorus (μg/l). His data support the hypothesis that nitrogen is an important limiting nutrient in hypertrophic lakes. Canfield also suggests that aquatic macrophytes have an important impact in Florida lakes (see discussion in a following section).

Such studies support the general utility of simple models, even if they require some modification in specific situations, as useful eutrophication assessment and prediction tools.

The use of in-lake phosphorus retention coefficients with simple phosphorus models. The phosphorus retention characteristics of a lake have been used in conjunction with simple empirical models to predict the in-lake total phosphorus concentration. If the necessary data are available, the phosphorus retention coefficient (R) can be calculated (Dillon & Rigler, 1974) as:

$$R = 1 - (\text{annual outflow P load/annual inflow P load}) \quad (6.12)$$

If the data for Equation 6.12 are not available, Kirchner & Dillon (1975) have developed another method for calculating the phosphorus retention coefficient, as follows:

$$R = 0.426\, e^{(-0.271 q_s)} + 0.574\, e^{(-0.00949 q_s)} \quad (6.13)$$

The correlation coefficient for Equation 6.13 is 0.10.

These two estimates of R (Equations 6.12 and 6.13) can also be pooled to produce an estimate of the phosphorus retention coefficient which is usually more accurate than either estimate alone. Reckhow & Chapra (1983) provide details on the method of calculating this pooled estimate of R.

Using any of these methods to calculate an estimate of R, one can use the simple total phosphorus model of Dillon & Rigler (1975), as follows:

$$TP = (L_p/q_s) \cdot (1 - R) \tag{6.14}$$

One can then use first-order error analysis to obtain an estimate of the prediction error of Equation 6.14, as discussed by Reckhow & Chapra (1983; Reckhow, 1983). Salas (1982) provides information on calculation of the phosphorus sedimentation coefficient for warm-water lakes.

It is emphasized that if it is necessary to partition a waterbody into several sub-basins (e.g. an elongated reservoir; see Chapter 8), then one should make use of the phosphorus retention coefficient in calculating the movement of phosphorus from an 'upstream' sub-basin to the adjacent 'downstream' basin. Use of the phosphorus retention coefficient will allow one to make more realistic estimates of the amount of phosphorus entering a downstream sub-basin from the sub-basin immediately upstream (i.e. the phosphorus not retained in the upstream sub-basin becomes the phosphorus load to the downstream sub-basin).

Simple phosphorus models and other eutrophication-related water quality indicators. The relationship between the phosphorus load and the in-lake phosphorus concentration is the focus of most of the simple eutrophication models identified above. However, it is reiterated that in-lake phosphorus concentrations, *per se*, do not constitute the perceptible water quality deterioration associated with lake and reservoir eutrophication. Instead, such water quality characteristics as water clarity, chlorophyll *a* concentration, and dissolved oxygen concentration in the hypolimnion are symptoms of unacceptable conditions in a waterbody, and are easily perceived by the public. Some of the available models for dealing with these latter water quality characteristics (which themselves are a function of the in-lake phosphorus levels) are presented by Vollenweider (1975, 1976a), Carlson (1977), Rast & Lee (1978), Ryding (1980) and OECD (1982). Examples of some of these models, taken from the individual regional projects, as well as the summary compilation of the OECD Cooperative Programme for Monitoring of Inland Waters (OECD, 1982), are provided in Table 6.1.

As shown in Table 6.1, the OECD study has resulted in the development of predictive models for a range of lake and reservoirs, located primarily in the temperate zones of the earth. As noted earlier in Chapter 4, the Pan American Center for Sanitary Engineering and Environmental Science (CEPIS) is currently coordinating a study to develop a similar quantitative framework for warm water lakes and reservoirs in the tropical/subtropical regions. Annex 1 illustrates the use of these various water quality characteristics in assessing possible limitations on different water uses.

Many investigators have examined the cross-sectional relationship between chlorophyll and in-lake phosphorus concentrations. For example,

Table 6.1 OECD relationships between annual phosphorus load and several lake and reservoir water quality parameters (modified from OECD, 1982)

OECD Project	Derived Relationship	n	r	SE
I. Annual mean total phosphorus concentration ($\mu g/l$):				
Combined OECD Study[1]	$1.55X^{0.82}$	87	0.93	0.192
Shallow Lakes and Reservoirs[2]	$1.02X^{0.88}$	24	0.95	0.185
Alpine Lakes[3]	$1.58X^{0.83}$	18	0.93	0.212
Nordic Lakes[4]	$1.12X^{0.92}$	14	0.86	0.252
U.S.A. Study[5]	$1.95X^{0.79}$	31	0.95	1.60
II. Annual mean chlorophyll concentration ($\mu g/l$):				
Combined OECD Study	$0.37X^{0.79}$	67	0.88	0.257
Shallow Lakes and Reservoirs	$0.54X^{0.72}$	22	0.87	0.238
Alpine Lakes	$0.47X^{0.78}$	12	0.94	0.189
Nordic Lakes	$0.13X^{1.03}$	13	0.82	0.329
U.S.A. Study	$0.39X^{0.79}$	20	0.89	0.261
III. Annual maximum chlorophyll concentration ($\mu g/l$):				
Combined OECD Study	$0.74X^{0.89}$	45	0.89	0.284
Shallow Lakes and Reservoirs	$0.77X^{0.86}$	21	0.88	0.276
Alpine Lakes	$0.83X^{0.92}$	11	0.96	0.191
Nordic Lakes	$0.47X^{1.00}$	13	0.77	0.373
U.S.A. Study	–	–	–	–
IV. Annual mean Secchi depth (m):				
Combined OECD Study	$14.7X^{-0.39}$	67	-0.69	0.237
Shallow Lakes and Reservoirs	$8.47X^{-0.26}$	26	-0.55	0.237
Alpine Lakes	$15.3X^{-0.30}$	18	-0.74	0.171
Nordic Lakes	–	–	–	–
U.S.A. Study	$20.3X^{-0.52}$	22	-0.82	0.196
V. Areal hypolimnetic oxygen depletion(g $O_2/m^2.day$):				
Combined OECD Study	$\sim 0.1X^{0.55}$	–	–	–
Shallow Lakes and Reservoirs	–	–	–	–
Alpine Lakes	–	–	–	–
Nordic Lakes	$0.085X^{0.467}$	–	–	–
U.S.A. Study	$0.115X^{0.67}$	–	–	–

$X = \{[L(P)/q_s]/(1 + \sqrt{t_w})\}$ = 'flushing corrected average annual phosphorus inflow concentration' (OECD, 1982); see Equations 6.10 and 6.11 for definition of terms.

n = number of data points; r = correlation coefficient; SE = standard error of estimate; dash (–) indicates data not available.

References: [1]OECD (1982); [2]Clasen (1980); [3]Fricker (1980a); [4]Ryding (1980); [5]Rast & Lee (1978).

V.H. Smith & Shapiro (1981) compared a cross-sectional (global) model to lake-specific time series models, and found substantial differences in their predictive capabilities for some of the lakes in their sixteen lake data set. Thus, if sufficient data exist, it usually is preferable to develop a lake-specific phosphorus-chlorophyll model. Using different rationale, a similar conclusion was reached previously by Rast & Lee (1978). However, if lake-specific data are unavailable, one can use the cross-sectional models developed by OECD (1982) as initial predictors of algal biomass.

The dissolved oxygen concentration in hypolimnetic waters is another significant eutrophication indicator. The transition between oxic and anoxic conditions in the hypolimnion is associated with fundamental (usually detrimental) changes in the chemistry and biology of a lake or reservoir. Rast & Lee (1978), Ryding (1980) and OECD (1982) provide simple models for predicting the hypolimnetic oxygen depletion rate, based on areal phosphorus loading. OECD (1982) combined the data from these two studies to produce a model which relates the hypolimnetic oxygen depletion rate to the phosphorus load, as follows:

$$\text{HODR} \cong 0.1\,[(L(P)/q_s)/(1 + \sqrt{t_w})]^{0.55} \tag{6.15}$$

where: HODR = areal hypolimnetic oxygen depletion rate
(g O^2/m^2.day).

Vollenweider & Janus (1982) developed a cross-sectional model which relates the volumetric oxygen demand and the chlorophyll biomass. Using data primarily from the international OECD (1982) eutrophication study, this model is expressed as follows:

$$\text{VHOD} = 0.732(\text{Chl } a)^{0.633} \cdot (z_e^{0.891}/\bar{z}^{0.878}) \tag{6.16}$$

where: VHOD = volumetric hypolimnetic oxygen demand
  (g/m^3.month);
Chl a = average chlorophyll a concentration (μg/l); and
z_e = depth of euphotic zone (m) ($=4.6/[0.3 + 0.02$ Chl $a]$).

The standard error for Equation 6.16 is 0.131 (base 10 log VHOD) and the r^2 value is 0.907.

OECD (1982) also provided a simple model for predicting areal hypolimnetic oxygen depletion rate as a function of the annual primary production. Reckhow (1978) used a cross-sectional data base of 55 north temperate lakes to develop a model for classifying lakes as having either oxic or anoxic conditions in the hypolimnion. Most of these models can be used to predict the oxygen depletion conditions of the hypolimnion, given projected changes in the external total phosphorus load to the waterbody.

The discussion in the several previous sections on simple eutrophication models is not comprehensive. As noted earlier, there are a large number of empirical eutrophication models in the literature which have been used in a variety of limnological settings. It is not possible to illustrate all possible examples and applications here. Rather, the reader is referred to the individual references, especially Rast & Lee (1978) and OECD (1982), for further details on the availability and uses of such simple models.

Simple phosphorus models applied to reservoirs. Based on a previously cited study of 107 U.S. reservoirs operated by the U.S. Army Corps of Engineers (see Table 4.5), reservoirs generally receive a higher areal phosphorus loading than natural lakes, yet exhibit lower in-lake concentrations of total phosphorus and chlorophyll *a* than natural lakes. This disparity has been related to the shorter water residence times, higher sedimentation rates and decreased light transparency of reservoirs (see K.W. Thornton *et al.*, 1981, 1982; Kennedy *et al.*, 1985). A related question, therefore, is whether or not simple phosphorus models are appropriate for reservoirs, as well as the natural lake systems from which many were initially derived.

The available evidence suggests that such models are adequate for making initial assessments of eutrophication in reservoirs. Mueller (1982), for example, used several simple phosphorus models of the type identified above to assess reservoirs in the western United States. He reported that, although the standard errors were relatively large, the models could be valuable tools in reservoir planning and management when the degree of uncertainty was taken into account. Canfield & Bachmann (1981) assessed the predictive capability of several simple phosphorus models for 704 natural and man-made lakes in the United States, Canada and northern Europe. Their assessment showed that the models were equally good in natural and man-made lakes for predicting in-lake total phosphorus concentrations, although they were less precise for predicting chlorophyll and Secchi depth values in man-made waterbodies. Nevertheless, they report that many man-made lakes exhibited the same relationships established for natural lakes. They suggest that the deviations that did exist appeared mainly to involve man-made lakes with high levels of non-algal particulate materials.

J.A. Thornton (1979) and J.A. Thornton & Walmsley (1982; Walmsley & Thornton, 1984) have assessed the predictive capability of the simple phosphorus mass balance models developed in the OECD study (1982; also see Table 6.1 and R.A. Jones & Lee, 1984) for sub-tropical reservoirs in southern Africa. Their studies concluded that the models could generally be applied to their sub-tropical reservoirs.

Walker (1981, 1982, 1985) conducted an extensive assessment of simple empirical eutrophication models, as applied to 299 reservoirs operated by the United States Army Corps of Engineers. The results of his studies showed that such models can be adapted for use in reservoirs. He reported the error variance for the prediction of chlorophyll was generally more than twice that for predicting the in-lake phosphorus concentration and

Secchi depth. This is believed to be due more to uncertainties in evaluating the biological responses of reservoirs (e.g. effects of light, nutrient kinetics) than to uncertainties in reservoir nutrient retention characteristics. Nevertheless, he indicated that simplified eutrophication models are valuable for providing preliminary indications of reservoir surface water quality. He also suggested that organic nitrogen was highly correlated with chlorophyll levels in reservoirs and, thus, was useful as a trophic state indicator. However, this conclusion was restricted to reservoirs located in the mid and southern latitudes of the United States.

Walker's work (1981, 1982, 1985) supports the notion that simple empirical models are valuable tools for assessing eutrophication of reservoirs, as well as of natural lakes. He also suggested, however, that some modifications may be necessary to account for the effects of the types of factors identified above, thereby increasing their predictive capability. The reader is referred to the reports of Walker (1982; 1985) for further details of this study.

The results of the OECD Shallow Lakes and Reservoirs Study (Clasen & Bernhardt, 1980); show that the predictive capabilities of the simple phosphorus loading models developed for the reservoirs in the study (see Table 6.1) are similar to those developed in the other regional projects. In particular, the statistical characteristics of the Shallow Lakes and Reservoirs project models are similar to those calculated for the Nordic (Ryding, 1980), Alpine (Fricker, 1980a) and United States (Rast & Lee, 1978) studies. The results of these various studies provide strong support for the use of simple, empirical phosphorus models for both man-made and natural lakes.

It is emphasized that the available data suggest that these simple models may not be appropriate for assessing reservoirs with extensive longitudinal nutrient and other eutrophication-related gradients (e.g. dendritic reservoirs or very elongated reservoirs dominated by one major tributary input). With the appropriate alterations, however, one can still use these simple models in a satisfactory manner to assess eutrophication in such situations. Kerekes (1982) and Lee & Jones (1981a), for example, show how one can use the longitudinal gradients and/or distinct embayments of a reservoir to partition a reservoir with an elaborate surface structure into a series of 'sub-basins'. Each sub-basin, in turn, can be assessed individually with the use of a simple model. Necessary adjustments in sampling programmes to account for nutrient and other chemical and biological gradients are discussed in Chapter 8.

Simple phosphorus models and macrophytes. It is generally accepted that aquatic macrophytes can have a significant role in the eutrophication of a lake or reservoir. Wetzel & Hough (1973) and Wetzel (1975), for example, have discussed the interrelationships of macrophytes, attached algae and phytoplankton in the total primary productivity of a lake. The excessive growth of macrophytes can be especially severe in the warmer climate of the tropical/sub-tropical regions. The simple phosphorus models of the type identified above, however, were developed with the assumption that the primary trophic response of a waterbody to its nutrient inputs would be manifested in the form of phytoplankton, rather than aquatic macrophytes. Consequently, these simple models do not appear to be useful in their present form for assessing eutrophication in waterbodies which exhibit extensive macrophyte growths.

The development of models related to macrophyte problems is relatively sparse to date. Carignan & Kalff (1980) previously demonstrated that the bottom sediments of a lake supply the majority of the phosphorus used by rooted aquatic weeds, based on studies in southern Quebec (Canada). They reported that up to an average of 72% of all the phosphorus taken up by rooted macrophytes under hypertrophic conditions can be supplied by the sediments, rather than the water column. Carignan (1981) subsequently developed an empirical model for predicting the phosphorus uptake by roots of submersed macrophytes. However, while such models are useful for assessing the role of macrophytes in phosphorus cycling in lakes and reservoirs, they do not provide a practical tool for eutrophication management purposes.

A study which attempts to address the role of aquatic macrophytes as they affect the prediction of chlorophyll a concentrations in lakes is that of Canfield et al. (1983, 1984; Canfield & Jones, 1984). Their study showed that the chlorophyll a concentration in Lake Pearl (Florida, United States) increased as the percentage of the total volume of the lake infested with aquatic macrophytes decreased. As pointed out in a previous section, because of the phosphate-rich soils, many Florida lakes exhibit nitrogen limitation and are hypertrophic. Based on their findings, Canfield et al. (1983, 1984) used data from 32 Florida lakes to develop a predictive equation for chlorophyll a concentrations in lakes with significant macrophytes growths. This equation relates chlorophyll a, total phosphorus and nitrogen concentrations and macrophytes, as follows:

$$\log \text{Chl } a = 1.02 \log \text{TN} + 0.28 \log \text{TP} - 0.005 \text{ PVI} - 2.08 \quad (6.17)$$

where: Chl a = chlorophyll a (μg/l);
TN = total nitrogen concentration (μg/l);
TP = total phosphorus concentration (μg/l); and
PVI = total lake volume infested with macrophytes (%).

The r^2 value for this equation is 0.86. Canfield et al. (1983, 1984) suggest that PVI values greater than 15% constitute major weed problems, in terms of public perception, in Florida lakes. They also reported an increase in the predicted Secchi depth, based on the chlorophyll – Secchi depth relationship developed by J.R. Jones & Bachmann (1978).

Canfield et al. (1983, 1984) emphasize, however, that inclusion of PVI in their regression analysis only increased the coefficient of determination of their previous chlorophyll a prediction equations from 0.82 to 0.86. They suggest that other factors (e.g. zooplankton grazing, light limitation by inorganic turbidity and hydraulic flushing) may be of equal or greater importance in determining in-lake chlorophyll levels. Further study outside their limited Florida lake data set is required to assess the generality of their relationship.

Several researchers have presented simple methodologies for determining the biomass of aquatic macrophytes in lakes. For example, Maceina et al. (1984), also studying several Florida lakes, reported that a fathometer represented a potentially valuable tool for predicting the biomass of aquatic macrophyes, particularly if fathometer tracings previously have been correlated with actual biomass samples. One can use such information to estimate whole-lake vegetation coverage and percent volume infestation. Maceina et al. also caution, however, that this estimation technique is not equally reliable for all waterbodies, due primarily to differences in the type of aquatic vegetation present in a waterbody.

Downing & Anderson (1985) have studied the accuracy, precision and cost-efficiency of quantifying standing macrophyte biomass, using various quadrant sizes ranging from $1-100 \, m^2$. Their study showed that all five quadrant sizes studied yielded equivalent biomass estimates. They suggest the use of small quadrant samples with greater replication, especially when sampling macrophyte beds of low standing biomass, with a potential 30-fold decrease in sampling costs.

Because of the need of macrophytes for adequate sunlight energy for photosynthesis, an inverse relationship between the extent of macrophyte infestation in a waterbody and the degree of water transparency is implied. Rast & Lee (1978), for example, suggested the extent of macrophyte encroachment into a waterbody can be used as a measure of trophic condition. Subsequently, Canfield et al. (1985) have used data from lakes in Finland, and in Florida and Wisconsin (United States), to assess the relationship between water transparency (measured as Secchi depth) and the maximum degree of colonization of the waterbody by aquatic macrophytes. They developed the following predictive equation:

$$\log MDC = 0.61 \log(SD) + 0.26 \qquad (6.18)$$

where: MDC = maximum depth of macrophyte colonization (m); and
SD = Secchi depth (m).

The r^2 value for this equation is 0.56. Canfield et al. (1985) caution, however, that their model has a 95 per cent confidence interval of 46–236% of the calculated MDC value. They suggest this phenomenon is due to differing light requirements for different types of plants colonizing the lake bottom.

Their model tends to underestimate MDC values in very clear waters, and overestimates MDC values when Secchi depths are low. Because of this problem, Canfield et al. (1985) also developed regression equations for their individual data sets, as shown in Table 6.2.

While advising caution in their use, Canfield et al. (1985) suggest that their models do provide reasonable first approximations, given the variability observed in lakes. One can also attempt to relate the MDC values back to the simple phosphorus models identified earlier, via the common variables of Secchi depth and chlorophyll a.

More recently, Duarte & Kalff (1986) tested the hypothesis that morphometric characteristics are a major factor controlling submerged macrophyte biomass in the littoral zone. Based on data from a number of stations in Lake Memphremagog (Quebec, Canada and Vermont, United States), they derived a relationship for predicting the maximum biomass of submerged macrophyte communities as a function of the slope of the littoral zone. Using a slope of 5.33% as a critical value between gentle and steep slopes, they derived two predictive equations as follows:

Slope $<5.33\%$:
MSMB (maximum submerged macrophyte biomass; g fresh wt./m^2)
$= -29.8 + 1.403$ slope$^{-0.81}$ (6.19)

Slope $>5.33\%$:
MSMB $= 13.2 + 3.434$ slope$^{-0.8}$ (6.20)

For Equation 6.19, n = 28, $r^2 = 0.60$ and S.E. = 121, while for Equation 6.20, n = 16, $r^2 = 0.79$ and S.E. = 186. Duarte & Kalff (1986) also considered the sediment organic matter of the littoral zone in deriving a further predictive equation. They cautioned, however, that their predictive equations appear to be appropriate only for non-turbid, temperate zone lakes. Their equations tend to overestimate the submerged macrophyte biomass in highly-turbid, temperate zone lakes, while underestimating the biomass in tropical/sub-tropical lakes. In these latter cases, solar irradiance and the degree of light penetration appear to be more important than slope in determining the maximum submerged macrophyte biomass.

Although the above-cited examples are encouraging, work to date is not sufficient to allow development of generalized relationships for macrophyte-

Table 6.2 Relationship between macrophyte colonization and Secchi depth for lakes in Finland, Florida and Wisconsin (from Canfield et al., 1985)

Region	n	Equation	r^2
Finland	27	log MDC = 0.51 log (SD) + 0.18	0.82
Florida	26	log MDC = 0.42 log (SD) + 0.41	0.71
Wisconsin	55	log MDC = 0.79 log (SD) + 0.25	0.57

infested lakes and reservoirs, such as the simple phosphorus models of the OECD (1982) international eutrophication study. It is clear that further work is necessary in this important area, particularly for lakes and reservoirs in the warm, tropical/sub-tropical regions.

Simulation waterbody models

In contrast to the simple statistical (usually correlation or regression) models of the previous section, the models discussed here are characterized by the inclusion (to varying degrees of detail) of a structural, mathematical description of the important physical, chemical and biological processes in the lake or reservoir system. Major advantages of these 'dynamic' models are:

1. They allow for more temporal and spatial detail;
2. They generally consider a number of relevant variables (e.g. phytoplankton, orthophosphate, detritus, zooplankton, etc.), rather than just one or two; and
3. They are potentially more accurate than empirical, statistical models if they can be properly calibrated.

The increased spatial and temporal resolution is often a desirable property, especially in describing peak algal bloom events, which usually cause most of the nuisance problems associated with cultural eutrophication. They can also be more 'general' than empirical, statistical models, if properly constructed, in that they allow one to attempt to consider situations outside the range of the data used to derive them. Such models can also permit one to attempt to assess the relative magnitude and impact of phosphorus loads from different sources in a drainage basin (e.g. internal loading versus stormwater inputs).

The disadvantages of dynamic models are related to the same items which constitute their advantages; namely, that the greater detail sought requires a large (sometimes exhaustive) data base. Moreover, the potential accuracy of such models can be diminished simply because the mechanisms of some of the processes incorporated into the models are often not satisfactorily known. They are also not as readily transferable from one lake to another, due to their inclusion of lake-specific 'interface' terms and the need for sophisticated computer systems.

Dynamic eutrophication models typically contain three types of terms:

1. Hydrologic and hydrodynamic characteristics;
2. Descriptions of chemical and biological transformations; and
3. Input, output or exchange of materials through the boundaries (i.e. 'interface' terms); these latter components are usually driven by hydrologic and hydrodynamic factors.

These terms are usually mathematically expressed in a set of differential equations which attempt to describe, for any instant in time, the relevant processes driving important parameters. Solution of the set of differential equations gives the expected or predicted behavior of the waterbody, based on the input variables and driving forces characterizing the waterbody.

The rate of change of these various terms, and the processes they describe, usually depend on extraneous or forcing variables (e.g. temperature, solar radiation, wind speed and direction, etc.). This means, of course, that relatively detailed records of such variables must be available if one is to attempt to successfully use a dynamic model.

Space limitations preclude a detailed discussion of specific dynamic eutrophication models here. Rather, this section focuses on the nature and acquisition of the above-defined informational needs. The reader is referred to the specific studies highlighted in this section for further details regarding dynamic eutrophication models. Straškraba & Gnauck (1983, 1985) also provide a good review of simulation models.

Hydrologic and hydrodynamic components of dynamic models. Consideration of hydrological throughflow is important because it determines the flushing rate of a lake or reservoir. This latter parameter can constitute a significant loss factor for both algal biomass and nutrients. It can be difficult, however, to determine how much hydrodynamic detail is required in a given situation. The most important internal transport mechanisms for biomass and nutrients in a lake or reservoir are wind-induced mixing and heat-generated buoyancy. The latter factor causes thermal stratification, most notably in deep lakes. For modeling purposes, therefore, a minimum of two water layers is usually distinguished – the epilimnion on top and the hypolimnion at the bottom of the waterbody. For simplicity, these two layers are usually considered to be completely mixed (Snodgrass & O'Melia, 1975; Imboden & Gächter, 1978; DiGiano et al., 1978). The extent of exchange of nutrients and biomass between the two water layers can be estimated on the basis of measured temperature profiles (DeBruijn et al., 1980). In practice, one usually assumes the waterbody is fully mixed, in order to circumvent the difficult problem of internal mixing and transport. Fortunately, if a lake-wide average value, rather than single station values, is sufficient to answer the management question, then a fully mixed model is quite appropriate (Thomann et al., 1975). In such situations, this simplification works reasonably well because the non-linearity of algal growth is fairly weak; that is, excessive algal growth in areas of high nutrient levels in a waterbody is roughly balanced by less than average growth elsewhere in the waterbody (this may be due to instability of the horizontal differences, varying with time from area to area). However, this approach also results in a loss of spatial detail.

If the data show that strong nutrient gradients exist, a multi-sectional model of the waterbody may be required (van Straten, 1981; see also Chapter 8). A problem with multi-section models, however, is determining the exchange of materials between the lake sections. If no gradients of a conservative substance (e.g. chloride) exist, there may be no way to obtain estimates of exchange other than with two- or even three-dimensional hydrodynamic circulation models, which can be very complex. In principle, the best solution to such a problem is to use a fully transport-oriented eutrophication model. Unfortunately, only a few examples of such models exist in the literature (e.g., Lam & Jaquet, 1976). A review of the role of hydrodynamics in spatially-detailed eutrophication modeling of shallow lakes is presented by Shanahan & Harlemann (1982). In a more recent publication, Shanahan & Harlemann (1984) discuss another approach for reducing the complexity of coupled hydrodynamic-transport-eutrophication models.

Chemical and biological transformations. Chemical and biological reactions and transformations usually form the core of a eutrophication model. Although a simplification of reality, Figure 6.3 illustrates the most important cycles of carbon and nutrients in a segment of a waterbody. Fortunately, however, simplification is a basic goal of model building (Imboden, 1982). The choices made in each modeling effort on how to translate the generally-accepted principles outlined in Figure 6.3 into a practical model are influenced (or should be influenced) to a large degree by the specific modeling goal.

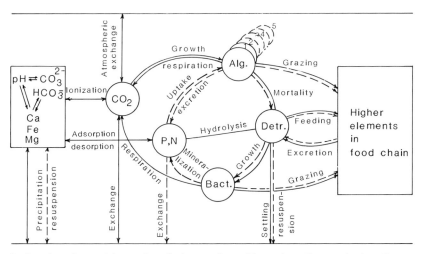

Explanation of terms: (–) = carbon; (– –) = nutrients; Alg. = algae; Detr. = detritus; Bact. = bacteria; P = phosphorus; N = nitrogen; CO_2 = carbon dioxide; CO_3^{2-} = carbonate; HCO_3^- = bicarbonate; Ca = calcium; Fe = iron; Mg = magnesium.

Figure 6.3 Schematic diagram of main carbon and nutrient cycles of interest in the eutrophication process (from van Straten, 1986)

A large number of variants exist, both with respect to the number of dependent variables ('state variables' or 'components') considered (Scavia & Robertson, 1979; Canale, 1976), and the mathematical formulation of the sub-processes. A review of the latter is provided by Swartzman & Bentley (1979). Differences in the number of state variables among models arises from a more or less subjective judgement about the degree of detail desired (or required) to yield sufficient realism. In practice, however, the selection of a specific model frequently is dictated by the availability of data.

Questions to be considered in selecting a dynamic model are:

1. Whether or not to include only a single nutrient (e.g. phosphorus) or several nutrients (e.g. phosphorus, nitrogen and/or silica);

2. Whether or not to consider grazing of phytoplankton by zooplankton and higher trophic levels;

3. Whether or not to describe chemical components in detail (e.g. orthophosphate, condensed phosphorus, particulate organic phosphorus, dissolved organic phosphorus, etc.) versus simplifications (e.g. only dissolved biologically-available phosphorus and phytoplankton phosphorus); and

4. Whether or not to include a single (or few) algal species versus multispecies modeling.

The answers to such questions will depend in part on the nature of the problem being assessed and the availability of sufficient data. They will certainly affect the selection of the eutrophication model eventually used in a given situation.

Most eutrophication models (whether statistical or dynamic) concentrate on the biological cycle, rather than on the chemical component. The literature contains few examples of dynamic models which consider in-lake chemical processes in detail (e.g., see DiToro, 1976, 1980; DeRooij, 1980; Lum et al., 1981). This is somewhat surprising since the carbon dioxide balance and pH (both of which are influenced by algal growth and mineralization processes) are heavily affected by the presence of calcium and magnesium compounds. The rise in pH which normally accompanies extensive algal blooms can cause the precipitation of calcium carbonate and, consequently, the co-precipitation of phosphorus (Koschel et al., 1983). This mechanism can cause a significant loss of phosphorus to the hypolimnion and lake sediments. Similarly, phosphorus can be precipitated with iron and calcium when river water comes into contact with lake water of a higher pH. It is expected that the increasing attention being paid to the global acid rain problem will yield results also applicable to eutrophication modeling.

Interface terms. As noted earlier, these items refer primarily to the input, output or exchange of materials through boundaries. An important component of this category is the external nutrient load to the waterbody. Since the external load constitutes the link between a lake or reservoir and its catchment area, considerable monitoring efforts are often needed to evaluate interface terms, regardless of whether one is using a dynamic or statistical eutrophication modeling approach.

Another class of interface terms concerns exchange processes, including the exchange of gases (oxygen, nitrogen, carbon dioxide) in the waterbody with the atmosphere, the transport of constituents across the thermocline in stratified lakes, and the exchange of nutrients between the water column and sediments. The sediment-water nutrient exchange process is extremely complex (see Chapter 5). Physico-chemical processes, such as adsorption or chemosorption on calcium and iron compounds, are affected by the oxygen conditions and pH of the water. These latter parameters, in turn, depend both on the biological processes in the water column and the transport and transformation processes within the sediment (Sly, 1982).

In most lake eutrophication models, the sediment component (if considered at all) is extremely simplified. There is little doubt that sediment-water interactions (and our lack of knowledge about them) are largely responsible for the less-than-satisfactory performance of dynamic models for long-term predictions. For example, the straight line in Figure 6.4 illustrates the

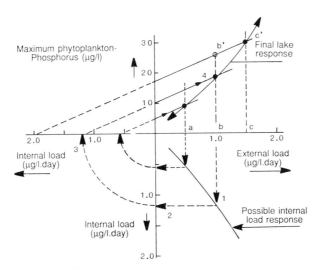

Explanation of terms: Line c′–b′ = initial response of phytoplankton to a load reduction from c to b. Due to the load reduction (1-2-3), the final response lines slowly shift parallel downward, and finally end in the more favorable intersection point; see text for discussion.

Figure 6.4 Phytoplankton response to external phsophorus load reduction (from van Straten, 1986)

predicted response of a waterbody to phosphorus load reductions, using a model with a simplified sediment segment. This model suggests that one cannot lower the algal biomass below a certain level, because the internal phosphorus source would go on indefinitely. In reality, however, this situation is unlikely to happen, since internal phosphorus loading (where it occurs) can vary considerably from year to year (e.g. see Ryding, 1985). The 'true' behavior is represented by the reduced nutrient release likely to occur over the long term, following reduction of the external load. Thus, the true situation for eutrophication management is less pessimistic than predicted with these so-called 'short memory' models. The difficulty, of course, is that no one can be sure, *a priori*, about the time needed for sediment 'restoration'. Solution of this latter problem will require more fundamental research on nutrient behavior in sediments.

Approaches for selecting a dynamic eutrophication model. The first approach is based on the assumption that a realistic representation of a lake or reservoir system is best obtained by putting together mathematical formulations for a large number of ecological sub-processes. This approach strives for completeness and, in a sense, the resultant model is believed to be universal. Thus, in principle, the model can be applied to a large variety of situations. An example is the model CLEAN (Park *et al.*, 1974), later expanded to CLEANER (Park, 1979), which now comprises about 40 different state variables. The constituting sub-process formulations and associated parameters (several hundred in number) are derived from laboratory experiments or other independent investigations. Using this approach, ideally one should be able to predict the future, even for large deviations from the present situation, because 'everything' relevant is included in the model. In practice, however, the expectations usually fall far short of reality. This is because, despite careful experimentation, at present it is impossible to know all relevant sub-processes and process parameters without error. This 'complete' model will reflect these uncertainties. Because the number of inevitable sub-process errors is usually large, their effect on the performance of the model as a whole is practically unpredictable. Thus, the final answers remain unreliable and uncertain. The usual 'solution' of calibrating the model against field data is impractical here because, with so many variables, it is unlikely that the appropriate field data will be available.

The alternative approach is to start simply and update the model as knowledge and experience are gained. This approach usually starts with use of the field data, and a limited amount of *a priori* knowledge about those processes believed to be dominant in the behavior of the lake or reservoir system. Such models tend to be simple initially; updates are made only if deviations between observations and model outputs are unacceptably large in relation to the goals of the study. Thus, these types of models are 'tailor-made', rather than universal. New hypotheses are generated as a result of deviations, and only those improvements which find support in the field data are used to refine the model. With this approach, parameter estimation (or calibration) and structure identification are closely related. Beck (1982, 1981) provides an example of this procedure.

Frequently, measurement data are not accurate or complete enough to allow for deterministic modeling. In such cases, it is better to carry the fundamental uncertainties along in the generation of hypotheses (Hornberger & Spear, 1980) or in the predictions (Fedra *et al.*, 1981; van Straten, 1981). A relatively simple model structure also allows for a simplified analysis (Verhagen, 1976; van Straten, 1983). The basic advantages of this approach include both the transparency of the structure, and the explicit treatment of uncertainties. However, a fundamental disadvantage with this approach is that the predictive power often is restricted to waterbodies similar to those upon which the model was based. This can limit the practical application of these models.

Thus, with respect to the development and use of complex eutrophication models, one is faced with the dilemma (Beck, 1981) of choosing between universal models (which cannot be calibrated) or well-calibrated, simpler models (with a smaller range of predictive power). For the present, the answer to this dilemma is to attempt to combine the best features of these two extremes. In fact, such development has already begun (e.g. see Beck & van Straten, 1983).

Despite these inherent limitations, the use of dynamic mathematics is often the only choice for solving more complex management questions, especially in judging the effectiveness of alternative options in some cost-benefit or optimization analyses. For example, the year-to-year variance in the occurrence of algal blooms is lower with phosphorus retention reservoirs than with tertiary treatment, even though the average phosphorus concentration in the reservoir may be higher (Somlyódy & van Straten, 1986). A proper trade-off requires a proper evaluation of these kinds of effects. There are virtually no alternatives to mathematical modeling in these situations. An example of a dynamic model (SALMO) used in the decision making process is presented in Chapter 11.

CHAPTER 7

ESTIMATING THE NUTRIENT LOAD TO A WATERBODY

As discussed in Chapter 3, practical experience suggests that the most effective long-term measure for the control of eutrophication of lakes and reservoirs is the reduction of the external nutrient inputs. To determine the most efficient way to accomplish this goal, it is necessary to identify the most significant nutrient sources to a waterbody, as well as their relative contribution to the total nutrient load. One also must identify which sources can be most readily controlled. Without such information, elementary decisions regarding the effective management of eutrophication are difficult to make.

MAJOR PHOSPHORUS AND NITROGEN SOURCES

Many sources, located both outside and inside the drainage basin can contribute nutrients to a waterbody. Potential external nutrient sources (Figure 7.1) include effluent discharges from municipal and industrial point sources (e.g. sewage treatment plants) and diffuse or non-point sources such as land runoff and deposition from the atmosphere. Important internal sources include nutrient regeneration from the bottom sediments, and ground water seepage (i.e. subsurface flow).

This chapter provides an overview of the phosphorus and nitrogen sources in the drainage basin which have been shown to be significant in stimulating eutrophication. It is recognized that all these nutrient sources are not measured routinely in all countries. In such cases, one should attempt to use at least rough estimates of the nutrient loads from these sources, recognizing that these estimates may have a large degree of uncertainty. The following sections provide guidelines for such situations.

Point sources

The nutrient load from municipal and industrial point sources includes the direct discharge of sewage or effluents from a treatment facility. The

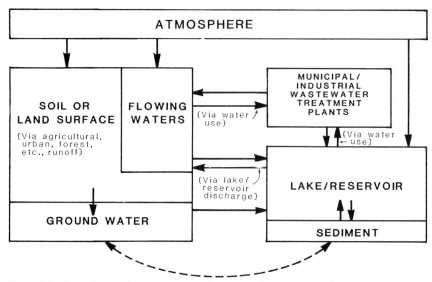

Figure 7.1 Potential nutrient transport pathways to a lake or reservoir

quantity of pollutants discharged depends on such factors as the wastewater flow rate, the degree and type of waste treatment used, and the constituents in the influent wastewaters. The use of detergents containing large amounts of phosphorus, for example, can result in an increased phosphorus content in municipal wastewaters. Industrial discharges to municipal sewage treatment plants may also contribute to the nutrient content of sewage effluents.

Certain urban and rural pollution sources can reach a recipient water in a more or less concentrated form and, therefore, also can be regarded as a form of point source input (although others may consider these inputs as non-point sources). Among such agricultural sources, private waste disposal (septic) systems, urine tanks, manure stacks and silage juice from silos may contribute substantial amounts of nutrients if improperly designed, if malfunctioning, or if located in areas with soils unsuited for immobilizing nutrients. Urban runoff through piped sewer systems also can contribute large amounts of nitrogen and phosphorus to a waterbody, especially following storm events when the sewer system receives large quantities of urban drainage. Private septic systems also exist in some urban areas.

Diffuse (non-point) external sources

Nutrients are deposited via atmospheric inputs directly onto the surface of a lake or reservoir from precipitation, dry fallout and/or by turbulent transfer to the water surface. Atmospheric nutrient deposition can include both particulate and soluble nutrient forms, the latter form usually being dominant. The origin of nutrients in atmospheric deposition can be diverse. The proximity of stack emissions and local amounts of rainfall, for example, can affect the atmospheric deposition of potential pollutants to a waterbody.

Both urban and rural factors can affect the input of nutrients from land drainage (runoff) to lakes and reservoirs. Urban storm water drainage can become a relatively more significant proportion of the nutrient source to a waterbody as municipalities reduce point source nutrient inputs by constructing adequate municipal wastewater treatment plants. Urban storm water can influence lakes and reservoirs by polluting the ground water, by disturbing treatment processes in wastewater treatment plants during periods in which they receive a high hydraulic load, by overflows from combined sewer systems, or by direct drainage from separate sewer systems.

Rural sources of phosphorus and nitrogen include land drainage from agricultural and forested areas. Significant agricultural diffuse sources include animal operations, such as from animal feedlots, leaking manure stacks and urine tanks and farm wastewaters (e.g. from milk rooms and machinery washing areas).

Septic tank wastewater disposal systems, both for household and farm sewage, as well as pit latrines, may also be significant nutrient sources in areas where they occur. The magnitude of the phosphorus and nitrogen load from septic tanks and pit latrine systems will depend on such factors as the ground water hydrology of the local area and the soil absorption capacity for the nutrients (e.g. the aqueous chemistry of phosphorus in calcareous systems).

A potentially significant factor affecting nutrient loads in rural land drainage is the materials usage in the watershed, especially the spreading of inorganic fertilizers, animal manure, and 'night soil'. The use of fertilizers (with their high phosphorus and nitrogen content) can increase the nutrient loads in land runoff, compared to that from unfertilized soils. Practices which do not allow for the proper incorporation of inorganic fertilizers into the soil (i.e. making fertilizers prone to being washed from the soil), or which involve the application of fertilizers at excessive rates, can cause short-term, elevated nutrient concentrations in runoff. Similarly, improperly applied animal manure or 'night soil' (e.g. being spread on frozen ground) also may cause short-term, elevated nutrient losses.

A large part of the nutrient load generated from agricultural land is related to the natural nutrient content of the soil. Other important factors include the soil texture, soil chemistry, soil type (mineral or organic), and physiography (drainage density, slope). The single most important characteristic in many regions appears to be the soil texture (soil particle size distribution). Sandy (coarse-grained) soils produce the smallest quantity of water runoff, since such soils allow for rapid water infiltration. By contrast, clay (fine-grained) soils often drain poorly, promoting water runoff. Any land use that exposes soil to the erosive forces of rainfall increases nutrient loss. Intensive farming on rural land or construction sites in urban land areas are examples of intense land uses which expose the soil, resulting in increased nutrient losses via erosion. On some intensively-farmed lands, the soils are artificially drained. Tile drainage usually reduces soil loss and phosphorus pollution, but may increase nitrogen losses. This is because the soil retention capacity for nitrogen is less than that for phosphorus. Thus, the water collected in drainage fields may be high in nitrogen.

Streambank erosion may be appreciable for some streams. The nutrient contribution, however, is often small compared to sheet and rill erosion. Streambank materials are likely to have a lower nutrient content than topsoil.

Forest land is not usually a major nutrient source per unit area of land. However, intensive forest practices such as clear cutting and scarification can increase the exposure of soil, thereby resulting in an increased nutrient runoff to a waterbody.

Diffuse internal sources

A certain portion of tributary flows is derived from ground water seepage. Ground water (sub-surface flow) can also seep directly into a lake or reservoir. The concentrations of phosphorus in ground waters are usually low. As a result, ground water is often ignored as a potential phosphorus source. However, nitrate nitrogen is often found in large concentrations in ground water, especially in agricultural areas.

Once nutrients enter a lake, they may be recycled many times between the sediments, aquatic plants and water column. Over the long term, sediments normally act as a net sink for nutrients. However, as a result of release from sediments, some of the nutrients can be transferred back into the water column (see Chapter 5). For example, wind-induced mixing of bottom sediments, or anoxic conditions in the upper sediment layer and in bottom waters will favor this nutrient release or 'regeneration'.

Nutrient recycling is particularly important in shallow lakes (i.e. mean depth < 5 m), in which the external nutrient inputs have been reduced to very low levels. In each case, the frequent resuspension of nutrient-enriched bottom sediments and the release of nutrients in shallow lakes can maintain high nutrient concentrations in the water column, thereby delaying recovery processes, especially in formerly highly-eutrophic waters. Consequently, if long-term results are desired, merely limiting external nutrient inputs may be insufficient to improve the water quality conditions of shallow, nutrient-enriched lakes over the short term. Calculation of the internal loading of phosphorus from the sediments to the water column is discussed in a later section of this chapter.

Nitrogen fixation, whereby microorganisms convert nitrogen gas (free nitrogen) to organic forms of nitrogen, can be a potentially significant source of nitrogen for some lakes and reservoirs. This process is most important in eutrophic lakes containing high phosphorus levels and large populations of blue-green algae, including nitrogen-fixing algae.

Biological availability of nutrients

Total phosphorus and total nitrogen, which consist of both the soluble (dissolved) and particulate forms, are often the focus of eutrophication control programmes. The soluble fraction is generally considered to be easily utilized by algae. This fraction is often termed the 'biologically available' or bioavailable fraction (see Chapter 4). In contrast to the soluble fraction, only a portion of the particulate nutrients may be bioavailable. In such cases, a large proportion of the total nutrient load may actually be in a biologically unavailable particulate form (see Lee et al., 1980). Therefore, the potential bioavailability of particulate-associated nutrients is very important with regard to the effective management of eutrophication, especially the selection of the significant nutrient sources to be controlled. Based on PLUARG (1978a) studies in the United States Great Lakes Basin, approximately one-third (on average) of the phosphorus associated with suspended sediments at the rivermouth was found to be in the biologically-available form.

Phosphorus in municipal sewage treatment plant effluents is largely bioavailable at the point of discharge (e.g. 75->90% for municipal wastewater treatment plants in the Great Lakes Basin). Such phosphorus is primarily in a soluble form easily used by aquatic plants (DePinto et al., 1980; Young et al., 1982). Thus, point sources that discharge directly to lakes and reservoirs are usually the most important sources of bioavailable phosphorus to consider in eutrophication control programmes.

The bioavailability of phosphorus from other sources is less well known (PLUARG, 1978a; Lee et al., 1980). The bioavailability of phosphorus from shoreline erosion is low (approximately three to ten per cent). Twenty-five to 50% of the phosphorus in atmospheric deposition can be bioavailable. Ground water inputs are likely to contribute mostly dissolved (bioavailable) phosphorus to a waterbody, although the concentrations are normally low.

As eutrophication management strategies are refined, it will become more important to consider the bioavailability of nutrients. Cost-effectiveness, in terms of the cost per unit amount of bioavailable nutrient that can be prevented from entering a lake or reservoir, is an important consideration for developing eutrophication control programmes. For this reason, nutrient sources high in bioavailable phosphorus and other dissolved nutrients are especially attractive as nutrient control targets. Furthermore, shallow lakes are likely to be more sensitive than deep lakes to potentially bioavailable forms of nutrients, because of resuspended sediment particulate material (which is often more bioavailable to aquatic plants). A simple method for estimating the available phosphorus load to a waterbody is discussed by Cowen and Lee (1976) and Cowen *et al.* (1978).

QUANTIFYING THE NUTRIENT LOAD

After the major nutrient sources in the drainage basin have been identified, the next step is to quantify the annual loads to the waterbody. Quantitative information on the annual total nitrogen and phosphorus load is often sufficient for developing eutrophication control programmes. Nevertheless, ideally, it is desirable to measure or estimate the annual inputs of total and particulate phosphorus, soluble orthophosphate, total nitrogen, ammonia, nitrate and, if possible, organic nitrogen. Algal bioassay techniques can be used to determine the bioavailability of the nutrient loads (see Chapter 4).

Exceptions to the use of the annual nutrient load are waterbodies which have rapid flushing rates (i.e. short water retention times). A related exception is when most of the annual nutrient input enters a lake or reservoir immediately prior to, or during, the growing season. Such a load can have a greater impact on the growing season biological productivity than can the annual load. These exceptions are discussed further in Chapter 11.

Methods for quantifying nutrient inputs from the different sources in the drainage basin are identified in the following sections. It is important to point out that the necessary precision for the nutrient load estimates depends on the intended use of the information. For example, determination of the major components of the annual nutrient load often requires only approximate estimates. On the other hand, a detailed analysis of the cost-effectiveness of alternative eutrophication control options usually requires more precise nutrient load estimates. Therefore, when attempting to quantify nutrient inputs, the following questions should be addressed:

1. What is the information to be used for?

2. How precise must it be?

Direct measurement of the nutrient loads is preferred, since this approach usually provides the most accurate estimates. If it is not possible to make direct measurements, however, the concept of the unit load is very useful. Unit loads represent the quantity of nutrients generated per unit area and per unit time for each type of nutrient source in the drainage basin. For non-point nutrient sources, for example, unit loads (often called unit area loads) are usually expressed as kilograms of nutrient contributed per hectare of drainage basin area per year (kg/ha.yr). For point sources, unit loads are a function of the number of individuals contributing the nutrients, and are often expressed as kilograms per capita per year (kg/capita.yr). The concept of unit area loads can also be applied to septic systems and feedlots.

Unit load values are most useful for long-term estimates of nutrient inputs (e.g. average annual nutrient load). They are less useful for estimating short-term nutrient inputs (e.g. see discussion on rapidly flushed waterbodies in a following section of this chapter). The use of unit areas loads to estimate the annual nutrient load is described by Rast & Lee (1983).

Although unit loads can be very useful for estimating and comparing nutrient inputs, accurate use of such information requires that one understands the factors that can affect the values obtained, especially the variations that can occur in such estimates. Unit load values can be affected by any factor that contributes to the movement of nutrients from their sources in a drainage basin to the waterbody itself.

The actual transport of nutrients to a waterbody is largely influenced by the quantity of water movement in the basin. The greater the flow, usually the greater the quantity of nutrients transported. In fact, high flows lasting only a few months or less can transport the majority of the annual nutrient load from a watershed to a waterbody within that period of time. A precautionary note is that the values of unit area loads can have a broad range, sometimes spanning several orders of magnitude. However, experience suggests that the unit load concept, nevertheless, can be of considerable value in estimating the relative contribution from various sources, and in comparing data obtained from a specific watershed (e.g. see Rast & Lee, 1983).

Point source loads

The most accurate estimates of nutrient loads are usually those of point sources (e.g. nutrient discharges from both municipal and industrial sources are relatively easy to measure). Many sewage treatment plants and factories monitor their effluent discharges on a routine basis. In such cases, calculations of nutrient loads may be available directly from the discharger or from the appropriate monitoring or regulatory agency. Alternatively, multiplying average annual flows and nutrient concentrations can be used to calculate annual loads. When nutrient data are lacking, a reasonable estimate of the nutrient concentration often can be made using knowledge of nutrient concentrations discharged from similar plants with similar influent waters and treatment processes.

Municipal wastewater treatment plant unit loads can also be estimated from per capita inputs. By knowing the number of people served by a municipal plant and the annual average per capita nutrient contribution in the final effluent, an annual nutrient load can be estimated. Table 7.1 provides examples of average per capita nutrient loadings from municipal sewage treatment plants in the North American Great Lakes Basin. The total municipal load can be calculated as the product of the per capita contributions multiplied by the number of individuals served by the plant.

Table 7.1 Phosphorus unit loads for U.S. sewage treatment plants in the North American Great Lakes Basin (modified from Sonzogni et al., 1980)

	Total phosphorus unit loads* (kg/capita.yr)	
Lake basin	Large plants	Small plants
Superior	0.2	0.5
Michigan	0.3	0.5**
Huron	0.2	0.2**
Erie	0.3	0.6
Ontario	0.2	0.7

*Unit loads are expressed in terms of population equivalents (kg P/capita.yr); large plants have flows equal to or greater than 3800 m^3/day and practice phosphorus removal; smaller plants have flows less than 3800 m^3/day and generally did not practice phosphorus removal.
**A large but undetermined proportion of the small plants in the drainage basin practice phosphorus removal.

This method must be used with caution, however, since the effective per capita phosphorus and nitrogen contributions can vary considerably from location to location, depending on the basic foodstuffs in the region and the food habits of the local population. In areas in which phosphate detergents are used, the effective per capita contribution can be higher than in areas in which such detergents are not used. A summary of general per capita loads for general various types of waste treatment facilities in a drainage basin is presented in Table 7.2.

Small scale waste disposal (septic) systems and urban runoff during storm events may be considered point sources under some conditions. Septic systems can usually trap nutrients in the soil so that they do not reach receiving waters. However, if such systems fail and discharge effluents directly to a watercourse, the above-noted per capita technique can be used to calculate annual loads of untreated sewage. Due to potentially large, rapid and irregular variations in urban storm water discharges during storm events, special monitoring schemes should be carried out for such storm water drainage systems. The monitoring scheme should be developed to reflect the flow variations as closely as possible.

Diffuse (non-point) external loads

Non-point source nutrient loads include land drainage from urban, agricultural and forested areas. As discussed below, estimates of the non-point nutrient load can be obtained by direct measurement of runoff from such land use areas, by measurement of the nutrient load of tributaries draining the area, or by indirect calculations based on unit area loads.

Table 7.2 Characteristics of nutrient load based on waste treatment facilities in catchment area (modified from Technical Standard, 1982)

Type of load	Annual population equivalents[1] in terms of		
	BOD_5	Total N	Total P
1 Inhabitant; no treatment	1	1	1
1 Inhabitant; septic tanks	0.7	0.8	0.8
1 Inhabitant; mechanical treatment	0.7	0.8	0.7
1 Inhabitant; biological treatment	0.2	0.4	0.4
1 Inhabitant; land treatment or advanced treatment with P-precipitation	0.1	0.3	0.1
1 Livestock unit[2]: Diffuse nutrient loss after agricultural utilization (dry farming)	1	5	1
1 Livestock unit[2]: Diffuse nutrient loss after agricultural utilization (liquid manure)	2	12	5
1 Livestock unit[2]: At pasture farming on permanent grassland[3]	0.5	3	0.6
NOTE: Wastewaters from industrial and agricultural production should be similarly evaluated.			
Nutrient flux from crop production/ha of agricultural area[4] (without the influence of organic manure from livestock production)	–	9	0.5

[1]One population equivalent = 54 g BOD; 13 g N; 2 g P. Use the P and N values for P- and N-limited waterbodies, respectively. For waterbodies which are not nutrient-limited, use the BOD_5 population equivalents.
[2]One livestock unit = 1 cattle or horse, 6 pigs, 14 sheep or 150 heads of poultry.
[3]Six months on the pasture, slight slope only; 6 months dry feeding, no access to a waterbody.
[4]The annual nitrogen loss depends on the types of crops and related amounts of fertilizers applied, as follows: for grassland, approximately 10 kg/ha; clover, 30 kg/ha; grain, 40 kg/ha; root crops, 70 kg/ha; vegetables, 80 kg/ha.

Direct rivermouth measurements. The most reliable method for estimating nutrient loads to lakes and reservoirs is direct measurement of these parameters at the mouth of the tributary to the waterbodies. The nutrient load calculated in this manner would include the nutrient input from non-point sources, plus the point sources, in the tributary drainage basin. The latter would be the 'indirect point sources'. An estimate of the non-point source nutrient input can be obtained by subtracting the indirect point source nutrient load from the total load measured at the tributary mouth. In cases where streams and rivers are monitored on a routine basis (e.g. water supply or pollution surveillance), relevant information for making such calculations can be obtained from an appropriate agency or institution.

In collecting rivermouth data, accurate information on flows usually is most critical. In contrast, necessary information on nutrient concentrations can be collected less frequently than the flow information. This is because the concentrations of dissolved nutrients usually do not vary as much as the flows, either from year-to-year or within an annual cycle. However, the quantities of particulate forms of nutrients in a tributary can vary significantly with flow. For tributaries subject to high flow runoff events, it is important to have flow measurements taken during the high flow conditions. Studies in both temperate and tropical regions indicate that a very large portion of the total annual load of nutrients may be contributed during short-term, high-flow runoff events. In North American and European streams, for example, a large portion of the annual nutrient input to waterbodies usually occurs during the spring thaw period. Storms at certain times of the year also may produce high flows and high nutrient loads. LaBaugh & Winter (1984), Winter (1981) and LaBaugh (1985) discuss how uncertainties in the hydrologic measurement can affect the calculation of accurate phosphorus loads for lakes and reservoirs.

Several different calculation techniques are traditionally used to calculate annual nutrient inputs, based on a series of flow and concentration measurements taken over a year (R.V. Smith & Stewart, 1977; Verhoff et al., 1979, 1980; Fricker, 1980b; Westerdahl et al., 1981; Whitfield, 1982; Yaksich & Verhoff, 1983). These techniques include:

1. the average short-term load;

2. the product of the average short-term flow and average short-term concentration; and

3. the product of the weighted mean concentration and the annual flow.

Each of these techniques is useful, depending on the flow and concentration data available. However, for nutrients such as phosphorus, whose concentration in a tributary may increase or decrease with the flow, a procedure known as the flow interval method may provide a more accurate estimate. With this method, an average short-term load, adjusted to reflect the variability of the flow over an annual cycle, is calculated. This latter method is particularly useful when the available data base for a tributary includes daily flow measurements over an annual cycle and nutrient concentration measurements over a range of flows.

With the flow interval method, short-term tributary flows are grouped according to their velocities, resulting in a series of flow intervals. The average flow from each interval is multiplied by the average nutrient concentration measured during any days within the interval to provide an average short-term load for each interval. The annual load then is calculated as the sum of the average separate loads for each interval multiplied by the number of events of the year that the flow falls within an interval. The number of intervals chosen is arbitrary, but depends on the data available. This flow interval method decreases the bias involved in having only a limited number of nutrient measurements available over the annual cycle.

It has been suggested as a minimum effort (Rast & Kerekes, 1982) that one can calculate reasonably accurate nutrient loads based on the average daily flow record for the tributary, and tributary nutrient concentrations taken at biweekly or shorter time intervals. One should also consider whether or not a given nutrient load estimate reflects the 'average' hydrologic conditions (i.e. the long-term average hydrologic input), or else is atypical (i.e. a 'wet' or 'dry' year). As a general rule, the frequency of tributary sampling should reflect the variability of the streamflow.

Indirect measurements. As previously stated, an alternative method for estimating non-point source loads is to use unit area loads based on nutrient export coefficients. This method is of value if it is not possible to measure consistently the tributary inputs, or if no other nutrient load estimates are available. It can also provide an independent evaluation of the accuracy of nutrient load estimates made in other ways.

This approach is based on the observation that, under average hydrological conditions in a watershed over the annual cycle, a given land use activity in the watershed will export a relatively constant load of nutrients per unit land area, to the receiving waters draining the land. The nutrient load, therefore, can be calculated as the product of the areal extent of the various land use activities in the watershed multiplied by the appropriate nutrient export coefficient. The same general approach can be used to calculate the input of nutrients directly to the surface of a lake or reservoir from atmospheric deposition. It is stressed, however, that this approach is less accurate than direct measurement.

A number of studies have been conducted on nutrient export coefficients from various land use categories in different regions of the world, including those of Vollenweider (1968), Loehr (1974), Sonzogni & Lee (1974a, b), Uttormark *et al.* (1974), Omernik (1976, 1977), Wanielista *et al.* (1977), PLUARG (1978a), Rast & Lee (1978, 1983), Duncan & Rzoska (1978) and Reckhow *et al.* (1980).

The use of export coefficients to estimate nutrient loads, however, should be done with caution. This is because land use activities and climatic patterns vary considerably around the world. Most of the above-cited studies, for example, were conducted in temperate regions of the world. However, there are probably no nutrient export coefficients applicable on a global scale. Nutrient export coefficients should be viewed primarily as a method of obtaining rough estimates of annual nutrient loads. Rast & Lee (1978, 1983) discuss methods for assessing the accuracy of phosphorus and nitrogen annual load estimates based on nutrient export coefficients (unit area loads).

If a given set of nutrient export coefficients is not representative for a given watershed, and more detailed information is required, it is possible to measure the nutrient runoff from the watershed directly and calculate export coefficients specific for the land use activities in the watershed. Such site-specific coefficients would nearly always be more accurate than the general values reported in the literature.

Atmospheric inputs. Both wet and dry fallout onto the surface of a lake or reservoir add nutrients directly to the waterbody. The inputs from such sources would be proportional to the lake surface area. The quantity of atmospheric fallout onto the land surface of the drainage basin is already included as part of the rural and urban runoff loads. The relative importance of atmospheric nutrient inputs depends on the magnitude of the inputs from all other sources in the drainage basin.

Atmospheric nutrients can originate well outside the drainage basin of an individual lake or reservoir. Large atmospheric nutrient loads are usually characteristic of waterbodies in close proximity to urban areas, likely a result of combustion emissions. In agricultural areas, the erosion of soils by winds is likely to be a major source of phosphorus. The volatilization of ammonia nitrogen from feedlots and manure storage areas adds to the nitrogen content of the atmosphere. In contrast, atmospheric deposition in forested regions usually contains fewer nutrients than seen in agricultural or urban regions.

Unit loads analogous to the unit area loads for land runoff can be used to estimate atmospheric inputs. Estimates of the annual atmospheric nutrient load per hectare of lake surface are presented in Table 7.3. The human activity in the area can have an effect on the atmospheric unit loads. Deposition rates are generally lowest for forested and non-agricultural rural locations and higher in the urban and agricultural regions.

Rural and urban inputs. A number of sophisticated mathematical models are available for predicting nutrient loads in runoff from arable land. Such models, however, usually require extensive hydrologic data. Therefore, unit

Table 7.3 Atmospheric deposition of total nitrogen and total phosphorus in watersheds grouped according to dominant land use[a] (modified from Hendry et al., 1981)

Dominant land use	Total nitrogen (kg/ha.yr)	Total phosphorus (kg/ha.yr)
Coastal	5.8	0.31
Urban	7.2	0.48
Rural (non-agricultural)	6.2	0.27
Rural (agricultural)	8.8	0.66
Observed range in literature[b]		
Forested lands	0.99–11.3	0.07–0.54
Rural-agricultural lands	10.5–38.0	0.12–0.97
Urban-industrial lands	4.7–24.0	0.26–3.67

[a]Data from Florida, United States.
[b]Based on a number of different studies; modified from Reckhow et al. (1980).

area loads provide an alternative and less complex approach for estimating nutrient loads from rural and urban areas to surface waters.

Unit area loads for urban lands are generally of the same magnitude or higher than rural cropland loads. Table 7.4 contains total phosphorus unit area loads for combined sewered areas, separate sewered areas, unsewered areas, small urban areas, and urbanizing land. As shown in Table 7.4, the degree of industrialization is a factor in the loss of phosphorus from urban lands. Factors such as urban animal populations and tree density also can affect urban unit area loads. Developing urban lands (e.g. construction sites), characterized by large areas of exposed soil, produce large phosphorus runoff loads. However, while unit area loads from construction sites can be very high, the area devoted to such practices in a drainage basin is often relatively small. Therefore, the load from this source will not necessarily be the dominant portion of the total load from an urban area.

Data on nitrogen unit area loads from urban land are scarce. Values ranging from 1.9 to 14 kg/ha.yr have been reported in the literature. The nitrogen unit area loads for developing urban land are usually several times greater than the loads from other urban sources. Figure 7.2 illustrates an approach used in Sweden for obtaining a rough estimate of the nitrogen and phosphorus loss from urban areas to a lake or reservoir, based on the percentage of impermeable surface area and the annual rainfall.

As an example of sub-tropical values, J.A. Thornton & Nduku (1982) reported total phosphorus unit area loads of 14, 280 and 13–53 mg/m^2.yr for 'natural bush', industrial and residential areas, respectively, for the Lake McIlwaine region in Zimbabwe. Total nitrogen unit area loads for these respective land uses are 31, 584 and 85–209 mg/m^2.yr.

Countryside households. On-site septic tanks can serve as wastewater disposal systems for household sewage and various kinds of farm wastewaters. Septic tank systems may or may not be effective in trapping and preventing nutrients from entering a tributary or a lake via ground water.

Table 7.4 Total phosphorus unit area loads (kg/ha.yr) for urban land[a] (modified from PLUARG, 1978a)

Type of urban area	Degree of industrialization		
	Low	Medium	High
Combined sewered areas	9	10	11
Separate sewered areas	1.25	2.5	3.0
Unsewered areas	1.25	–	–
Small urban areas (sewer systems not differentiated)	2.5	2.5	2.5
Urbanizing land (e.g., construction sites)	25	25	25

[a] Based on data from the North American Great Lakes Basin

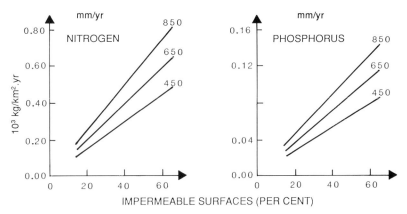

Figure 7.2 Estimation of urban unit area loads based on annual precipitation and impermeable surface area (from Malmquist, 1982)

The success of a septic tank system in preventing nutrient loss depends basically on the soil retention coefficient (S.R.), which provides an estimate of how well a septic tank system can immobilize nitrogen and phosphorus in the soil. The coefficient can range from 0–1.0. For example, if it is assumed that half of the nutrients entering a septic tank eventually reach a body of water, an S.R. value of 0.5 would describe the nutrient retention capability of the system (i.e. half of the nutrients entering the system are retained by the system). If no nutrients reach the waterbody, then S.R. = 1.0.

The soil retention coefficient is dependent upon many factors, including the absorption capacity, natural drainage and permeability of the soil. It also depends on the age of the system (i.e. how long it has been used for infiltration purposes). However, estimates of S.R. still must be based primarily on knowledge of the soil conditions in the region in which each septic tank is located, as well as past experiences in other similar watersheds.

Calculations of nutrient loads from septic tanks can be based on per capita loads, as discussed earlier, multiplied by a factor which describes the ability of the nutrient to move through the soil (calculated as $1 - S.R.$). As a general rule for average septic systems, S.R. values for nitrogen and phosphorus are often in the order of 0.20–0.55 and 0.25–0.40, respectively.

Farm wastewaters. Activities associated with farming may also contribute significantly to non-point nutrient pollution in the drainage basin. Many farm activities related to animal production can produce polluted wastewaters. For example, in some countries, wastewaters from farm households and manure stacks, and silage juice from silos, are often connected to septic tank systems.

Milk rooms and the washing area for milking machinery and implements are two main sources of untreated wastewaters which are often discharged to the nearest watercourse. As a rough estimate, for example, the effluent from a farm with 50 ha of arable land, and 30 milk cows plus replacement stock and one employee, is equivalent to about 13 per capita nitrogen and phosphorus load units (see Table 7.2).

Rural cropland. Cropland, which represents an intensive rural land use, can be a major nutrient source for lakes and reservoirs. Ideally, detailed monitoring data should be used to estimate the nutrient loads from cropland areas. However, such monitoring is expensive, and usually not justified for routine work. Fortunately, considerable information is available on nutrient losses from various types of rural land in different geographical locations. Not withstanding the precautions presented earlier, such data can be normalized on a unit area basis to produce unit area loads that can be extrapolated to similar areas.

Table 7.5 summarizes typical unit area loads for total phosphorus from rural cropland, as a function of the type of surface soil. These unit area loads are typical of total phosphorus losses for cropland in temperate regions with similar climate. However, it is reiterated that unit area loads should be used primarily as a guide, rather than as an absolute methodology for calculating nutrient loads.

Wherever possible, locally measured unit area loads should be used to estimate non-point nutrient loads. When developing unit area loads for specific types of land use, an area-weighted average is desirable. For example, if a cropland in a rural area is comprised of 80 per cent row crops and 20 per cent mixed farming, the annual total phosphorus load from these types of cropland can be estimated separately, and then summed to provide a total load value for the drainage basin.

Unit area loads for total nitrogen (the sum of inorganic and organic nitrogen) tend to be about an order of magnitude greater than the total phosphorus unit area loads. Several international studies have reported nitrogen losses from agricultural lands ranging between 10–20 kg/ha.yr.

Rural non-cropland. This category of land usage comprises pasture, grass-

Table 7.5 Total phosphorus unit area loads (kg/ha.yr) for rural cropland (modified from Sonzogni *et al.*, 1980)

	Surface soil texture				
Land use	Sand	Coarse loam	Medium loam	Fine loam	Clay
Cultivated fields – row crops (low animal density)	0.25	0.65	0.85	1.05	1.25[a]
Cultivated fields – mixed farming (medium animal density)	0.10	0.20	0.30	0.55	0.85

[a] Unit area loads can be higher for soils with very high clay content.

land, wetlands, forests and other idle lands, which usually have a stable ground cover. Consequently, these types of land uses contribute less nutrients to receiving waters than do croplands. Drainage from rural non-cropland areas is not normally a major nutrient source in a drainage basin, unless the drainage basin consists mainly of this type of land use activity. Since nutrient inputs from these sources essentially reflect 'background' conditions, there are few remedial measures available to reduce the nutrient load from such areas. Typical total phosphorus unit area loads for non-croplands are summarized in Table 7.6. As noted, the unit area loads generally increase as the soil becomes more clayey in character.

Information on nutrient losses from wetlands is scarce. Available information suggests that the net phosphorus loss from wetlands over the annual cycle is approximately zero (see Lee et al., 1975). This implies wetlands remove phosphorus from inflowing waters during some parts of the year and release it during other times. However, more intensive use of wetlands (e.g. muck farms) may result in the wetland becoming a sink for phosphorus, thereby removing it from surface waters. Reliable estimates of nitrogen losses from wetlands are not well developed, although there is some evidence that wetlands generally serve as a net nitrogen sink.

The degree of animal activity in the drainage basin also can affect the unit area loads for rural non-cropland. Unconfined animal production in pasture and rangeland does not usually result in significantly increased unit area loads if the activity does not disturb the vegetative cover. In contrast, runoff from land used for manure disposal and for feedlots usually produces high unit area loads (Table 7.7; see also Table 7.5). Intensive livestock operations, such as beef cattle and dairy feedlots, can contribute large quantities of nutrients to receiving waters. In Africa, nutrient inputs from wildlife and birdlife in lakeside game preserves can be substantial, especially in nutrient-poor areas (e.g. see Jarvis, 1982; Kilham, 1982a; Balon & Coche, 1974). The impact of such sources, however, is site dependent. On a unit area basis, the nutrient loads from such sources can be quite high.

Livestock access to streams can also result in the stirring up of bottom

Table 7.6 Total phosphorus unit area loads (kg/ha.yr) for rural non-cropland (Modified from Sonzogni et al., 1980)

Land use and intensity	Surface soil texture					
	Sand	Coarse loam	Medium loam	Fine loam	Clay	Organic
Pasture/range – dairy	0.05	0.05	0.10	0.40	0.60	—
Grassland/idle land	0.05	0.05	0.10	0.15	0.25	—
Forest/wooded	0.05	—	—	—	0.10[a]	—
Wetlands	—	—	—	—	—	0.20

[a] Unit area loads may be higher in certain unique forested areas with clay soils.

Table 7.7 Unit area load values for land used for feedlots and manure disposal (modified from Loehr, 1974)

Land use	Unit area load (kg/ha.yr)	
	Total nitrogen	Total phosphorus
Land receiving manure[a]	4–13	0.8–2.9
Feedlots[b]	100–1600	10–620

[a]Crop or unused land used for manure disposal.
[b]Runoff from confined, non-enclosed animal holding and feeding areas.

sediments by walking in the streams. Grazing livestock can also denude stream shorelines, accentuating erosion and increasing the nutrient input.

In some areas, wastewaters are applied to agricultural land as a treatment and disposal method. Experience in North America suggests that annual unit area phosphorus loads from non-cropland receiving municipal wastewater spray irrigation approximates that from cropland.

Overall, unit area loads for most non-croplands are usually relatively low. Except for feedlots or manured lands, one usually can assume rural non-cropland unit loads to be about one-tenth that of cropland loads.

Diffuse internal loads

Ground water. To date, ground water inputs directly to a lake or reservoir generally have not been considered an important nutrient source. However, the actual significance of this potential nutrient source has not yet been studied in detail, and it may be greater than previously assumed. It is clear, for example, that such factors as the geology of an area and the presence of ground water pollution can affect the nutrient content of ground waters.

Release of nutrients from bottom sediments. Regeneration of nutrients from bottom sediments ('internal loading') can contribute significantly to the total nutrient load to a waterbody. This occurs most commonly under anoxic conditions in hypolimnetic waters, although others (e.g. see Lee *et al.*, 1977, and J.A. Thornton & Nduku, 1982) have reported significant nutrient releases also can occur under oxygenated conditions. Because of incomplete knowledge regarding the exchange of nutrients between the sediments and the water column, the available procedures for estimating the magnitude of internal nutrient loading to a lake or reservoir are limited. Laboratory experiments are helpful for identifying the mechanisms and processes involved in the movement of nutrients between sediments and water, and for obtaining an estimate of the gross nutrient release rate from sediments. However, results obtained from such laboratory experiments may not be directly extrapolated as predictive tools, primarily because of the very specific conditions prevailing during the tests (Ryding, 1985).

At present, the only reliable method for estimating the phosphorus flux between the sediments and the water column involves detailed phosphorus mass balance calculations made at monthly or smaller intervals. This procedure is preferable to laboratory studies of the type indicated above, because of the variability of the latter. For example, comparison of estimates of internal nutrient loading based on laboratory experiments with intact sediment cores with estimates based on mass balance calculations using field studies often show great variability (Ryding, 1985).

Mass-balance studies consist primarily of quantifying the phosphorus input from:

1. The total phosphorus content in the water column for the entire lake or reservoir;
2. All relevant phosphorus inputs to the waterbody; and
3. All relevant phosphorus outflows.

The general calculation for any given time interval is:

$$L_{int} = L_{out} - L_{ext} \pm TP \tag{7.1}$$

where: L_{int} = internal phosphorus load during time interval;
L_{out} = phosphorus loss (outflow) from the waterbody during time interval;
L_{ext} = external phosphorus load (inflow) to the waterbody during time interval; and
TP = difference in total phosphorus content in the water column during time interval.

For a given time interval, the load values can be calculated as the product of the phosphorus concentration and the water volume during this interval (i.e. phosphorus load = phosphorus concentration x volume of water). The internal phosphorus load estimates usually are more accurate when measurements are made at frequent intervals (e.g. at least monthly). In determining the total phosphorus content in the waterbody, one should sample at multiple depths to obtain accurate estimates. The phosphorus mass in any given layer or strata in a lake or reservoir is calculated as the product of the phosphorus concentration in the water layer and the volume of the layer (i.e. phosphorus mass in water layer = phosphorus concentration in water layer x volume of water layer). The total phosphorus content in the waterbody is the sum of the values for the individual water strata.

Calculated at monthly intervals (or shorter, if appropriate information is available), a negative value of L_{int} in Equation 7.1 indicates a net deposition of phosphorus during the month, while a positive value indicates a net release (i.e. internal loading to the waterbody during the month).

Table 7.8 presents an example of a phosphorus mass balance calculation. In this example, from Rotsee (Vollenweider, 1976b), phosphorus mass

Table 7.8 Mass balance approach to estimating internal phosphorus loading to a waterbody (modified from Vollenweider, 1976b)

Date	Total P mass in lake (t)	Total P inflow to lake between dates (t)	Total P outflow from lake between dates (t)	Calculated P mass in lake on date (t)	Measured P mass in lake on date (t)	Difference between calculated and measured P mass[a] (t)
20/03/69	3.06					
(29)[b]		+0.09	−0.41	2.74	1.54	−1.20
18/04/69	1.54					
(34)		+0.11	−0.26	1.39	1.45	+0.06
22/05/69	1.45					
(19)		+0.06	−0.08	1.43	1.64	+0.21
10/06/69	1.64					
(30)		+0.10	−0.14	1.60	1.55	−0.05
10/07/69	1.55					
(35)		+0.11	−0.07	1.59	1.32	−0.27
18/08/69	1.32					
(28)		+0.09	−0.04	1.37	1.61	+0.24
11/09/69	1.61					
(29)		+0.09	−0.04	1.66	1.81	+0.15
10/10/69	1.81					
(27)		+0.09	−0.07	1.83	1.21	−0.62
06/11/69	1.21					
(29)		+0.09	−0.20	1.10	2.02	+0.92
05/12/69	2.02					
(41)		+0.13	−0.51	1.64	2.42	+0.78
15/01/70	2.42					
(34)		+0.11	−0.49	2.04	1.82	−0.22
18/02/70	1.82					
(33)		+0.11	−0.34	1.59	1.50	−0.09
23/03/70	1.50					
Total period of time = 368 days		Annual total P input to lake = +1.18 t	Annual total P outflow from lake = −2.65 t		Sum of P fluxes to sediments (negative values) = 2.45 t	
			Difference between P input and outflow = −1.47 t		Sum of P fluxes from sediments (positive values) = 2.36 t	

[a] Calculated as (measured P mass in lake) − (calculated P mass in lake).
[b] ( ) = number of days between sampling dates.

balance calculations were made at approximately monthly intervals between March, 1969–March, 1970. The phosphorus content in the lake at the end of each monthly interval is calculated on the basis of the measured inflow and outflows during the interval. This calculated difference between these two numbers determines whether or not there is an internal phosphorus load to the lake during the monthly interval. If more phosphorus leaves the lake through the outflow than enters through the inflow, one would conclude that the lake acted as a source of phosphorus (i.e. internal loading from the sediments). Conversely, if less phosphorus flows from the lake than entered it, the lake would be retaining phosphorus (i.e. phosphorus sedimentation).

Several useful points of information are contained in Table 7.8. First, over the annual cycle (at least for the study period), the lake acted as a net source for phosphorus. Approximately 1.18 t of phosphorus entered the lake, compared with the lake outflow of approximately 2.65 t. Thus, lake sediments supplied a total of approximately 1.47 t of phosphorus to the lake over the annual cycle. Furthermore, based on the monthly phosphorus fluxes, it can be seen that the lake functions as a phosphorus source during some months and as a phosphorus sink during other months. In this example, these periods correspond roughly to the thermal stratification period and the period during which the lake was completely mixed, respectively.

The necessary data to make such calculations as presented in Table 7.8 are usually already being measured in routine eutrophication monitoring programmes. Consequently, such calculations are relatively easy to make. In fact, one also can attempt to make calculations of the phosphorus flux between the epilimnion and hypolimnion with the same general mass balance approach (see Vollenweider, 1976b).

It is emphasized that this procedure will allow one to make estimates only of the net phosphorus flux. Nevertheless, the information gained with this procedure is likely to be very useful for estimating the magnitude of the internal loading of phosphorus, relative to the external sources. Nitrogen deposition or release can also be roughly approximated, by using conversion factors based on the N:P ratio of the material residing in the surface layer of the sediment.

As a practical observation, a high internal loading of phosphorus during the summer months (especially if the internal load is the main phosphorus input to the waterbody during this period) can delay expected improvements in the water quality of a lake or reservoir. This is especially the case for eutrophication control programmes based on reducing the external phosphorus load.

Nevertheless, it is usually desirable to first bring the external phosphorus load down to a satisfactory level before applying in-lake treatment methods to a waterbody exhibiting excessive internal loading (see Chapter 9). Otherwise, the in-lake control measures can be ineffective in the eutrophication control effort.

Further details on mass balance calculations of the internal loading of phosphorus to a lake or reservoir are provided in other sources, including the reports of Serruya (1975), Vollenweider (1976b), Larsen & Malüeg (1980), Jacoby et al. (1982) and Nürnberg (1984).

Nitrogen fixation. There still is a great deal of uncertainty over the relative importance of nitrogen fixation to the total nitrogen input to a lake or reservoir. This process, whereby microorganisms convert nitrogen gas in the atmosphere to organic forms of nitrogen, can be a nitrogen source in some lakes, particularly eutrophic lakes with large populations of blue-green algae. One study of a eutrophic North American lake suggested that nitrogen fixation accounted for almost 10 per cent of the total nitrogen load (Torrey & Lee, 1976). In both, Lake George and Reitvleidam in Africa, nitrogen equivalent to up to 50% of the inflow has been measured (Horne & Viner, 1971; Ashton, 1981). An accurate estimate of this input usually requires sophisticated experimental work.

Shoreline erosion. This nutrient source may be significant for lakes with shorelines that are long, relative to their total drainage basin. However, although shoreline erosion can contribute large quantities of total phosphorus to a lake, a large proportion of this phosphorus is usually tightly bound to sediments. Consequently, it does not usually represent a significant source of biologically available phosphorus. Thus, shoreline erosion is generally ignored as a significant nutrient source (PLUARG, 1978a).

Waterfowl. Wastes from waterfowl can be a significant nutrient source for small lakes used by large numbers of birds. For example, geese and ducks can consume vegetation and grain on land and then transport nutrients from such vegetation to lakes. An example of such potential nutrient inputs to lakes and reservoirs is provided below in Table 7.9.

Summary comments on nutrient load quantification

Observed ranges in unit area loads (primarily for the temperate zone) are compared in Figures 7.3 and 7.4. It is clear from these summaries that human activities in a drainage basin and land use intensity can significantly impact the nutrient contributions from non-point sources.

A SIMPLIFIED APPROACH FOR ESTIMATING THE ANNUAL NUTRIENT LOAD TO A LAKE OR RESERVOIR

A general overview of the factors to be considered in watershed nutrient load models was presented previously in Chapter 6. This section provides an example of a simplified approach for estimating the nutrient load to a lake from a hypothetical watershed. It is based primarily on the technique developed by Monteith et al. (1981) for choosing between point and non-

Table 7.9 Nutrient load characteristics based on direct utilization of a waterbody (from Technical Standard, 1982)

Type of load	Annual population equivalents[a] in terms of		
	BOD_5	Total N	Total P
1. Duck breeding with free access to the water (100 ducks/day)	14	8–12	16
2. Duck farming on slatted floors with swimming troughs (100 ducks/day)	11	7	10
3. Geese breeding with free access to the water (100 geese/day)	42	24–36	48
4a. Trout cultivation in channels per 1000 kg true rate of stocking/day[b]	85	110	110
4b. Average load per day at an annual production of 1000 kg fresh weight of fish in net cages in channels	30	30	30
5. Recreation loads caused by primary body contact (100 bathing persons/day; seasonal mean)	2	2	2

[a] One population equivalent (P.E.) = 54 g BOD; 13 g N; and 2 g P. Use P and N values for P- and N-limited waterbodies, respectively. For non-nutrient-limited waterbodies, use P.E. values of BOD_5.

[b] Since approximately 50% of the nutrients are bound to particles, it should be possible to reduce the load by almost 50% using settling basins.

point control strategies.

In evaluating nutrient loads (and land use impacts) originating from various point and non-point sources in the drainage basin, one important problem is determination of the proportion of the various sources contributing to the composite nutrient load measured at a downstream section (e.g. the river mouth to a lake or reservoir). The composite downstream nutrient load at the rivermouth is not simply the algebraic sum of all nutrient inputs in the watershed, since various chemical, physical and/or biochemical processes can take place within the watercourse during the period of downstream travel of the nutrients. These processes can markedly affect the water quality profile along a watercourse, as well as the nutrient load profile (see Figure 7.5).

Thus, the means for providing a link between the various sub-system inputs and converting them to a total input (i.e. total nutrient load) are generally missing. In many simplified (or even complex) nutrient transport models, this problem is 'solved' by introducing a single factor or term referred to as the retention rate, delivery rate, transport rate, etc. Unfortunately, such terms oversimplify and/or confuse the linkage between the nutrient sources in the drainage basin and the total nutrient load which ultimately reaches a lake or reservoir. In selecting an appropriate eutrophication control strategy in a real situation, however, consideration of this linkage can be very important.

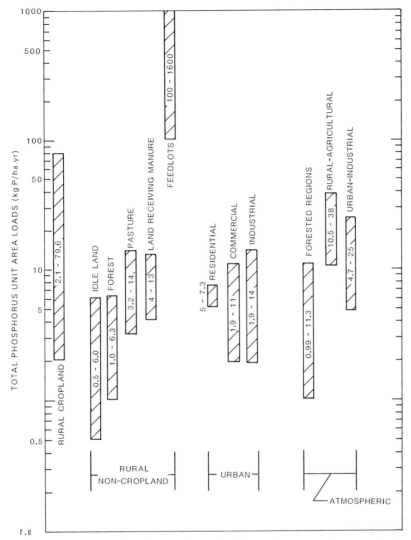

Figure 7.3 Range of total phosphorus unit area loads for various non-point sources (updated from Loehr, 1974)

Many complex watershed models are difficult to use simply because of the lack of appropriate input data. Consequently, more simplified techniques, based primarily on extrapolations from unmonitored areas, are a practical choice. This section discusses the use of a simplified approach for estimating the annual phosphorus input to a lake.

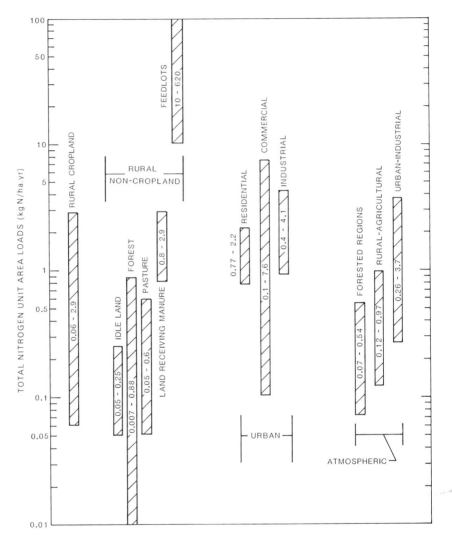

Figure 7.4 Range of total nitrogen unit area loads for various non-point sources (updated from Loehr, 1974)

Specific steps in use of simple watershed model

The first step of a simplified approach (Monteith *et al.*, 1981) for estimating the total nutrient input from the drainage basin to a waterbody is to define the boundaries of the drainage area. The drainage basin is further subdivided into sub-basins, generally based on the drainage of smaller feeder streams to the main tributary. Hydrologic or other physiographic maps can be used

Estimating the nutrient load to a waterbody

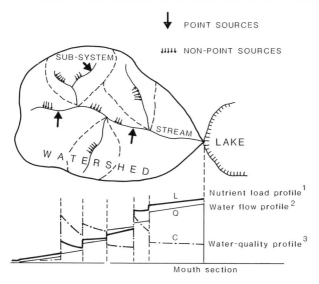

[1] Relative nutrient load to lake; increasing load indicated in upward direction.
[2] Relative streamflow; increasing streamflow indicated in upward direction.
[3] Relative water quality in stream; degraded water quality indicated in upward direction.

Figure 7.5 Schematic of drainage basin, showing load, flow and water quality profiles (from Jolánkai, 1984)

to define tributary boundaries and determine the sub-basin areas. If adjacent sub-basins have homogenous surface soil texture and hydrology, they can be grouped together. An example of a hypothetical tributary drainage basin divided into such sub-basins (dashed lines) is presented in Figure 7.6.

After establishing the sub-basin boundaries, each important nutrient point source and the location of its discharge are identified. Urban areas also are identified and their areas estimated. The urban areas can be classified, based on whether they are served by separate storm sewers, combined sewers, or are unsewered. The remaining land areas can be considered rural land, which can be further sub-divided into cropland and non-cropland, countryside households, farming activities, etc. Such additional subdivisions allow more detailed estimates of nutrient loads to be made. For accounting purposes, each point and non-point nutrient source in the drainage basin is assigned an identification number. Nutrient contributions from these various sources can then be estimated on the basis of unit area or per capita loads, or else measured directly at the appropriate locations.

The tributary is then divided into several river stretches, separated by 'points of entry'. The point of entry represents the point at which all the nutrient sources within a given sub-basin enter the main channel. Thus, the point of entry usually corresponds to intersections of sub-basin streams

Control of eutrophication

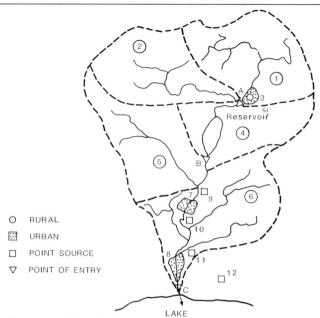

Figure 7.6 Schematic of hypothetical drainage basin (modified from Johnson *et al.*, 1978; Monteith *et al.*, 1981)

with the main channel, or to positions upstream and downstream of river features such as reservoirs or low gradient flood plains which could prevent nutrients from moving downstream.

Any 'upstream' lakes or reservoirs in a watershed must also be investigated. An upstream waterbody can act as an effective nutrient trap, retaining a large part of the phosphorus which enters it. In such cases, the phosphorus load would not enter the downstream waterbody of concern. Therefore, the total phosphorus load to the downstream waterbody may be overestimated. The same situation can occur when significant quantities of the phosphorus load enter an upstream sub-basin or arm of the main waterbody. It is possible to correct a phosphorus load estimate for such retention effects by calculating the phosphorus retention coefficient for the upstream lake or sub-basin.

In such cases, the phosphorus retention coefficient, $R(P)$, defines the fraction of the total phosphorus load entering a waterbody which is subsequently retained in the waterbody (by sedimentation). Therefore, a phosphorus retention coefficient of 0.6 indicates that 60% of the phosphorus load entering a waterbody is retained in it. A convenient calculation of the phosphorus retention coefficient, $R(P)$, is as follows (Larsen & Mercier, 1976):

$$R(P) = \sqrt{t_w}/(1 + \sqrt{t_w}) \tag{7.2}$$

where: t_w = water residence time (yr);
[= lake volume (m^3)/annual inflow volume (m^3/yr)]

The phosphorus load entering an upstream lake or reservoir can be multiplied by the appropriate value of the phosphorus retention coefficient [calculated as $1 - R(P)$] to define the quantity of the nutrients which actually leave the upstream waterbody.

The relative distance of specific nutrient sources in a drainage basin from a receiving waterbody also should be considered. In setting nutrient control priorities, attention should be directed first to the large sources of 'biologically-available' phosphorus (see earlier discussion on this topic) in the watershed located closest to the waterbody of concern. Nutrients from such sources would likely reach the waterbody more rapidly than from more distant sources.

After all the nutrient sources in a watershed have been identified and quantified, the total input to a lake or reservoir at the tributary mouth can be determined. It is noted that a lake or reservoir may have several discrete watersheds draining into it. Each individual watershed should be analyzed separately.

The total nutrient load to a lake or reservoir is the sum of the inputs from individual watersheds, plus the direct point sources. This can be expressed as:

$$L = L_T + L_{DS} + L_{DD} + L_A + L_G + L_S \qquad (7.3)$$

where: L_T = total tributary load (summed up for all the rivers entering the lake or reservoir);
L_{DS} = direct sewage load;
L_{DD} = direct diffuse load (urban and rural);
L_A = atmospheric input;
L_G = ground water seepage; and
L_S = sediment release of nutrients in lake.

The tributary load (L_T) will include various load components, such as indirect sewage discharges to a given river or one of its tributaries, as well as non-point sources (e.g. agricultural and urban). To illustrate the derivation of L_T, a conceptual physical layout of the hypothetical watershed in Figure 7.6 is presented in Figure 7.7. Based on the information in Figure 7.7, the individual nutrient inputs in the watershed can be added together to get an estimate of the total nutrient load to the waterbody (Table 7.10).

The use of transmission coefficients (the fraction of the nutrient load transmitted between any two points of entry of the river stretch) helps account for nutrient losses within the river system itself. If the transmission coefficient is equal to one, the nutrient delivery to the receiving waterbody is 100 per cent, implying no nutrient loss occurs within the river stretch. Values less than one indicate the degree to which nutrients are retained within the stretch of the river.

The concept of 'effective transmission', the quantity of phosphorus transmitted from the point of entry to the river mouth is also shown in Figure 7.7. It is emphasized that the 'effective transmission' is distinct from

Control of eutrophication

Source	Point of Entry	Area (ha)	Special Features
1 (a) Cropland (b) Non-cropland	A A	5,000 10,000	Clay soils, row crops Clay soils, forest
2 (a) Cropland (b) Non-cropland	A A	9,000 4,000	Clay soils, row crops Clay soils, woodland
3 Unsewered	A	500	Little industry
4 (a) Cropland (b) Non-cropland	B B	15,000 3,000	Loam soils, mixed farms Loam soils, pasture (dairy)
5 (a) Cropland (b) Non-cropland	C C	12,000 7,000	Sandy soils, mixed farms Sandy soils, forest
6 (a) Cropland (b) Non-cropland	C C	20,000 2,000	Loam soils, row crops Loam soils, dairy farms
7 (a) Combined storm sewer (b) Unsewered	C C	500 500	Some industrialization
8 (a) Separate storm sewer (b) Combined storm sewer	C C	1,000 1,000	Heavily industrialized Heavily industrialized
9 Municipal	C	--	Primary treatment Population Served: 10,000
10 Municipal	C	--	Activated sludge Population Served: 30,000
11 Industrial	C	--	Electroplating
12 Municipal	C	--	Activated sludge Population Served: 100,000

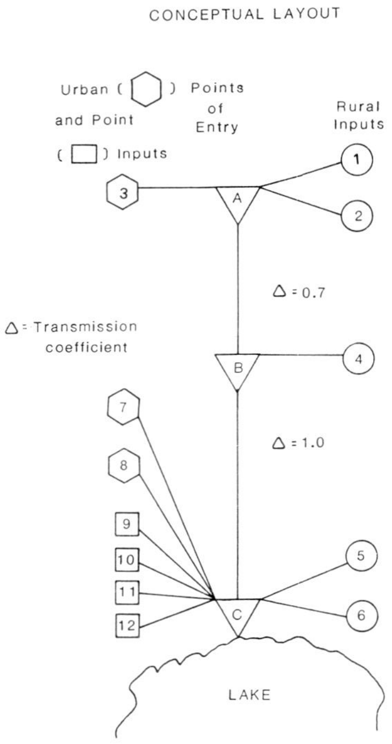

Figure 7.7 Physical layout and features of hypothetical drainage basin (modified from Monteith *et al.*, 1981)

142

Estimating the nutrient load to a waterbody

Table 7.10 An example of a calculation of contributions from various sources within a hypothetical watershed[a] (Modified from Sonzogni et al., 1980; Monteith et al., 1981)

Source	Load at point of entry (kg/yr)	Effective transmission	Load adjusted to tributary mouth (kg/yr)
1(a)	6,250	0.7	4,375
(b)	1,000	0.7	700
2(a)	11,250	0.7	875
(b)	400	0.7	280
3	625	0.7	440
4(a)	4,500	1.0	4,500
(b)	300	1.0	300
5(a)	1,200	1.0	1,200
(b)	350	1.0	350
6(a)	17,000	1.0	17,000
(b)	200	1.0	200
7(a)	5,000	1.0	5,000
(b)	625	1.0	625
8(a)	3,000	1.0	3,000
(b)	11,000	1.0	11,000
9	8,000	1.0	8,000
10	27,000	1.0	27,000
11	0	1.0	0
12	90,000	1.0	90,000
			TOTAL: 181,845

[a] See Figure 7.7 for the characteristics of the hypothetical drainage basin.

the 'transmission coefficient', the latter being the quantity of phosphorus transmitted between any two points of entry. Every stretch of river between any two points of entry has a transmission coefficient associated with it. By contrast, the effective transmission is the product of all transmission coefficients at a given entry point. In Figure 7.7, for example, the effective transmission at point of entry A is equal to 0.7 (i.e. 0.7 x 1.0). If the transmission coefficient between points of entry A and B was 0.9, the effective transmission of nutrient sources entering at point A would be 0.63 (i.e. 0.7 x 0.9). Monteith et al. (1981) provide further discussion of the transmission coefficient and effective transmission.

The basic approach outlined here has been used successfully in several waterbodies. Prominent examples include the North American Great Lakes (PLUARG, 1978a; Johnson et al., 1978; Sonzogni et al., 1980; Monteith et al., 1981) and Lake Balaton (Somlyódy & van Staten, 1986). Additional guidelines can be obtained from these references.

NUTRIENT ESTIMATES FOR WATERBODIES WITH RAPID FLUSHING RATES

The approach discussed above may have to be adjusted for waterbodies with rapid flushing rates. This is necessary because the use of annual phosphorus load estimates in simple eutrophication models (see Chapter 6) can produce erroneous results for rapidly flushed waterbodies. Such models assume that the entire phosphorus load to a waterbody during a given calendar year can affect its biological response (e.g. algal biomass) during that year. However, in a rapidly flushed waterbody, a certain portion of the phosphorus load entering the waterbody at the beginning of the calendar year may be flushed out of the waterbody before the onset of the growing season. Thus, this portion would likely have no impact on algal growth during the growing season. As a result, one may erroneously conclude that the annual phosphorus load did not have as significant an effect on algal biomass as expected.

In this situation, Ryding & Forsberg (1979) suggest that one consider only the 'hydrological relevant P-load' when using simple phosphorus models of the type developed during the OECD (1982) eutrophication study. Ryding & Forsberg identify the hydrological relevant P-load as the phosphorus load entering the waterbody during the growing season, plus the load entering the waterbody during the period corresponding to one flushing time prior to the growing season. This latter period is calculated as the lake discharge (m^3.yr.): lake volume (m^3) ratio.

Ryding & Forsberg (1979) used a monthly discharge:volume ratio of 1.0 to screen for waterbodies with rapid flushing rates. They showed that for Lake Glaningen in Sweden (hydraulic residence time, $t_w = 0.21$ yr), one would only include the phosphorus load during one month prior to the growing season (plus the growing season input) to calculate the hydrological relevant phosphorus load for 1975. In contrast, for Lake Malmsjön ($t_w = 1.0$–1.7 yr), the hydrological relevant phosphorus load for 1975 included the phosphorus input for the last two months of 1974. Lake Malmsjön had a high phosphorus input for the last two months of 1974. Although this input was of no consequence during the 1974 growing season, it was important in assessing the biological response of the lake during the 1975 growing season. Using data from eight shallow, phosphorus-enriched lakes in Sweden, Ryding & Forsberg (1979) also showed that, for 80% of the lakes with t_w values less than 0.4 years, the hydrological relevant phosphorus load was about half the annual load. Overall, both the annual and hydrologically relevant methods of calculating the phosphorus load gave similar results with hydraulic residence times between 0.2–0.9 years.

R.A. Jones & Lee (1979) used a similar rationale in calculating the biologically relevant phosphorus load for Lake Lillinonah in Connecticut (United States). Walker (1985) also discusses calculation of the nutrient load based on a time period other than the annual cycle.

Kerekes (1975) provides an example of the effects of rapid flushing rates on lake metabolism. He reported that the phosphorus concentrations in undisturbed, oligotrophic lakes in Kejimkujik National Park in Nova Scotia (Canada) were proportional to the hydraulic load (expressed as flushing rate) up to a certain point. After a flushing rate of about seven times per year was reached, the phosphorus concentration exhibited a plateau value. This plateau value did not change appreciably as the flushing rate increased further, even though the phosphorus input might be increasing as the hydraulic load was increasing. Kerekes interpreted this to mean that, when the flushing rate increased to seven times per year or greater in his study lakes, an inflow-outflow steady state was achieved, and water leaving the lake was carrying out solutes at the same rate they were being supplied to the waterbody.

These examples indicate that both the hydraulic flushing rate and the phosphorus residence time must be considered when assessing the predicted response of a rapidly flushing waterbody to a eutrophication control programme, especially one based on reduction of the external nutrient load.

It also is noted that the hydraulic residence time (t_w) is inversely related to the flushing rate (i.e. $t_w = 1/$flushing rate). As used in this document, the hydraulic residence time denotes the time necessary to replace a volume of water equivalent to the volume of the lake/reservoir basin. Sometimes called the 'filling time', it does not represent the time necessary to replace every original water molecule in the waterbody, but rather the time necessary to replace a water volume equivalent to that of the waterbody.

In the OECD (1982) international eutrophication study, the water residence time was calculated as the quotient of the waterbody volume divided by the annual inflow volume (i.e. t_w = waterbody volume (m^3)/annual inflow volume (m^3/yr)). This formulation assumes that the precipitation and evaporation rates at the lake/reservoir surface, as well as the ground water inflows and outflows, are approximately equal over the annual cycle. The degree to which these assumptions are true in a given situation often is unclear. Others have calculated the water residence time as the waterbody volume/annual outflow volume. Ideally, the water residence time should be calculated as the waterbody volume divided by the net sum of all water inputs and outflows (i.e. t_w = waterbody volume/net hydraulic load). Obviously, the most accurate calculations are made on the basis of frequent, precise measurements of most or all of the significant water inflows and outflows of a waterbody. LaBaugh and Winter (1984), Winter (1981) and LaBaugh (1985) discuss how uncertainties in the hydologic measurements can affect the accuracy of annual nutrient load estimates for a waterbody.

Palmer (1975) earlier suggested calculating the water residence time in coastal regions on the basis of the measured concentration gradients of conservative substances (e.g. dissolved solids). Using measurements of conservative mass exchanges over the annual cycle, the coastal water residence time was calculated as the volume of the conservative substance in the coastal region divided by the mass exchange under steady-state conditions. This approach may be useful for calculating the water residence time for sub-basins or 'arms' of a dendritic waterbody. However, on a practical level, it does not appear to be useful as a general methodology for lakes and reservoirs.

Overall, the results of the OECD (1982) study suggest that calculation of the water residence time as the waterbody volume (m^3)/annual inflow volume (m^3/yr) is sufficiently accurate for use in most cases. The major exceptions are waterbodies with large, irregular inflows or outflows over the annual cycle, and waterbodies with no specific input or output tributaries (e.g. 'terminal' lakes).

RELIABILITY OF NUTRIENT LOAD ESTIMATES

After the nutrient sources in the drainage basin have been identified and an estimate made of the annual phosphorus and nitrogen loads to a lake or reservoir, the reliability of the calculated nutrient load estimates should be evaluated. This is a difficult task, since there are no absolute standards available for identifying major errors. As pointed out earlier, the most accurate estimates of nutrient loads are normally those made with data from actual monitoring of the systems under study. By contrast, unit area or per capita loads will only provide approximations of such inputs. Nevertheless, the latter often represent a practical choice for assessing the accuracy of total nutrient load estimates, especially if the intent is to determine the relative importance of various nutrient sources in the drainage basin.

Even if a nutrient source is actually monitored, however, it is rare that it is sampled sufficiently to produce accurate estimates of nutrient loads within ± 25 per cent. In some situations, it may be desirable to estimate an upper and lower range of total loads. If the ranges do not overlap, the relative importance of nutrient sources can then be established.

Experience and knowledge of the local area under study is very important in making accurate nutrient load estimates. Thoughtful application of all the factors that can affect the magnitude of the loads, as discussed earlier in this chapter, will help reduce uncertainty in the load estimates. Even so, error and bias can occur when extrapolating unit area loads from one watershed to another, or even from one time period to another in the same watershed. It is virtually impossible to define and quantify all uncertainty components in nutrient load estimates. Therefore, the best practical advice is to remain aware of potential sources of error in estimating unit area nutrient loads and to interpret results accordingly. Rast and Lee (1978, 1983) and Reckhow et al. (1980) discuss methods for assessing the accuracy of nutrient load estimates, based on the characteristics of the drainage basin and (or) the lake basin.

CHAPTER 8

GUIDELINES FOR SAMPLING A WATERBODY

Determination of the trophic state and general water quality represents the core of any assessment or classification of lakes and reservoirs. As noted earlier, the objectives of eutrophication management programmes should be clearly defined, and primary attention should be given to obtaining the information considered essential to the implementation of these objectives. The reliability of eutrophication control programmes can be no better than the reliability of the data bases used in developing the programmes.

Many eutrophication measurement programmes normally focus on nutrient inputs from the land and the atmosphere, and how the nutrients and other related materials partition themselves among the biotic and abiotic components of the aquatic environment. The goal of this chapter is to provide general guidelines for obtaining the necessary in-lake information for development of effective eutrophication management programmes.

WHAT TO SAMPLE

In addition to the primary causative variables of eutrophication (mainly nutrients), water quality parameters which reflect the impacts of eutrophication also have to be measured and assessed.

It is necessary to collect analytical data for several reasons, including:

1. Assessment of the 'average' condition of a waterbody during a specific time interval (e.g. seasonal or annual);
2. Classification of a lake or reservoir (see Annex 1);
3. Assessment of the load/response relationship of a lake or reservoir (see Chapter 6);
4. Selection of appropriate control measures for external (drainage basin and atmospheric sources) and internal (sediment regeneration and ground waters) growth-limiting nutrients (see Chapter 9); and

5. Prediction of changes in trophic state and/or water quality by means of mathematical models (see Chapters 6 and 11).

In selecting appropriate analytical procedures for determining the values of water quality parameters related to eutrophication, several factors should be considered, including:

1. The required rapidity of the analysis;
2. The required sensitivity and detection limits;
3. The constraints on accuracy and precision; and
4. The total number of analyses to be done per year.

A water sample, whether measured in the laboratory on in the field, should be indicative of the actual condition of the waterbody at the time and point of collection. This type of sample can be considered a representative sample of the waterbody for the component or components of interest. Furthermore, the sample should provide a true description of the temporal and spatial variations in eutrophication-related water quality for the duration of the sampling programme. This latter characteristic describes a valid sample. Satisfactory sampling should be both valid and representative. Thus, a non-representative sample cannot be valid.

Because of this requirement, the following factors must be considered in designing an adequate eutrophication sampling programme:

Validity of the samples	*Representativeness of the samples*
Validity of the sampling sites	The necessary sample size
Validity of the sampling frequency and timing	A network of single samples at random versus integrated samples
	Sample collection (discrete intervals versus continuous hauls)
	Sample transportation and storage

A detailed discussion of these factors and how to assess them is provided by Mancy & Allen (1982).

Essential criteria for assessing the eutrophication-related water quality in a lake or reservoir are outlined in Table 8.1. In addition, the minimum parameters to be measured at the sampling site itself include:

1. temperature and dissolved oxygen profiles;
2. pH;
3. specific conductance; and
4. Secchi depth.

Table 8.1 Primary parameters for assessing the eutrophication status of a waterbody

	Parameter	Units[a]
I.	*Morphometric conditions:*	
	Lake surface area	km^2
	Lake volume (average condition)[b]	10^6 m^3
	Mean and maximum depth	m
	Location of inflows and outflows	–
II.	*Hydrodynamic conditions:*	
	Volume of total inflow (including ground water) and outflow for different months	m^3/day
	Theoretical mean residence time of the water (renewal time, retention time)	yr
	Thermal stratification (vertical profiles along longitudinal axis, including the deepest points)	–
	Flowthrough conditions (surface overflow or deep release, and possibility of bypass flow)	–
III.	*In-lake nutrient conditions:*	
	Dissolved reactive phosphorus; total dissolved phosphorus; and total phosphorus	µg P/l
	Nitrate nitrogen; nitrite nitrogen; ammonia nitrogen; and total nitrogen	µg N/l
	Silicate (if diatoms constitute a large proportion of phytoplankton population)	mg SiO$_2$/l
IV.	*In-lake eutrophication response parameters:*	
	Chlorophyll *a*; Pheophytin *a*	µg/l
	Transparency (Secchi depth)	m
	Hypolimnetic oxygen depletion rate (during period of thermal stratification)	g O$_2$/m^3.day
	Primary production[c]	g C/m^3.day; g C/m^2.day
	Diurnal variation in dissolved oxygen[c]	mg/l
	Dissolved and suspended solids[c]	mg/l
	Major taxonomic groups and dominant species of phytoplankton, zooplankton and bottom fauna[c]	–
	Extent of attached algal and macrophyte growth in littoral zone[c]	–

[a]The terminology and units proposed by the International Organization of Standardization is recommended for expressing the parameters.
[b]A bathymetric map and a hypsographic curve also is necessary in many cases.
[c]Can provide additional information on the trophic conditions of a waterbody; recommended if resources are adequate or if special situations require more detailed information (also see Table A.3 in Annex 1).

Dissolved oxygen may also be determined in the laboratory, after appropriate preservation in the field. Although many samples can be stored for long periods if properly preserved, the shorter the time interval between the collection of a sample and its analysis, the more reliable the results will be.

Microbial activity can change the nitrate/nitrite/ammonia balance, as well as the concentration of dissolved reactive phosphorus and dissolved total phosphorus, in a water sample. Ideally, therefore, filtration of the water sample at the sampling site should be done to allow the most accurate determination of these parameters. If this is not possible, one usually is able to obtain reasonably accurate values if the time between sample collection and analysis does not exceed a couple of hours. Transportation of such samples should be in a cooler, or appropriate cold storage container, and samples should be kept in the dark. This precaution is also valid for biological samples.

A collection of acceptable sample preservation techniques and analytical methods for measuring the necessary water quality parameters is provided by Golterman (1971), Wetzel & Likens (1979) and American Public Health Association *et al.* (1985).

As a general rule, analytical results also should include information on the statistical characteristics of the data, including its accuracy, precision, etc. It also is important to attempt to provide an estimate of the potential error associated with both the sampling strategy and the analytical results. Ekedahl *et al.* (1982) discuss the topic of analytical error associated with laboratory analysis. In addition, Chapters 6 and 7 provide some discussion of uncertainty and potential error related to eutrophication modeling and estimation of the annual nutrient load, respectively.

NECESSARY TEMPORAL AND SPATIAL RESOLUTION FOR DATA

If sampling techniques are not carefully selected, analytical results may be partially or completely invalid for their intended uses. In particular, samples must be collected and handled so that, upon measurement, the values of the parameters of interest are the same as those in the waterbody at the time of sampling (see discussion of representative samples above). This topic is discussed further by Mancy & Allen (1982) and Wilson (1982).

Where to sample

Selection of sampling sites and frequency of sample collection depend largely on the morphometry and hydrodynamic properties of a waterbody (OECD, 1982). Therefore, it is important to know the position of all major inflows and outflows, and any related 'short-circuiting' of water and associated materials entering the waterbody. The short-circuiting of inflow materials can occur when major inflows and outflows of a waterbody are located relatively close to each other. In this situation, inflow waters and the materials carried in them may exit through a nearby outflow tributary before they have the opportunity to mix with the main volume of the waterbody. The possibility of water short-circuiting can be assessed with appropriate measurements of temperature and/or specific conductance, as discussed in the following sections.

All major tributaries must be considered, including measurement and/or calculation of both the concentrations and masses of nutrients and other parameters of interest entering the waterbody. Within the waterbody itself, selection of sampling locations should make allowances for the possibility of a heterogeneous distribution of water quality. Standing waters can be thermally or chemically stratified throughout portions of the year. Thus, particulate materials of differing densities from that of the lake water tend to be distributed heterogeneously. This is especially the case for suspended silt in the water column and for blue-green algae with gas vacuoles. Winds can often blow such algae to one side of a lake or reservoir.

In principle, winds also can cause lateral heterogeneity in the distribution of dissolved materials. The vertical or horizontal distribution of such materials can be assessed easily with rapid field tests for temperature, dissolved oxygen, electrical conductance and turbidity.

Spatial consistency in the siting of sampling stations in the waterbody is important, both for assessment of changes in the values of measured water quality parameters over time and for insuring a reference point against which water quality in other parts of the waterbody can be assessed. Thus, as a practical matter, one normally would not change the location of a sampling station for a given parameter or group of parameters once a sampling programme has begun. If one desires to sample another location in a waterbody, it is better to add a new sampling station, rather than change the location of an existing station. The exception would be cessation of a sampling programme at a given site, coincident with the initiation of a new sampling programme.

During the stratification period, the location of the thermocline can be determined by vertical temperature profiles at 0.5 m intervals. The point at which the temperature shows the maximum change (decrease) per depth interval corresponds to the location of the thermocline. A temperature change of $1°C$ or greater per meter of depth can be used as a rule-of-thumb in temperate lakes and reservoirs. Lee & Jones (1980) suggest that for temperate zone waterbodies with surface temperatures of $20°C$ or greater, a temperature decrease of $0.5°C$ or more per meter of depth signifies a stratified waterbody. Any significant increase in specific conductance with depth might signify salt-related density stratification.

The required number of vertical samples depends on the purpose of the measurement. For non-stratified waterbodies, it usually is adequate to take samples approximately 0.5 m below the surface, 0.5 m above the bottom, and at mid-depth.

Ideally, for stratified waterbodies, samples should be taken approximately 0.5 m below the surface, middle of the epilimnion or bottom of the euphotic zone, approximately 1.0 m above and below the thermocline and approximately 0.5 m above the bottom.

Shallow waterbodies may not be sufficiently deep to sample all five depths. In such cases, one should consider both the relative proportion of the epilimnion and hypolimnion, and the depth of the euphotic zone in selecting the sampling depths. However, shallow lakes and reservoirs may not exhibit a sustained stratification during the growth season, especially in windy areas. In such cases, an integrated sample (e.g. collected with a hose lowered through the water column, thereby obtaining water from each water layer) may provide adequate information on average in-lake nutrient and chlorophyll levels.

If resources are inadequate to sample all five depths, the two thermocline samples might be omitted, although some valuable information may be lost (e.g. see following discussion on metalimnetic chlorophyll maxima).

An exception to these general sampling depth guidelines are waterbodies which exhibit maximum chlorophyll levels in the thermocline or the hypolimnion. Several researchers have reported this phenomenon (e.g. Eberly, 1959; Ichimura et al., 1968; Baker & Brook, 1971; Watson et al., 1975; Fee, 1976; Pick et al., 1984; Moll et al., 1984). The occurrence of non-epilimnetic maximum chlorophyll values has been suggested as a relatively common characteristic of Alaskan lakes (personal communication, Paul Woods, U.S. Geological Survey, 1985).

The possibility of non-epilimnetic chlorophyll maxima can be assessed by initial reconnaissance sampling of a waterbody for chlorophyll concentrations, optical density and/or dissolved oxygen concentrations during the growing season. Examination of water column profiles usually will indicate the presence of sub-surface peaks of such parameters. Small sampling intervals will be necessary, since chlorophyll maxima can be concentrated in very small water layers in some cases.

It also is mentioned, however, that it is not clear how knowledge of the occurrence of non-epilimnetic chlorophyll maxima should be used in assessing the overall water quality or trophic status of a lake or reservoir. Pick et al. (1984), for example, suggest that the importance of metalimnetic peaks in regard to the total primary productivity of a waterbody may be greatly exaggerated by the measurement of chlorophyll a. Thus, from the perspective of publicly perceptible deterioration of surface water quality conditions, sub-surface chlorophyll peaks appear to be less important than surface water quality in regard to selection of appropriate measures for the eutrophication control.

If a waterbody exhibits a high degree of lateral heterogeneity in water quality, the necessary number of discrete water samples to adequately characterize the waterbody may be prohibitive. One can attempt to compensate for this heterogeneity by mixing portions of individual water samples in quantities proportional to the heterogeneity, thereby obtaining a single composite sample for the waterbody. This approach would not be appropriate, however, for lakes or reservoirs which exhibit distinct longitudinal gradients of eutrophication-related water quality. An example would be a reservoir with an elongated shape (see Figure 4.1). In such a case, depth measurements should be taken at the outflow, at the inflow and at selected intermediate positions. The area of greatest depth should always be included into the sampling programme. Particular care should be taken when sampling under ice cover.

Samples taken from the hypolimnion are necessary to assess hypolimnetic oxygen depletion and phosphorus regeneration from the bottom sediments (i.e. internal nutrient loading) during the stratification period. The latter parameter often is determined with mass balance calculations of phosphorus inputs and outputs and the in-lake phosphorus concentration (see Table 7.8). Composite samples containing water taken from both the epilimnion and hypolimnion, proportional to the volume of each layer, usually are required to make such calculations. Alternatively, one can obtain discrete samples from these two water layers and make appropriate adjustments in the calculations to account for their respective volumes. Chapter 7 provides further discussion of the mass balance approach for calculating the internal phosphorus loading to a waterbody.

It should be recognized that selective withdrawal of water from the hypolimnion of reservoirs during the period of thermal stratification must be taken into account in calculating the hypolimnetic oxygen depletion rate during this period. Theoretically, since hypolimnetic withdrawal from reservoirs potentially could remove oxygen-depleted waters at a more rapid rate than in natural lakes, the hypolimnetic oxygen conditions may be better in reservoirs than in natural eutrophic lakes. However, because of selective discharge of oxygen-depleted waters or waters with high concentrations of nutrients and heavy metals,the potential for degraded water quality downstream from a reservoir is increased, compared to natural lakes.

With regard to spatial location of sampling stations, Lee & Jones (1980) suggest that, if a waterbody has no significant arms or sub-basins, and is fairly well-mixed horizontally, a single sampling station at the deepest part of the waterbody is usually adequate to characterize the eutrophication–related water quality. However, if the chlorophyll or nutrient concentrations vary by more than a factor of ± 10 along the length of an elongated waterbody, or if the specific conductance varies by greater than 40-60 $\mu S/cm$ ($25°$ C), a single sampling station usually will not provide an adequate description of the 'average' water quality conditions of the waterbody (e.g. see Figure 4.1). In such cases, it will be necessary to establish additional sampling stations along the length of the waterbody. The number of additional stations should be sufficient to take into account the longitudinal variability in water quality. Determination of the necessary number of sampling stations to account for longitudinal gradients of water quality in reservoirs is discussed further in following sections of this chapter.

Another approach to account for longitudinal gradients in water quality is to treat the waterbody as a series of serially-connected sub-basins. This approach essentially partitions the waterbody into a series of connected 'sub-basins', each with its own unique water quality characteristics. Each sub-basin can be examined individually for its trophic status and water quality characteristics. Kerekes (1982) previously has demonstrated the utility of this approach in assessing phosphorus load–trophic response relationships in reservoirs. This concept also is discussed by Walker (1985) and Frisk (1981).

When to sample

Eutrophication-related water quality parameters measured in surface water samples, or samples taken in the upper portion of the epilimnion, can be highly variable over time in both natural lakes and man-made impoundments. In contrast, the amplitude of variations in hypolimnetic water quality at diurnal (or even weekly) intervals is much lower. Thus, samples from the lower layers of a stratified waterbody reflect water quality characteristics of a longer period of time, compared to epilimnetic samples. However, because of the possibility of hypolimnetic oxygen depletion, and related nutrient and heavy metal releases from the bottom sediments, the measured water quality in the hypolimnion can be worse than in surface water samples.

When diurnal variations in water quality do occur, and are relevant to the sampling programme, sampling times have to be chosen carefully to reflect this variation. For example, in waters with high chlorophyll levels, the dissolved oxygen concentration in the surface waters is normally minimum at sunrise and maximum at noon. Thus, if samples are always collected at the same time of day and do not reflect this diurnal variation, one may obtain biased results for the mean value of the dissolved oxygen. To determine diurnal variations in dissolved oxygen, initial investigations over a 24-hour period are needed. Minimum requirements include measurements at pre-dawn, midday and pre-sunset.

To obtain adequate data for accurate water quality assessment and trophic classification of a waterbody, the following minimum sampling frequency for the essential criteria (see Table 8.1) is recommended:

1. Samples should be collected monthly from November to March, and approximately biweekly from April to October (encompassing the period of thermal stratification) in northern temperate climates. The same regime would apply over the corresponding growth and non-growth months in southern temperate climates;

2. In tropical/sub-tropical regions, samples should be collected biweekly

from the start of the rainy season until three months after it is over, as well as during the period of thermal stratification. Samples should be collected monthly at other times of the year;

3. In both cases above, sampling also should be done during any overturn periods. In addition, if algal blooms occur between the above-noted sampling intervals, samples should be taken during the bloom periods.

As a practical matter, if one has to choose between temporal and spatial resolution, temporal resolution may be the logical choice, since one station often is sufficient to describe the average conditions in a lake or reservoir. The exception would be reservoirs with longitudinal gradients in water quality (discussed further in the following section).

In temperate lakes with a water residence time of many years, average lake concentrations are close to spring overturn concentrations. In contrast, in irregularly-flushed lakes and in reservoirs, the spring concentration may be substantially different from the annual average concentrations. In these latter cases, full year measurement cycles usually are required to obtain reliable information (Janus & Vollenweider, 1981).

For waterbodies which exhibit substantial year-to-year fluctuations in hydrological conditions, calculations based upon a single year cycle may be inadequate. For example, in warm climates with dry and wet seasons, the variability in both volume–flow and nutrient concentrations is often very high (e.g. see Chapter 7 for discussion of rapidly flushing waterbodies).Thus, as a general rule, a minimum sampling programme covering three consecutive years is recommended.

The magnitude of the internal phosphorus load (as phosphate release into the euphotic layer) can be approximated with a relatively high frequency of vertical sampling (e.g. biweekly) in the epilimnetic and hypolimnetic waters. Special attention should be given to the period of thermal stratification (Janus & Vollenweider, 1981). Calculation of the internal phosphorus loading in a waterbody was discussed previously in Chapter 7. Ryding & Forsberg (1977) and Vollenweider (1976b) also provide guidelines for calculating the internal phosphorus load to a waterbody.

Sampling strategies in waterbodies with longitudinal water quality gradients

As noted previously, longitudinal gradients of water quality in a waterbody usually prohibit the use of a single sampling station to characterize the 'average' water quality in the waterbody, especially reservoirs. Consequently, a relevant concern with such waterbodies is to determine how many sampling stations are needed to characterize the water quality adequately and where should they be located in the waterbody.

Based on a study of DeGray Lake, a reservoir in southern Arkansas (United States), K.W. Thornton *et al.* (1982) provide an approach for addressing this concern. Previous monitoring data indicated this reservoir had considerable longitudinal and vertical variation in water quality. The data showed the water quality gradient was most pronounced at the upper end of the reservoir (nearest the tributary inflow), and progressively decreased as one moved closer and closer to the downstream (dam) end of the reservoir.

To address this problem, K.W. Thornton *et al.* (1982) partitioned the reservoir into fifteen transects, averaging five stations/transect. The reservoir was sampled at 0, 2, 4, 6, and 10 m, and at 5 m intervals thereafter to the bottom, for total phosphorus, turbidity and chlorophyll *a* in July, 1978, and in January and October, 1979. Analysis of variance and Duncan's multiple range test (Cochran, 1963; Snedecor & Cochran, 1967) were used to detect significant differences between the means of the parameter values between the various transects. General linear statistical models were used to characterize the transect means, assuming a normal distribution of the data, homogeneity of variance, etc., usually required for statistical models.

If the model showed that the slope of the mean value of a parameter in a transect was not significantly different from zero, then a minimum of one station was needed to characterize the transect area. If a linear function was needed to account for a significant portion of the variance among the transect means, a minimum of two stations was needed to characterize the transect area. If a quadratic function was necessary, a minimum of three stations were needed, etc. Once the analysis of variance was completed, the transects were compared in order to identify areas of similarity and overlap.

In the 15-transect Degray Lake example, K.W. Thornton *et al.* (1982) showed that transects 1–5 had similar means for all variables over all dates. Thus, one station was sufficient to characterize the entire area represented by all five transects. The means of transects 6–13 required a linear model to account for the variance among the transect means. Thus, a minimum of two stations was necessary to characterize the water quality in the area represented by these transects. Transects 14 and 15 were either distinct or required a linear model. Thus, these two transects each needed a separate sampling station. Based on this analysis, DeGray Lake needed a minimum of five sampling stations to characterize its longitudinal water quality gradients. Thus, sampling stations on transects 3, 10, 12, 14 and 15 (Figure 8.1) would be adequate to characterize the water quality of DeGray Lake.

In addition to the number of sampling stations, the number of samples to be collected at each station also depends on the data variability of the desired water quality parameter, as well as the desired precision of the parameter estimate. Based on these factors, K.W. Thornton *et al.* (1982) also provided a general formula for random sampling in DeGray Lake, as follows:

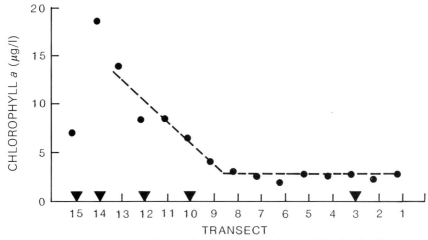

Figure 8.1 Representative sampling station locations for characterizing longitudinal water quality gradients in DeGray Lake (from K.W. Thornton et al., 1982)

$$n = t^2 s^2 / d^2 \qquad (8.1)$$

where: n = number of samples;
 t = appropriate value from student's t distribution;
 s^2 = sample variance (based on existing data or preliminary surveys); and
 d = desired precision about the mean value.

They suggested an initial t value of 30 degrees of freedom to begin the analysis procedure. When applied in an iterative manner, Equation 8.1 will produce a value for n converging on the appropriate sample size, usually after 3–5 iterations.

For the period when the reservoir is stratified, a different random sampling design can be used. A general formulation for stratified sampling is:

$$n = (\sum w_i s_i)^2 / (d^2/t^2) \qquad (8.2)$$

where: n = total number of samples;
 w_i = weighting factor for stratum (e.g. ratio of volume of epilimnion to total lake volume);
 s_i = standard deviation of samples in stratum; and
 t = appropriate value from student's t distribution.

Within each individual stratum (e.g. epilimnion or hypolimnion), the number of necessary samples can be calculated as:

$$n_i/n = w_i s_i / \sum (w_i s_i) \qquad (8.3)$$

where: n_i = number of samples in the stratum i; and
 n = total number of samples.

One point to emphasize here is that, while simple random sampling requires only a single mean and variance (see Equation 8.1), stratified sampling requires an estimate of the mean and variance of each stratum sampled.

CALCULATING THE COSTS OF SAMPLE COLLECTION

K.W. Thornton et al. (1982) provide a methodology for calculating the costs of obtaining necessary samples, based on the type of analyses presented above. Cale & McKown (1986) also present an approach for calculating the costs of collecting the necessary number of samples in a waterbody to achieve a desired data precision. This approach is essentially a comparison of the maximum expected variance of the data with the desired data reliability. Although both approaches overlap to a certain degree, that developed by Cale & McKown is discussed here because of its greater emphasis on calculation of sampling costs.

Cale & McKown have approached the problem of desired data precision versus costs so that one can attempt to answer the following types of questions about a water quality sampling programme:

1. Will the funds available for a sampling or monitoring programme be sufficient to achieve data of a desired precision?
2. If not, what is the precision of the data that can be obtained for the available funds?
3. If the initial funding level is insufficient to obtain the desired data, how much additional funding will be necessary to achieve this precision?

Data quality is expressed in terms of a desired statistical confidence level. In scientific/technical work, a 95% confidence level (i.e. $p = 0.05$) often is used as the standard of data acceptability. However, as Cale & McKown point out, this value is actually an arbitrary criterion which has gained general acceptance simply because of its repeated usage over time. It is both a reflection of the fact that improvements in data quality are less expensive and less difficult at lower significance levels than at higher levels and a standard of data reliability. In fact, in a given situation, data with a confidence level of 90% may be adequate for answering the question being asked. That is, one might conclude that the funds necessary to increase the confidence level of the data from 90 to 95% may not be worth the additional accuracy gained. In other cases, a particularly sensitive issue may require a data confidence level even greater than 95%. Unfortunately, there are no *a priori* guidelines for deciding unequivocally if whether or not incremental improvements in data reliability are worth the additional costs. In most cases, it is as much a socioeconomic and political decision as a technical decision.

As noted above, the approach of Cale & McKown for calculating sampling costs is basically a comparison of the maximum expected variance of the data being sought with the desired data reliability. The results of this comparison are then related to the associated costs. This approach consists of the following steps:

1. Based on the desired sampling frequency and the associated costs, determine the number of samples per trip (n) that can be collected with the available funds;

2. Based on literature values or pilot study results, calculate the largest standard error and standard deviation likely to occur for the desired number of samples (n);

3. Based on the results of Step (2), calculate the maximum variance likely to occur under the specific conditions of the sampling programme;

4. Calculate the half-width of the confidence interval (e.g. 95%) that will result from the variance calculated in Step (3); and

5. Compare the results of Step (4) to the desired data reliability. Based on this comparison and the data quality goals of the sampling/monitoring programme, one can decide whether or not it is reasonable to spend the necessary funds to obtain the desired data.

The costs of any monitoring activity can be characterized as either 'fixed' or 'variable'. Fixed costs can be further identified either as being relatively constant and independent of the monitoring programme (F_1), or as a specific function of the programme (F_2). Examples of F_1 costs include literature reviews, data analysis and development and refinement of monitoring programmes. F_2 costs are most directly related to the specific monitoring programme and include such items as costs of travel time to the study site, equipment setup and calibration, per diem costs, etc. By contrast, variable costs (v) are a direct function of the intensity of the monitoring effort and include the consumption of chemicals and supplies, collection of field samples and laboratory analyses of the samples. These latter costs are variable in that they are a function of the number of samplng trips, number of samples collected per trip, etc.

Given these definitions, the costs (C) of collecting a required number of samples (n) for a given number of sampling trips (M) over the duration of a lake or reservoir monitoring programme can be determined as follows:

$$C = F_1 + M(F_2 + nv) \qquad (8.4)$$

where: C = total costs (e.g. monitoring budget);
F_1 = fixed costs independent of monitoring effort (see text);
F_2 = fixed costs dependent on monitoring effort (see text);
v = variable costs (see text);

M = number of sampling trips; and
n = number of samples collected per sampling trip within budget limitation.

The values of M and n are calculated with the use of standard statistical procedures. The value of n can be calculated as:

$$n = (s^2 t^2_{1-\alpha/2})/d^2 \tag{8.5}$$

where: n = required number of samples to achieve desired precision of data;
s^2 = estimate of data variance, based on the preliminary sample;
t = Student's t-test distribution for $1 - \alpha$ confidence level and $n_o - 1$ degrees of freedom;
n_o = size of a preliminary sample; and
d = half-width of the desired data confidence interval.

The value of M can be calculated as:

$$M = (s_t^2 Z^2_{1-\alpha/2})/d_t^2 \tag{8.6}$$

where: s_t^2 = variance estimate of population change over time (e.g. chlorophyll concentration);
Z = appropriate value from a table of cumulative normal distribution, with confidence level of $1 - \alpha$;
d_t = half-width of the confidence interval for measuring a change in the population.

Equations 8.5 and 8.6 provide estimates of the values of n and M, based on the desired level of data confidence for the monitoring programme.

A hypothetical application of this approach is provided with the data in Table 8.2. In this example, it is assumed a lake manager wishes to initiate a lake restoration programme based on reducing the external phosphorus load to Lake X and that he has a total monitoring budget of $30,000. The objective of the control programme is to reduce the in-lake chlorophyll

Table 8.2 Hypothetical costs of monitoring programme for Lake X (modified from Cale & McKown, 1986)

Activity	Type	Estimated cost[a]/day	Samples/day
Field sample collection	v	$250	25
Laboratory analyses	v	$150	50
Travel costs	F_2	$200	
Literature research; data anlysis; report writing, etc.	F_1	$300	
Secretarial/graphics	F_1	$200	

[a] Such items as overhead expenses can be incorporated into this figure or listed as a separate item.

concentration from 25 µg/l to 10 µg/l. The manager wishes to sample Lake X biweekly during the growing season (May-September) and monthly during the rest of the year (October-April) for a period of three years. In addition, the manager desires that changes in the chlorophyll concentrations in Lake X as low as 2 µg/l between sampling trips be detected, and that the data obtained have a confidence level of 95 per cent (i.e. $p = 0.05$). It is also assumed that six days of professional and secretarial assistance are needed per year. Given this background, the basic question to be answered is whether or not the $30,000 monitoring budget is adequate to achieve these monitoring objectives.

Using the approach of Cale & McKown (1986), this problem is assessed as follows:

1. Based on the desired sampling schedule, 17 sampling trips can be made each year for the three year monitoring programme. Thus, $M = (17)(3) = 51$. The number of days of necessary professional assistance is six days each year for the three year monitoring programme. Based on the cost estimates in Table 8.2, the number of samples (n) that can be collected per sampling trip, within the $30,000 monitoring budget (C), is calculated with Equation 8.4, as follows:

$$\$30,000 = (18)(300 + 200) + 51[(200) + n(250/25 + 150/50)]$$
$$= 9,000 + 10,200 + 663\,n. \tag{8.7}$$

Solving Equation 8.7 for n, the budget will allow 16.29 samples to be taken during each sampling trip. For this example, the value of n is rounded off to 16 samples per sampling trip.

2. Based on a review of relevant literature (or pilot study results, if available), the manager determines that the maximum standard error (s.e.) usually associated with measuring the mean chlorophyll concentration in similar waterbodies is 5 µg/l, for a sample size of ten samples.

The standard error (s.e.) = standard deviation $(s)/\sqrt{n}$. Thus, the maximum standard deviation expected for this monitoring programme is calculated as $s = (\text{s.e.})(\sqrt{n}) = (5)(\sqrt{10}) = 15.81$. Since the monitoring budget of $30,000 allows approximately 16 samples/sampling trip (see step 1 above), the maximum expected standard error (s.e.) to be expected should be no larger than $(s)/(\sqrt{16}) = (15.81)/(4) = 3.95$.

3. The manager is concerned with detecting differences in mean chlorophyll concentrations between sampling trips (as contrasted with differences within a single sampling trip). Assuming the data are independently and normally distributed, the associated variance (V) of the difference in mean chlorophyll concentrations between any two sampling trips is calculated as:

$$V(\bar{x}_1 - \bar{x}_2) = s_t^2 = V(\bar{x}_1) + V(\bar{x}_2) = (s_1^2/n_1) + (s_2^2/n_2) \tag{8.8}$$

where $\bar{x}_1$ and $\bar{x}_2$ are the mean values of the 16 samples collected during any two different sampling trips. One can maximize the variance by maximizing each of the standard error terms (i.e. s.e. = $s/\sqrt{n}$) in the summation term on the right side of Equation 8.8. Since the maximum expected standard error is 3.95 (via step 2), each of the two summation terms will have a maximum value of 3.95.

The maximum variance (V) then can be calculated as:

$$V[\bar{x}_1 - \bar{x}_2] = s_t^2 = V(\bar{x}_1) + V(\bar{x}_2)$$
$$= [(s_1^2/n_1) + (s_2^2/n_2)] = (3.95)^2 + (3.95)^2 = 31.2 \qquad (8.9)$$

It is noted that one could calculate the average variance of the data in the same manner. However, this would not insure the desired data precision will be achieved. Thus, as a practical matter, this example uses the maximum variance as the criterion for insuring the monitoring data to be obtained are within the desired precision range.

4. Since the manager is concerned with differences in the mean chlorophyll values between sampling trips, the expected half-width of the confidence level (d) for the maximum variance calculated in step 3 is determined with the use of Equation 8.6.

Knowing that $M = 51$, $s_t^2 = 31.2$ and $Z_{1-\alpha/2}$ (i.e. the half-width of the 95% confidence level) $= Z_{1-0.05/2} = Z_{0.975} = 1.96$, rearrangement of Equation 8.6 allows one to calculate the value of d as follows:

$$d_t^2 = (s_t^2 - Z_{1-\alpha/2}^2)/M = [(31.2 - (1.96)^2/51] = 0.536$$

Thus: $d_t = \sqrt{0.536} = \pm 0.73 \,\mu g/l$. Since the calculated half-width of the desired 95% confidence level for the monitoring data is within the required precision of $\pm 2\,\mu g/l$, the monitoring programme, as designed, would allow the manager to obtain data of the desired reliability. In fact, the desired precision range is more than twice the value of the expected precision. Thus, in this case, one can even consider reducing the intensity of the monitoring programme and still be assured of achieving the desired level of data reliability.

In contrast to the above results, if one had a total monitoring budget of only $20,000 for the same sampling programme, the calculated value of d would have between 2.85. This value is nearly 50% greater than the desired precision of $\pm 2 \,\mu g/l$. Based on this result, one would conclude that the desired data precision could not be achieved within the $20,000 monitoring budget. Thus, one would have to decide between either eliminating the monitoring programme, accepting the calculated data precision as the best achievable under the financial constraints of the monitoring programme, or increase the monitoring budget to the level necessary to obtain data of the desired quality.

The necessary funds needed in a specific case to obtain data of a desired precision also can be calculated. The procedure uses the same steps as the above example, but in a different order. Basically, one uses the estimate of the maximum variance to 'back-calculate' the specific value of n that would allow the variance not to be exceeded. The total costs are then calculated with Equation 8.4.

Based on the data in Table 8.2, Equation 8.6 can be rearranged to calculate for the maximum variance (s_t^2) as follows:

$$s_t^2 = [(M)(d_t^2)/Z_{1-\alpha/2}^2]$$
$$= [(51)(2)^2/(1.96)^2] = 204/3.84 = 53.10 \quad (8.11)$$

Thus, since s is an estimate of the maximum variance of the change in the mean chlorophyll concentrations in Lake X, it cannot exceed a value of 53.10 if one wishes to insure the desired data precision of $\pm 2\,\mu g/l$ in this example. Equation 8.9 shows that the total variance is equal to the sum of the individual standard error terms (i.e. s.e. $= s/\sqrt{n}$). Thus, to insure that a total variance of 53.10 is not exceeded, one must insure that the summation of the standard error terms in Equation 8.9 does not exceed this value.

One way to accomplish this is to assume that the two standard error terms are equal and that their sum is $\leqslant 53.10$. Under this assumption, Equation 8.9 can be rewritten, as follows:

$$V = s_t^2 = 53.10 = 2(s_1^2/n_1) \quad (8.12)$$

Since $n_1 = n_2$, one can solve Equation 8.12 for the value of the maximum standard error (i.e. s.e. $= \sqrt{53.10/2} = 5.15$.

Once the maximum value of an individual standard error term is calculated, the standard error expression in step 2 can be rearranged to solve for the value of n. If one assumes the calculated standard deviation (s) in step 2 is the maximum value to be encountered,

$$n = (s/\text{s.e.})^2 = (15.81/5.15)^2 = 9.42 \quad (8.13)$$

Now that the value of n is known, the monitoring budget required to obtain data of the desired precision can be calculated with Equation 8.4, as follows:

$$C = F_1 + M[F_2 + (nv)]$$
$$= (18)(300 + 200) + 51[(200 + 9.42(250/25 + 150/50)] \quad (8.14)$$
$$= 9{,}000 + 10{,}200 + 6{,}245 = \$25{,}445$$

This calculated value is approximately half-way between the $30,000 budget (which will exceed the desired data precision) and the $20,000 budget (which will not achieve the desired precision), consistent with the monitoring realities of this example.

It is emphasized that one can also analyze a multi-parameter monitoring programme using this same type of analysis applied in an iterative manner. Initially, one would establish a priority listing of the monitoring parameters of interest, based on the most critical data needs. The first parameter in this list then would be subjected to the analysis illustrated above. Equations 8.11–8.14 would be used to determine the monetary requirements for obtaining data of the desired precision for the first parameter. The calculated costs for obtaining the first parameter in the priority list would be subtracted from the total monitoring budget. The same analysis then would be applied to the second parameter in the priority list, using the revised (reduced) monitoring budget as the appropriate value of C for Equation 8.4. This procedure can be continued for each succeeding parameter in the priority list until the available budget was expended. If this occurred before all the parameters in the priority list had been examined, one can either reduce the number of parameters to be monitored, reduce the desired data precision, increase the monitoring budget, or some combination of these options.

The reader is referred to the report of Cale & McKown (1986) for further details (also see K.W. Thornton *et al.*, 1982). Both sources provide detailed examples of the necessary calculations for making the types of assessments outlined above.

COMPILATION AND PRESENTATION OF DATA

Realistically, one cannot sample a waterbody continuously. Consequently, one should attempt to minimize any bias in the data caused by the sampling frequency. Lee & Jones (1980), for example, suggest that all data should be converted to describe weekly mean values. For the biweekly sampling interval suggested in the previous section, this can be done by calculating the mean value of two consecutive biweekly values. This calculated mean value can be considered representative of the conditions (for the water quality parameter being measured) for the week between the two sampling dates. The same general procedure can be used for the monthly sampling interval, except that the arithmetic mean value obtained from the two consecutive monthly samples would be used for each of the three weeks between the monthly sampling dates. Lee & Jones (1980) suggest that, while not foolproof, this procedure appears to be adequate for determining reliable mean in-lake nutrient and chlorophyll values for the simple, empirical OECD load-response models discussed previously in Chapter 6.

For in-lake chlorophyll and Secchi depth measurements, it is suggested that both annual and summer mean values be calculated, since the latter period usually represents the period of maximum eutrophication-related water quality degradation (Lee & Jones, 1980). Volume-weighted mean values (i.e. mean values adjusted to reflect the volume of the water column from which they were collected; see further discussion below) would appear to provide the most accurate estimate of the 'average' concentration of an in-lake parameter for the waterbody as a whole. However, for the period of thermal stratification, the arithmetic mean in-lake chlorophyll and nutrient concentrations in the surface waters (i.e. approximately the top two meters) usually provide more relevant information on the publicly perceptible symptoms of eutrophication than does a composite sample composed of water from the epilimnion and hypolimnion.

As a practical matter, it is obvious that the use of multiple sampling stations for waterbodies with significant longitudinal gradients of water quality will result in the accumulation of a larger set of data points than a waterbody with only one station. Thus, a primary concern is how to calculate one 'average' value for a water quality parameter which applies to the whole waterbody, even if the parameter is characterized by markedly different values along the length of the waterbody. There are two possibilities in such cases. To begin with, if it is necessary to partition a waterbody into separate 'sub-basins' (including distinct embayments separate from the main body of the lake or reservoir), one can treat the waterbody as a series of connected sub-basins, rather than as one single waterbody (see discussion in previous section). However, this approach may make it difficult to evaluate the effectiveness of potential eutrophication control measures for the waterbody as a whole.

An alternative approach is to apply a weighting factor to the data from each of the sub-basins. In this way, an 'average' value which takes into account the relative proportion of the sub-basins from which the data are obtained can be calculated. One can calculate the relative volumes of each sub-basin and apply this factor to the mean value of the parameter obtained. The 'average' value calculated in this manner will represent an integration of the relative proportions of each of the sub-basins, even though it represents the arithmetic summation of a series of data points, rather than a single measure. One also can attempt to weight the mean values according to the surface areas of the sub-basins, especially if the depth is not significantly different along the length of the waterbody. However, if the depth does change markedly from one end of the waterbody to the other, the use of surface areas as a weighting factor can produce erroneous results. In such cases, the use of volume-weighting can be attempted.

A precautionary note is that weighting techniques can produce a single value describing the 'average' nutrient concentration (or other water quality parameter) in a waterbody. However, it also must be recognized that this calculated number is an entirely artificial value, since it was derived from mathematical calculations (as contrasted with a measured value). Because it is a derived number (e.g. the total mass of nutrients in the waterbody divided by the total volume of water), one may never actually measure this 'average' value in the waterbody. Nevertheless, it is based on arithmetic manipulations of measured data and can be used to describe the relative conditions in a reservoir characterized by longitudinal gradients in water quality.

Another practical observation is that the arithmetic mean may not always be the best value to describe the 'average' in-lake condition of a given parameter (e.g. average in-lake chlorophyll or nutrient concentration). This is because the dynamic nature of a waterbody's metabolism is often characterized by a series of extreme conditions (e.g. high and low values). An algal bloom in a lake or reservoir, for example, can result in high chlorophyll levels for short periods of time, interspersed with low values over longer periods of time. In such cases, the 'average' chlorophyll concentration calculated as the arithmetic mean (i.e. Σ data values/number of observations) may not accurately depict the skewed nature of the data. Heyman et al. (1984), for example, showed that several in-lake eutrophication response parameters (e.g. chlorophyll and phosphorus) showed a non-symmetrical frequency distribution in 25 Swedish lakes. That is, the data did not exhibit the standard 'Bell-shaped curve' of a normal distribution. This may be the general situation for most lakes and reservoirs.

Thus, strictly speaking, the median value should provide a more accurate description of the average situation or conditions in the waterbody. The median value is the value which has an equal number of data points greater and lesser than itself. It can be calculated by plotting all the data for a specific water quality parameter on a probability scale, and selecting the 50% value. Alternatively, one can rank the data in an ascending or descending order and select the middle value.

As a practical observation, however, the difference in the 'average' condition described by the arithmetic mean versus the median value often will be insignificant when related to eutrophication assessment and control programmes. One can assess whether or not the 'average' condition based on the arithmetic mean is significantly different from that based on the median by directly comparing the two values. If one decides (statistically or subjectively) that the numbers differ by too great a margin, it is recommended that the median value be used to describe the average condition.

Another method of describing the average value of a given parameter is to use the geometric mean value. This value is calculated in the same manner as the arithmetic mean, except that the data are first converted to common logarithms. The arithmetic mean of the logarithms is calculated. The antilog of the arithmetic mean constitutes the geometric mean value. The geometric mean was used in the OECD (1982) international eutrophication study to calculate theoretical trophic boundary values for several common water quality parameters. Again, as a practical observation, some of the individual investigators involved in the OECD (1982) study indicated that the arithmetic mean value was usually adequate for use in the simple, empirical load-response models developed in the study.

Based on these suggestions regarding data compilation, one can readily apply the results obtained to various modes of data analysis (e.g. correlations and regressions). As shown in Chapter 11, one can use the mean values of total phosphorus and chlorophyll concentrations, as well as Secchi depth, to estimate the maximum or 'worse case' conditions likely to occur in a waterbody.

CHAPTER 9

AVAILABLE TECHNIQUES FOR TREATING EUTROPHICATION

GENERAL CONSIDERATIONS

As noted in Chapters 3 and 4, the effective control of lake and reservoir eutrophication is linked strongly to control of the basic causative factor; namely, the input of excessive quantities of aquatic plant nutrients. Based on the limiting nutrient concept (see Chapter 4), practical experience suggests an effective, long-term eutrophication control measure is to reduce the external phosphorus load to the waterbody. Alternatively, one may divert the phosphorus load around or away from the lake or reservoir. This latter method will protect the waterbody of concern, but can cause eutrophication problems in downstream rivers, lakes and reservoirs. Thus, the basic problem is not eliminated if nutrient diversion is the only eutrophication control method used; it is simply transferred to another location. The first portion of this chapter focuses on methods for reducing or eliminating the phosphorus load to lakes and reservoirs.

It is also recognized that reduction of the external phosphorus load may not be feasible in a given situation. In such cases, one may have to consider control programmes which attempt to treat the symptoms or impacts of eutrophication. These latter methods will not eliminate the basic problem, since they ignore the basic cause. Nevertheless, in some situations, it may not be possible to initiate necessary nutrient control programmes. In these cases, control programmes based on treating the symptoms of eutrophication may be the only control alternative, and do offer varying degrees of relief from the negative impacts of eutrophication. These latter measures also are discussed in this chapter.

CONTROL OF THE EXTERNAL PHOSPHORUS LOAD

If reduction of the external phosphorus load to a lake or reservoir is the primary eutrophication control measure to be used, it is important to know the major phosphorus sources in the drainage basin and the way in which the phosphorus enters the waterbody (see Chapter 7). Once these are known, one can compare the eutrophication control goals with the available control options and determine an appropriate control programme.

Direct reduction of phosphorus at the source

Phosphate elimination by chemical precipitation during the sewage treatment process. Municipal sewage can be treated in a mechanical–biological treatment plant, using chemical precipitation methods to eliminate the phosphate. Phosphates are precipitated from municipal wastewaters with use of aluminum or iron salts, or lime. In a mechanical–biological treatment plant (Bernhardt, 1983), the precipitant may be added (1) prior to the mechanical step ('pre-precipitation'); (2) during the biological treatment step ('simultaneous precipitation'); or (3) after the biological step in a separate flocculation and settling tank ('post-precipitation'). These procedures will reduce the effluent phosphorus concentration to about 1 mg/l, depending on the chemical dosage used and the flocculation behavior of the sewage.

Swedish treatment plants show better results with post-precipitation. Effluent phosphorus concentrations as low as 0.2–0.45 mg/l are common. This is the result of the optimization of this process. A special optimization measure is to recycle part of the precipitated sludge to the activated sludge basin. In this way, the dosage of chemicals can be reduced by 30–45%.

In Switzerland, total phosphorus concentrations in effluents of municipal wastewater treatment plants located in the catchment area of lakes may not be higher than 1 mg/l. Ambuhl (personal communication, EAWAG, 1985) suggested this concentration may be too high to control eutrophication effectively in Swiss lakes, indicating that effective control would require municipal wastewater treatment plant effluent phosphorus limitations of <0.2 mg/l. This level can be achieved with contact filtration of mechanically/ biologically-treated wastewaters. Contact filtration of wastewaters after mechanical-biological and chemical treatment is a good means of achieving a high quality effluent with respect to phosphate and suspended solids.

A listing of the approximate costs of sewage treatment of municipal wastewaters in Sweden is provided in Table 9.1. In addition, a comparison of the costs of various levels of wastewater treatment for the removal of phosphorus in the North American Great Lakes Basin is presented in Table 9.2.

It is noted that pilot plant studies in Brazil, utilizing combined biological and chemical treatment of wastewaters, have resulted in effluent phosphorus concentrations down to 0.1 mg/l (personal communication, H. Salas, CEPIS, 1984). McKendrick (1982, as cited in J.A. Thornton & Nduku, 1982) discusses the physico-chemical removal of nutrients from raw sewage and sewage effluents in sub-tropical Zimbabwe. Furthermore, the National

Table 9.1 Approximate costs for sewage treatment in Sweden, 1978 (from Forsberg & Ryding, 1981)

Number of Person Equivalents (P.E.)	Costs for post-precipitation plants (including sludge treatment)		Additional costs for deep-bed filtration	
	Capital costs*	Operating costs*	Capital costs*	Operating costs*
2,000	130	100	–	–
5,000	100	70	25	8
20,000	60	50	10	4
50,000	45	40	7	3

*The annuity used in calculation of capital costs is 10% for post-precipitation plants and 13 percent for deep-bed filters; Swedish crowns (krona)/capita.yr (1 Swedish crown = $0.11 U.S.)

Water Research Institute in South Africa has been assessing the use of ferric chloride (e.g. from mine wastes) as a nutrient precipitant to treat sewage (personal communication, J.A. Thornton, NIWR, 1985).

Restriction of detergent phosphates. It is possible to restrict the quantity of phosphates in detergents, assuming suitable phosphate substitutes are available. Ideally, such alternative compounds must not cause new environmental problems, or ones worse than existed with the use of phosphates. Phosphate substitutes should also not interfere with sewage treatment processes or with water treatment for drinking purposes.

The extent to which phosphate substitutes can be used in detergents cannot be determined in a routine manner. Such factors as the amounts of water to be used and the water management customs of a particular country or region, and the potential environmental impacts of phosphate substitutes must be taken into account. In the Federal Republic of Germany, for example, a study on the effects of NTA (a phosphate substitute becoming commonly used in detergents) on the aquatic environment, in sewage treatment and in the preparation of drinking water is being conducted as part of such an assessment. Local social customs must also be considered. In some countries, for example, a detergent may not be considered useful unless it produces suds.

Laws regulating the amount of phosphate in detergents do exist in some countries. For example, in Canada, a detergent phosphate limitation of 2.2 percent (by weight) is in effect. Switzerland does not allow any phosphates in detergents. The allowable levels of detergent phosphates in the Federal Republic of Germany depends on the water hardness and the type of washing treatment used in washing machines. In most states in the United States portion of the Great Lakes Basin, a maximum detergent phosphate content of 0.5 per cent (by weight) is allowed. Phosphate restrictions are either recommended or optional in some other countries.

Table 9.2 Municipal point source phosphorus treatment options, based on experiences in the North American Great Lakes Basin (from Phosphorus Management Strategies Task Force, 1980)

Process	Effluent quality (mg P/l)	Incremental costs[a]	Comments
A-S/Chemical Activated sludge (A–S)		Capital–$3,500 per 1000 m³/day capacity.	Peak flow clarifier over-rate 32.6 m³/day essential
plus		Chemicals–$5 per 1000 m³ treated. O&M–$80/yr per 1000 m³/day capacity.	
alum or ferric chloride,	0.6–1.0		
ferrous salts	0.6–1.0	These waste salts are low cost.	Ferric conversion in aeration section; poor quality can cause problem.
or lime	1.0		Used in primary section only; limiting pH may make it diffiuclt to achieve 1 mg/l.
plus polymer	–	Expensive for continuous use.	Temporary solution to periodic hydraulic overloads.
Physical/Chemical Primary			Peak flow clarifier overflow rate 32.6 m/day essential.
plus alum or ferric chloride,	1.0	Similar to A-S/Chemical.	
ferrous salts	N/A[b]	N/A	Cannot be used unless converted to ferric salts.
or lime	1.0	Similar to A–S/Chemical	Effluent needs pH adjustment; lime feeding is difficult.
plus polymer	–	Expensive for continuous use.	Temporary solution to periodic hydraulic overloads.
AWT/Chemical Advanced Waste Treatment *plus* alum or ferric chloride,	0.3–0.5		Using effluent filter and chemical.
ferrous salts	N/A	N/A	N/A
or lime	0.3–0.5		Using upflow clarifier and chemical.

(*continued*)

Available techniques for treating eutrophication

Table 9.2 *Continued*

Process	Effluent quality (mg P/l)	Incremental costs[a]	Comments
Lagoons/Chemical Aerated or Facultative *plus* alum or ferric chloride,	0.5–1.0		Continuous chemical feed produces a 1.0 mg/l effluent; batch dosage and seasonal discharge gives 0.5 mg/l.
ferrous salts,	N/A	N/A	N/A
or lime	1–0	N/A	Difficult to handle.
Biological Bardenpho A/O Process	1.0 1.0	Expected to be significantly lower than A–S/Chemical and could be equivalent to conventional A–S alone.	Additional development work required.
Biological/chemical PhoStrip	0.5	34–37% lower than A–S/Chemical	Lime treatment of waste sludge sidestream reduces chemical requirements and sludge production.
Land Application Slow Rate	0.04–0.5	Competitive with A–S/Chemical	Applicability dependent upon soil conditions, health risks, cost competitiveness and public acceptance.
Rapid Infiltration	0.02–6	″ ″ ″	
Overland Flow	1–10	N/A	Applicable only where impermeable soils are available.

[a] Care must be taken in comparing costs for various treatment options. Phosphorus removal often can be incorporated into existing facilities. Costs of abandoning existing facilities must be considered where this option is proposed.

[b] N/A = not applicable.

Land use controls. This method involves the restriction ('protected zones') or control of land use activities in a drainage basin which result in the runoff of nutrients to a lake or reservoir. This approach has been used in the Federal Republic of Germany for the protection of drinking water supplies, as well as in other countries. It may be one of the most effective overall methods for attempting to control nutrient inputs to lakes and reservoirs, since it does not allow activities in the drainage basin which would generate nutrients. A drawback to this approach, however, is the need for appropriate institutional and legislative frameworks (see Chapter 3). Furthermore, in many cases (especially developing countries), conflicts between using riverbank areas for green belts versus agricultural activities are usually resolved in favor of the latter, even if the legislative requirements for the protection of river courses already exist. This is because riverbank areas in such settings are usually well-watered fertile areas, suited (and probably essential) for agricultural purposes.

The utilization of proper land use practices (as contrasted with land use control) to control non-point source nutrient inputs to lakes and reservoirs is discussed below.

Treatment of tributary influent waters

Pre-reservoirs. A reduction of nutrients in tributaries and reservoirs can be achieved with the use of 'bioreactors'. These basins retain nutrient-rich water for a short period of time prior to its entering the main body of a reservoir, thereby accentuating the opportunity for algal growth in the basins. Pre-reservoirs (called pre-impoundments or cascade reservoirs in some countries) are a type of bioreactor, their original purpose being to prevent the main reservoir from becoming rapidly filled with silt.

The use of pre-reservoirs as bioreactors for phosphorus elimination has been investigated by Benndorf & Pütz (1987). The elimination of phosphorus in pre-reservoirs is related to the enhancement of bioproductivity. The phosphorus becomes fixed in the increased algal biomass in the pre-reservoir, thereby being largely retained in the pre-reservoir via sedimentation. Such reservoirs also remove phosphorus by a simple reduction in flow velocity, which allows adsorbed phosphorus to settle out of the water column. This latter process, in fact, appears to be the dominant mechanism in some waterbodies, as contrasted with biological removal of phosphorus (e.g. some southern African lakes). Thus, inflow phosphorus will accumulate at the bottom of the waterbody, assuming there is sufficient oxygen available to keep it immobilized in the sediments. The effluent water from the pre-reservoir, therefore, usually contains less phosphate than the influent water. Several pre-reservoirs can even be connected in series, to form a system capable of removing nearly all the phosphorus in the influent waters. Štěpánek (1980) quotes a 96% removal of phosphorus using the pre-reservoirs.

A sufficiently long water retention time in the production zone is a requirement for the effective biogenic phosphorus removal of pre-reservoirs. In middle European waterbodies, the depth of this zone ranges from 0–3 m, depending on the light intensity. The retention time must be sufficiently long that algae are not prematurely washed out of the pre-reservoir.

The actual elimination rate of phosphate depends on the growth rate of the algae which, in turn, depends on the phosphate content of the influent waters, the light intensity and the water temperature. An optimal phosphate elimination rate is achieved when the water retention time in the bioproduction zone is greater than the critical retention time (i.e. the time necessary to achieve a significant algal biomass; see Chapter 5). The maximum influent phosphorus concentration should be <0.5 mg/l.

The use of pre-reservoirs has been developed to a substantial degree in the German Democratic Republic (Benndorf et al., 1975). Experience has shown that use of pre-reservoirs during the summer, under middle-European conditions, can reduce the orthophosphate content of the influent by 70–90%, primarily due to phytoplankton sedimentation. During the winter, the capacity decreases to 0–30%, due primarily to low light availability and water temperature, which restricts phytoplankton growth. As a result, the elimination process in winter consists mainly of sedimentation of abiotic sediment particles.

A variation of this approach is seen in the Small (Kis)-Balaton reservoir system (approximately 80 km^2) located at the mouth of the River Zala. This system is based on the concept of stimulating the growth of phytobenthos (particularly macrophytes), which remove nutrients from the water column. The first 20 km^2 segment of this reservoir system began operation in early 1985 (Sómlyody & van Straten, 1986; Joó, 1986).

Physical/chemical treatment of influent tributary waters before entering waterbody. Lakes and reservoirs can be treated by means of flocculation and filtration of phosphorus from the tributary inflow waters. The prominent example of this approach is the phosphorus elimination plant (PEP) in the Wahnbach Reservoir. This plant is designed for a maximum flow of 5 m^3/s, which is five times the long-term, average flow of the Wahnbach River.

The Wahnbach PEP has the following characteristics:

1. It is able to run for several weeks at full capacity (i.e. 5 m^3/s);

2. It can handle variation in flow capacity between 3,000 and 18,000 m^3/h;

3. It has the ability to operate for intervals of only a few hours within intervals of several days; frequent switching on and off without a decrease in the quality of the filtrate also is possible;

4. It shows no decrease in phosphate removal efficiency at water temperatures down to 0°C;
5. It can treat waters of high turbidity (up to 100 mg/l suspended solids content) at a maximal filtration rate of 16 m³/h without decreasing the duration of the filter runs to less than ten hours;
6. A decrease in the total phosphorus concentration to $<10\,\mu g/l$ is achieved;
7. Greater than 99% of the plankton grown during the summer months can be eliminated from the water (algal cells which break through the filter cause high concentrations of undissolved and dissolved organically bound phosphorus compounds in the filtrate, resulting in an undesirable phosphorus load in the reservoir);
8. Greater than 99% of inorganic, suspended, phosphate-rich material (which is flushed into the pre-reservoir from erosion of the arable land) is removed; and
9. Flocculation of phosphorus occurs with ferric (3^+) iron. The iron concentration in the effluent does not exceed 50 μg Fe/l.

The 'Wahnbach system' developed to accomplish these tasks consists of the following steps (Figure 9.1):

1. Precipitation of the orthophosphate ions present in the water by adding 4–12 mg Fe^{+3}/l in the acid pH range (i.e. pH = 6–7; average 6.4);
2. Destabilization of the colloids and suspensoids in the raw water, including the precipitated iron-phosphate-hydroxide compounds;

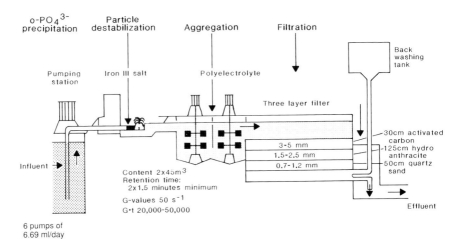

Figure 9.1 Diagram of the Wahnbach Reservoir phosphorus elimination plant (from Bernhardt, 1983)

3. Agglomeration by means of transport. The micro-flocs aggregate in a subsequent agglomeration step, forming larger, partly visible flocs. By adding a cationic polyelectrolyte, the flocs are made suitable for filtration. The specific cationic polyelectrolyte used in this method depends on the time of year and the dominant particulate substances present (i.e. algae, detritus or inorganic turbidity). It also depends on substances released into the water by the algae during the summer months, which can interfere with the flocculation process.

The filter of the phosphorus elimination plant consists of three layers of various granulations and densities. The filter has an area of 1100 m². The maximum water flow through is about 5 m³/s. Thus, the maximum water filtration rate is about 16 m/h.

The phosphorus elimination plant at the Wahnbach Reservoir was completed at the end of 1977, and the results of five years of operation are now available (Bernhardt, 1983). The quality of the filtrate and the average elimination are illustrated in Figure 9.2. About 95-99% phosphorus removal has been attained. The total five-year average phosphorus concentration in the filtrate was <5 µg/l. The decrease of soluble organic compounds has varied <60–80%, depending on the parameter of concern. More than 90% of the algae were eliminated during the algal growth periods. The decrease in turbidity has always been higher than 99%.

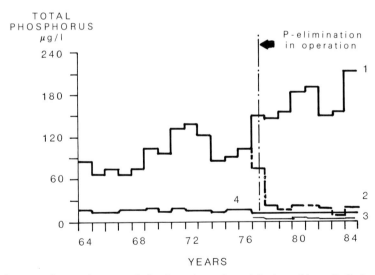

Explanation of terms: 1 = sum of all tributaries and precipitation without P-elimination; 2 = Sum of all tributaries and precipitation with P-elimination; 3 = PEP effluent; 4 = tolerable for oligotrophic conditions (e.g. Vollenweider, 1976a).

Figure 9.2 Quality of filtrate and average elimination for Wahnbach Reservoir (from Bernhardt, 1983)

Economic evaluation of the Wahnbach procedure has shown that the costs of phosphorus removal at the PEP were approximately 60% less than the costs of phosphorus reduction at individual sources in the drainage basin. In fact, even the maximum possible phosphorus reduction in the individual sources in the drainage basin would not result in a significantly greater water quality improvement, in comparison to the Wahnbach system. Thus, in terms of the costs per effort, the cost reduction with the PEP is even more significant.

Direct addition of phosphorus-precipitating chemicals to the influent waters. The external nutrient load to a lake also can be reduced by the direct addition of chemicals which precipitate phosphates to the influent water at the point the water flows into the lake. This procedure has been used successfully, for example, in the Federal Republic of Germany and in the Netherlands. It is most suitable for shallow reservoirs with a high phosphorus load (e.g. $10-50 \, g/m^2.yr$) especially in those cases where it is too expensive to pre-treat inflowing waters.

Rudolf & Uhlmann (1968) previously reported on the addition of iron salts to inlet waters as a control measure. Iron (III) sulfate has been applied to the inlet waters of De Grote Rug (a Dutch reservoir), in a concentration of $10 \, g \, Fe/m^3$ since 1976 (Bannink et al., 1980). No flocculation facilities are necessary because the divalent iron used is normally oxidized and precipitated in the reservoir itself. For waterbodies with short water retention periods (e.g. one month), application of trivalent iron or aluminum salts can give even better results. This has been seen in the Haltern Reservoir in the Federal Republic of Germany (Kotter & Patsch, 1980).

Filtration of tributary water through an aluminum oxide filter. A new method of phosphorus removal from small tributaries is the use of activated alumina columns. This method is useful mainly with small flows (<50 l/s) of phosphorus-rich waters which do not fluctuate greatly. An example is fish hatcheries fed by spring melt runoff in the upper part of streams of more or less constant size. The activated alumina is a technical grade aluminum oxide, from which products of varying interior surface sizes can be obtained. Products with an interior surface of $200-300 \, m^2/g$ have a high absorptive capacity. The absorptive capacity of this material for phosphate is higher than its capacity for all other substances present in natural waters (Bernhardt & Wilhelms, 1984).

This method has been in use for about $1-1/2$ years at the outflow of a fish hatchery in a lateral tributary of the Wahnbach reservoir (Figure 9.3), and has the following characteristics:

1. Operation without electricity;
2. A long running time for the aluminum oxide filter;

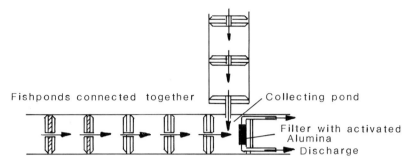

Figure 9.3 Arrangement of the activated alumina filtration plant in the collection pond of two rows of a fish-hatchery (From Bernhardt, 1983)

3. A decline in the orthophosphate concentration in the discharge of the fish culture to less than 50 µg/l, and of the total phosphorus concentration to less than 100 µg/l;
4. Low maintenance and service requirements;
5. Simple backwashing and treatment of the backwash;
6. Operational both during the summer period of extensive algal growth and during the winter period of low temperatures and ice on the ponds; and
7. Ability to replace individual aluminum oxide filters.

Canalization/diversion of wastewaters

Diversion of wastewaters. When the nutrient load to a waterbody comes mainly from very localized sources in the drainage basin, one can collect the nutrient-laden waters in sewer pipes and divert them, either to a conventional municipal wastewater treatment plant located below the lake or to a stream or lake below the lake of concern. However, simply diverting untreated waters to a downstream site is only a temporary measure since it merely moves the problem to another location, rather than treating it.

Nevertheless, this wastewater diversion can be effective in some cases. It has been used for the protection of many lakes in central Europe, particularly in the alpine and sub-alpine regions (e.g. Schliersee and Tegernsee), as well as the Soese and Innerste reservoirs in the Harz. It has also been applied to a number of lakes in Sweden and Denmark. Examples of the former include Lakes Norrviken, Uttran, Malmsjön, Glaningen, south Bergundasjön and Sätoftasjön (Ryding, 1981b), while an example of the latter is Lake Lyngbysø.

This technique has also been used in the United States, prominent examples being the lower Madison (Wisconsin) lakes (Sonzogni *et al.*, 1975), Lake Sammamish (E.B.Welch *et al.*, 1980) and Lake Washington

Control of eutrophication

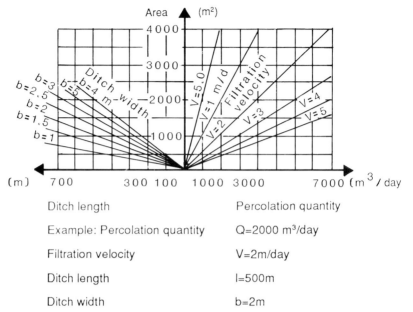

Figure 9.4 Calculation of the length of a seepage trench (from Bernhardt, 1983)

(Edmondson, 1972; Edmondson & Lehman, 1981).

Seepage trenches. Seepage trenches (as well as pit latrines) operate on the principle that phosphorus is removed when water passes through soil. The process is most effective when fine-grained sandy clays are present in the soil. The phosphorus becomes bound in the upper soil layers.

Seepage trenches are useful at sites where the quantities of effluent are low, and where the fluctuations in runoff are limited. It requires a drainage basin sufficiently small that extensive storm waters do not cause flood overflows from tributaries draining the basin. Seepage trenches can be used for tributaries having a flow not exceeding 100 l/s, and in a modified way below individual pollution sources (e.g. farms, pastures) if the slope is adequate and the soil composition allows seepage.

Figure 9.4 illustrates a sample calculation of the necessary length for a seepage trench. It is based on the quantity of seepage, the permeability of the filter material and the selected width of the trench. Ideally, a trench should not be more than 2–3 m wide, so that the upper filter layer can be removed as necessary.

It is advisable also to have a pre-basin to control maximum seepage. Maintenance costs for this type of system are comprised mainly of cleaning the pre-basin about every five years, and general operating expenses. Overall, this system is an economic, natural and satisfactory way of eliminating phosphorus, under the conditions described. It has been used successfully for years by the Wuppertal Municipal Works for protecting the Kerspe Reservoir (Grau, 1977).

IN-LAKE EUTROPHICATION CONTROL METHODS

In contrast to the previous section, which concentrated on treating a basic cause of eutrophication (i.e. external nutrient inputs), this section discusses methods for treating the in-lake symptoms of eutrophication. These latter methods, however, are usually not as effective over the long term as external nutrient control measures, and may have to be applied repeatedly. Nevertheless, the in-lake methods are effective for at least some period of time, and may even be the most reasonable approach in situations where it is too costly or otherwise unfeasible to build municipal wastewater treatment plants, etc. (e.g. developing countries). They also offer supplementary control measures in cases where the primary control programme is inadequate to achieve the control goals.

Dunst *et al.* (1974) provide a pre-1974 survey of previous experiences with many in-lake control methods, while Annex II provides a listing of more recent lake/reservoir reports which discuss the practical experiences with these methods. A recent compilation of in-lake treatment methods also is provided by Cooke *et al.* (1986). This latter reference discusses the range of available in-lake treatment methods, including their relative costs and effectiveness. Because of these various reports, the material discussed in this section is primarily descriptive. The reader is referred to Cooke *et al.* (1986) and the other cited references for further details.

Major in-lake control measures include the following:

1. *Nutrient inactivation* – this method involves the addition of phosphorus-precipitating chemicals (e.g. iron or aluminum salts) directly to a lake or reservoir, or to a tributary to the waterbody. These chemicals inactivate or immobilize the phosphorus. Drawbacks are the possible toxic effects of the added chemicals on biota, and the temporary nature of this treatment.

2. *Flow augmentation/flushing* – this method involves the transport of additional water (usually of low nutrient content) to a lake, thereby increasing its flushing rate. The increased flushing rate reduces the opportunity for biomass accumulation, while the increased water volume dilutes the in-lake nutrient levels. Flushing is particularly important in reservoirs where withdrawal of water from selected depths is possible. Because of this possibility lower water layers with a high nutrient content can bypass the algal-rich upper water layer by undercurrents. Also, direct dilution or reduction of algal crops can be achieved by flushing. A drawback is the need for large quantities of low nutrient content waters.

3. *Hypolimnetic aeration* – this method involves the introduction of oxygen to the hypolimnetic waters in a manner that preserves the thermocline. Maintaining oxygenated conditions in the hypolimnion will reduce the release of phosphorus and other reduced materials from the sediment into the water column. The ability of this method to effectively limit algal growths over a long period of time, however, has not yet been demonstrated. Further, the technology for this method has not yet been accepted on a global scale.

4. *Circulation* – this method is similar to hypolimnetic aeration, except that the hypolimnetic aeration is sufficiently vigorous that the thermocline is not preserved. The primary goal is to induce a mixing of the waterbody, thereby causing its destratification. The advantages and drawbacks of this method are similar to those of hypolimnetic aeration. A technically-related method is epilimnetic circulation. However, the aim of this latter method is not to add oxygen to water; rather, it is to prevent algal growth by circulating the algae out of the zone of light penetration for extended periods.

5. *Selective removal of hypolimnetic waters* – this method involves the withdrawal of nutrient-rich waters from the hypolimnion. This withdrawal effectively reduces the hypolimnetic volume, as well as the overall nutrient content of the waterbody. This procedure is generally applicable, however, only to small, deep lakes and to reservoirs in which waters can be withdrawn or discharged from selected depths.

6. *Lake level drawdown* – this method involves lowering the water level in the lake so that some or all of the bottom sediments are exposed to the atmosphere. Lake drawdown is used mainly for control of macrophytes and attached algae, and can be accompanied by dredging or by application of sediment covers. Drawbacks are the destruction of susceptible biota and the need to maintain the lake at low water levels or empty for extended periods of time.

7. *Covering bottom sediments* – this method involves covering the lake-bottom sediments with plastic sheeting or particulate materials (e.g. fly ash) to prevent sediment–water nutrient exchange, and to reduce macrophyte growths. Drawbacks are the associated costs and the possible effects of particulate materials on biota.

8. *Sediment removal (dredging)* – this method involves the dredging of nutrient-rich sediment from the lake bottom. Removal of the sediments will reduce the internal loading of nutrients and other materials (e.g. toxic substances). This method has been effective in lakes which have experienced severe nutrient enrichment over a long period of time.

Drawbacks are the associated expenses, the potential effects of dredging on biota, and sediment disposal problems.

9. *Harvesting* – this method involves the cutting and removal of nuisance growths of macrophytes and attached algae from the waterbody. This method will provide immediate relief from conditions which impair swimming, boating and water-skiing. Drawbacks are the associated expenses, the need for repeated application, and vegetation disposal problems (also see Chapter 10);

10. *Biological control ('Biomanipulation')* – this method involves the use of specific organisms to control growths of algae and/or other components of the food web. Examples include the use of fish to control macrophytes, the use of zooplankton to control phytoplankton, and the use of the manatee to control water hyacinths. However, extreme caution is advised in introducing foreign or exotic species to a given waterbody, since they may severely upset its ecological structure (also see Chapter 10);

11. *Chemical control* – this method involves the application of specific chemicals to waterbodies to kill undesirable aquatic plants. A common algicide is copper sulfate. Several herbicides have been used to combat macrophyte growths. Drawbacks are the associated expenses, the temporary nature of this method, and possible toxicity effects on other biota.

Table 9.3 provides a summary comparison of specific water quality problems in a lake or reservoir and the in-lake control methods which have been shown to be effective in treating them in various cases. Several of these methods are discussed below in more detail. Further, more detailed descriptions of these methods are provided in other sources (e.g. PLUARG, 1978a; United States Environmental Protection Agency, 1980; Loehr *et al.*, 1980).

Sediment nutrient control. Procedures for reducing the internal loading (regeneration) of phosphorus in lakes include all measures designed to reduce or restrict the release of nutrients from the bottom sediments. They also include measures for the removal of the released nutrients from the waterbody before they can get into the trophogenic zone.

The most effective in-lake sediment nutrient control measure is to dredge them. This procedure has been used successfully in Sweden, for example, in several lake restoration projects. Restoration of Lake Trummen is the best known example (Björk, 1983). A suitable area for the deposition of the dredged sediments also must be available, however, in order to use this approach most effectively. Furthermore, if the sediments contain elevated

Table 9.3 Water quality problems treatable by in-lake restoration measures

Control measure	Water quality problem							
	Odors	Fish kills	Toxic algae	Interference with swimming	Reduced commercial fishing	Excessive macrophyte growth	Poor drinking water quality	Excessive algal blooms
• Dredging								×
• Hypolimnetic aeration	×	×			×		×	
• Nutrient inactivation						×		×
• Altered circulation	×	×		×	×			×
• Algicides			×					×
• Biomanipulation			×	×				×
• Dilution\flushing	×	×			×	×		
• Removal of hypolimnetic waters	×	×			×			
• Lake drawdown								×
• Harvesting			×	×		×	×	
• Covering sediments	×	×						×

levels of heavy metals, its deposition in other regions may be hindered or prohibited because of regulatory requirements.

The control of phosphorus release from sediments is very important in shallow lakes, primarily because the zone of maximum algal productivity in such waterbodies borders directly on the sediments. Thus, any nutrient released from the sediments can directly enter this zone and stimulate algal growths. Sediment phosphate release is also an important consideration in waterbodies with long water retention times. In these latter waterbodies, the internal loading can become a significant component of the annual nutrient load.

If nutrient-rich sediments cannot be removed, one can attempt to reduce or inhibit the release of nutrients from sediments by producing an oxidizing (i.e. non-reducing) environment at the sediment surface. This can be achieved easily in most deep lakes by aerating ('oxygenating') the hypolimnion. A primary purpose of hypolimnetic aeration is to counteract the release of divalent manganese and iron ions, as well as hydrogen sulfide, from the sediment. It will, however, also work to provide an oxidizing layer at the sediment surface.

Hypolimnetic aeration produces an oxygenated aquatic environment in the hypolimnion which, in addition to enhancing removal of the external nutrient load, can gradually descend to the deeper hypolimnetic waters and sediments. The effect may be inadequate, however, if the sediments contain large amounts of unoxidized organic and inorganic substances. To insure a thorough oxidation of the surface sediment layer, it is possible also to inject a nitrate solution into the sediment. The nitrate provides anaerobic sediments with oxygen, which causes oxidation of the reduced organic and inorganic components in the sediment and also reduces the release of phosphates. However, this method, developed by Ripl (1976) and Leonardson & Ripl (1980) is relatively new and requires further assessment before widespread use.

Organisms living in the sediments can affect the release of phosphate from sediments. Particularly important organisms are insect larvae and worms, which can cause an increased exchange between the water and sediment via their burrowing ('bioturbation') throughout the sediment surface layer.

A related topic is the recycling of sediment nutrients via bottom-feeding organisms. As one example, it is known that bottom-feeding fish excrete phosphorus and nitrogen compounds into the water column. Such phosphorus excretion can be a significant contribution to the total nutrient loading of a lake or reservoir in some cases. An opportunity to assess the magnitude of this input was provided by the Minnesota (USA) Department of Natural Resources, which decided to restructure the fish population in Lake Marion. This is a shallow (mean depth = 1.98 m), large (area = 172 ha) lake in southcentral Minnesota. Pre-treatment estimates of the fish population were made, using mark–recapture methods. After rotonone treatment to eliminate the fish, a shore census of the dead fish was also made. This latter method gave much higher values. Based on the latter values, annual nutrient inputs from fish excretion were calculated to be $0.088 \text{ g/m}^2.\text{yr}$ for phosphorus and $0.27 \text{ g/m}^2.\text{yr}$ for nitrogen. By comparison, the Minnesota Pollution Control Agency previously had calculated a total phosphorus loading rate of $0.084 \text{ g/m}^2.\text{yr}.$ to the lake from its primarily agricultural watershed. This means fish excretion provided about half the annual phosphorus load to the lake.

It has been suggested that sediment nutrient release can be reduced or prevented by covering the sediments with an inert material. This method is not effective, however, when the applied layer is only a few centimeters thick, since sediment-dwelling organisms are capable of mixing the applied layer with the underlying sediment layers. Theoretically, this biological mixing problem could be solved by using fly ash from power plants, because ash material forms a cement-like layer on top of the sediment. Furthermore, the ash is able to bind phosphates. Unfortunately, however, fly ash also contains toxic substances (e.g. boron, selenium, molybdenum, arsenic and mercury) which restrict its use.

One can also attempt to cover the bottom sediments with plastic sheeting. This procedure is considerably more elaborate than covering the bottom sediment with loose materials such as ash. Therefore, it is useful primarily for very small lakes. This measure also may be useful in extreme cases at river or lake banks (e.g. beach areas) for preventing the growth of macrophytes.

In deep lakes, the euphotic zone usually is far from the bottom sediments, so that any phosphorus released from the sediments is not immediately available for algal growth (provided the lake remains stratified). The released nutrients will become available to algae only when the water becomes mixed (e.g. at autumn overturn). It is possible to remove the water lying immediately above the sediments from such lakes and reservoirs with a pipe, using the principle of a syphon. This method was developed by Olszewski (1973) and has been used effectively in several places, including a few lakes in Austria (Pechlaner, 1976).

Alteration of the flushing rates. Algae can only accumulate to nuisance levels in a lake or reservoir when the algal growth rate is faster than the rate of water renewal (i.e. the flushing rate). In some cases, therefore, one can attempt to reduce the accumulation of algae in a lake or reservoir by increasing the water flow-through rate (i.e. decreasing the water retention time).

An increase in the flushing rate can be useful even if it does not directly decrease algal growths, primarily because the input of low nutrient content waters to a waterbody can reduce the nutrient concentration in the lake, as well as in the sediments.

As discussed in Chapters 7 and 11, a rapid flushing rate also can affect the length of time prior to the growing season that one must measure the phosphorus load relevant to algal growth during the growing season.

Biomanipulation. The theoretical maximum algal growth which could occur for a given nutrient concentration usually is only achieved briefly (if at all) in most lakes and reservoirs. The actual phytoplankton concentration is usually considerably less than the theoretical maximum value, due to such factors as the feeding activity of zooplankton on phytoplankton (i.e. 'zooplankton grazing'). The feeding activity of zooplankton increase when the zooplankton assemblage is dominated by large *Daphnia*. This condition usually occurs when the fish population in a waterbody is dominated by large planktonivorous fish. By contrast, the presence of overcrowded coarse fish (which utilize *Daphnia* more readily) results in zooplankton assemblages dominated by smaller species. The smaller zooplankton species are unable to control the phytoplankton densities effectively.

Effective control of phytoplankton by zooplankton grazing can be enhanced by stocking a waterbody with carnivorous fish. These fish prey on smaller fish, thereby allowing larger species to dominate the zooplankton. The increased zooplankton grazing rate helps keep phytoplankton populations relatively small. Extensive experience with this method has been gained in Czechoslovakia (where it originated), the German Democratic Republic, the United Kingdom, The Netherlands and the United States.

Predatory fish (pike and pike-perch) were increased in the hypertrophic Bautzen Reservoir in 1977, to reduce the numbers of small fish that feed on filtering zooplankton (mainly *Daphnia*). As a result, the numbers of small zooplankton-feeding fish were reduced from about 7000 fish/ha in 1977–8 to about 3000 fish/ha in 1981–2. During the same period, the mean size of the predatory perch increased from 8 to 80 g/fish (Benndorf *et al.*, 1984a). In contrast, this approach is not effective where large phytoplankton dominate. For example, in Harbeespoort Dam, *Microcystis* is the dominant alga. This alga forms colonies which are too large for the dominant zooplankton (*Daphnia*) to graze upon (NIWR, 1985).

An enhanced concentration of large zooplankton can lead not only to a decrease in algal populations, but also to a reduction in the number of small zooplankton (e.g. *Rotatoria*). This effect can be important in reservoirs primarily used for drinking water, since some of the small zooplankton cannot be removed easily using the usual method of treating water with oxidizing agents, particularly chlorine.

The use of algal viruses on a technical scale to control blue-green algae has not been very successful to date (see Saffermann, 1973). This observation likely applies to other algal parasites as well (e.g. the widely distributed fungi belonging to the Chytridiaceae). These fungi can cause the collapse of an algal bloom. However, they are only effective when algae are present in high concentrations and in unfavorable environmental conditions. Therefore, the use of algal parasites for controlling eutrophication is limited at present.

Another in-lake method for attempting to regulate algal growths is by controlling the amount of sunlight energy available for photosynthesis. Two approaches for attempting to induce light limitation of algal growth in a lake or reservoir are:

1. Decreasing light penetration into the water column; and
2. Mixing the waterbody, in order to move the algae to deeper, darker parts of the waterbody.

Previous suggestions for decreasing light penetration into the water column include:

1. Covering the waterbody with opaque sheeting or floats;
2. Adding light-absorbing pigments (e.g. foodstuff dyes) to the waterbody; and
3. Spreading plastic beads or soot on the waterbody.

A practical limitation to this approach is that only relatively small waterbodies can be effectively covered or dosed with light-shielding or light-absorbing materials. Furthermore, some algal species are capable of persisting for relatively long periods of time in the dark (Hagedorn, 1981).

Attempts to control algal levels by mixing, or manipulation of the internal water circulation of lakes and reservoirs have usually been made in connection with waterbody destratification. Numerous reports on lake and reservoir destratification (which concentrate primarily on available technology and manipulation of water circulation) have been published (e.g. see Pastorok et al., 1981, 1982).

The success in using water mixing to decrease algal levels has been varied. It is related to such variables as the ratio of the mixing depth to the depth of the euphotic zone (i.e. whether photosynthesis or respiration dominate algal metabolism) and to the effects of increased nutrient availability and temperature associated with lake/reservoir mixing. Steele (1973) and Lorenzen & Mitchell (1973) have discussed the use of this latter approach to decrease algal biomass levels. Kalčeva et al. (1982) and Straškraba (1986) also discuss the effects of epilimnetic mixing to reduce the algal biomass to desired levels. Further, the algal concentration itself can affect light penetration into the water column. This self-shading phenomenon will reduce the depth of the euphotic zone. Numerous reports on lake and reservoir destratification discuss the available technology for air injection and manipulation of water circulation (e.g. see Pastorok et al., 1982).

Use of chemicals. Massive blooms of phytoplankton, particularly blue-green algae, and algae which can cause unpleasant odors and tastes in drinking water (e.g. some Chrysophyceae), are particularly undesirable. It is possible, however, to attempt to change the composition of the algal flora by changing the environmental conditions in the waterbody, even without reducing the nutrient influx. This can be done, for example, by changing the pH of the water. A massive bloom of the chrysophyte, *Synura uvella*, was brought under control in the German Democratic Republic by raising the pH of the waterbody with use of lime (Pütz et al., 1983). It is also possible to change an algal population dominated by blue-green algae to one dominated by green algae by decreasing the pH value, since blue-green algae apparently prefer a higher pH. Thus, a decrease in the pH gives the green algae a competitive advantage. Further, blue-green algae also are susceptible to specific viruses when the pH is low. This phenomenon has been demonstrated experimentally by Lindmark & Shapiro (1982).

Algae have been controlled in many countries in the past by the application to a waterbody of chemicals toxic to algae (e.g. copper sulfate and other algicides). Many experiences are documented in the literature (e.g. Rodhe, 1949; Jayangoudar & Ganapati, 1965; Eunpu, 1973; Meadley, 1970; Potter, 1971; and Scott et al., 1981, as cited in J.A. Thornton & Nduku, 1982). However, such measures should be used only in extreme cases, so that one does not inadvertently contaminate a lake or reservoir with the chemicals. The killing of large quantities of algae with the use of these substances, and their subsequent decay, also can overload the oxygen balance of a lake. Furthermore, when such chemicals are applied over a long period of time at relatively low concentrations, it can result in the growth of strains of algae which are resistant to the chemicals. This was observed in a canal of a waterworks in Sydney, Australia, which received a continuous application of copper sulfate (personal communication, I. Smalls, NIWR, 1984; Scott et al., 1981, as cited in J.A. Thornton & Nduku, 1982).

In contrast to algae, the use of herbicides can be useful in tropical lakes covered with floating plants, such as *Eichhornia* and *Pistia*. In such cases, the reoccurrence of the plants can be inhibited for long periods of time with just one spraying operation. It is even more effective if the re-introduction of such plants to the waterbody can be prohibited. For example, spraying operations at Lake McIlwaine and Hartbeespoort Dam (Africa) to control *Eichhornia* seedlings are still ongoing (for more than 15 years in the former case). This is because the seeds of this plant remain viable for many years.

Harvesting. In contrast to planktonic algae, rooted aquatic plants (macrophytes) can be removed mechanically from a waterbody (see also Chapter 10). This method involves the selective cutting or mowing of dense growths of macrophytes, which are then collected ('harvested') mechanically and usually disposed of elsewhere. Various kinds of mowing machinery have been developed to cut the plant shoots (see Cooke et al., 1986).

This method is often very useful for alleviating, at least temporarily, extensive macrophyte growths in the littoral zones of lakes and reservoirs. It allows most water uses to continue with minor interferences, and does not pose much hazard to other types of aquatic organisms. As discussed in Chapter 10, the harvested vegetation may be a beneficial product in some cases (e.g. livestock food, raw material for commercial products), especially in developing countries. Further, harvesting can be used in conjunction with other in-lake control measures.

Disadvantages of this approach include the fact that it it usually energy- and labor-intensive. In some cases, the occurrence of littoral zone macrophyte growths can be so extensive that considerable labor and/or equipment is required for their removal, resulting in high operating costs. In addition, the capital costs can also be relatively high. Probably the biggest disadvantage, however, is that the harvested vegetation must be disposed of in a manner that does not allow the release nutrients as the vegetation undergoes decay. Harvesting usually must be applied repeatedly, from year-to-year, and often several times per growth season.

In some cases, however, it is not always necessary to use mechanical equipment to control littoral zone macrophyte growths. Many macrophyte species can be kept under control by herbivorous fishes in the waterbody. For example, the grass carp (*Ctenopharyngodon idella*), indigenous to East Asia, has gained considerable importance in such applications in temperate and sub-tropical areas.

Oxygenation of the hypolimnion. The incomplete, anaerobic breakdown of aquatic organisms (primarily algae) and their byproducts, during periods of oxygen deficiency (particularly during the period of thermal stratification), can result in their accumulation in the hypolimnion. Such undesirable substances include divalent manganese and iron ions, hydrogen sulfide and ammonia, as well as macromolecular organic compounds, some of which have reducing and complexing properties. Virtually all of these substances can be converted to harmless compounds when sufficient amounts of oxygen are present in the water. The exceptions are the compounds which are difficult to degrade biologically. The latter compounds can cause significant problems in the production of drinking water (e.g. precursors for the development of potentially carcinogenic trihalomethanes).

Compounds easily broken down biologically, as well as undesirable oxidizable inorganic compounds, can be assimilated with hypolimnetic aeration, especially if the thermocline is not destroyed. This is because undesirable substances can become concentrated in the hypolimnion during the thermal stratification period. Complete mixing of the hypolimnetic waters, accompanied by oxygen enrichment, can oxidize the undesirable reduced compounds, thereby rendering them harmless. Maintenance of the thermocline also insures that planktonic algae in the epilimnion are not transported into the hypolimnion. If such transport occurred, the suitability of the hypolimnetic water for drinking water purposes would be considerably decreased. Moreover, it also inhibits a transport of nutrients from the sediment–water interface to the epilimnion, which could occur if air bubbles resulting from a complete mixing of the lake waters rise to the surface. Various techniques have been developed to effect such hypolimnic aeration without disturbance of the thermocline (see Bernhardt, 1978; Pastorok *et al.*, 1981, 1982).

CONTROL OF NON-POINT SOURCE NUTRIENTS IN THE DRAINAGE BASIN

Strictly speaking, the control of nutrients in the catchment area prior to their entrance in receiving waters is simply a variation of the control of external nutrients. However, as used here, nutrient control in the catchment area refers to control of non-point or diffuse sources of nutrients. These nutrient inputs are characterized by a lack of a 'pipeline' mode of entry into receiving waters, as in the case of municipal wastewater effluents. They are ubiquitous in most drainage basins settled by man, diffuse in character and often difficult to quantify. Available evidence suggests that the nutrient inputs to lakes and reservoirs in many developing countries arise primarily from non-point sources in the drainage basin. Therefore, the effective control of eutrophication in many countries may be more related to the treatment of non-point nutrient sources than to point source programmes.

Furthermore, non-point sources are usually more directly traceable to the actions of individuals, as contrasted with the cumulative input of many individuals (which characterizes municipal wastewater treatment plants).

Major non-point nutrient sources and possible remedial measures

Major non-point nutrient sources have been discussed previously (see Chapter 7) in regard to estimating the annual nutrient load to a lake or reservoir. Consequently, non-point source control measures are only briefly reviewed here, with reference given to more detailed sources of information. Prominent examples include the large scale study of non-point source pollution in the North American Great Lakes Basin (PLUARG, 1978*a*) and the Clean Lake Programme Guidance Manual (United States Environmental Protection Agency, 1980).

Urban non-point nutrient sources. Urban runoff can have a significant impact on the quality of receiving bodies of waters by polluting the ground water, by disturbing treatment processes in wastewater treatment plants during periods of floods, by overflows from combined sewer systems and by direct loading of receiving waters from separate sewer systems.

A primary goal for treatment of such non-point nutrient sources is to treat the pollution at its source; namely, reducing the amount of nutrients and other pollution in the runoff waters. This can be achieved by such measures as improved street sweeping practices. A decrease in the use of pesticides and fertilizers in gardening activities can be of help. Domestic sewer systems also can be connected to site treatments, in the form of circulation tanks (which are effective in containing coarse particulate matter).

Introducing permeable surfaces in urban areas can reduce the amount of urban runoff reaching lakes and reservoirs. It can also reduce the impacts of floods on sewer systems. If this approach is not feasible, it can be advantageous to construct storage or detention ponds for the storage of runoff waters prior to their entrance to treatment plants. By momentarily storing plug-flows, this approach can reduce the great flow and qualify variations characteristic of urban runoff waters. The stored runoff water can then be treated simultaneously with sewage in the plant, at a rate which allows a continuous flow through the treatment plant, thereby optimizing precipitation treatment processes.

A listing of possible remedial measures for urban non-point sources of nutrients, as identified in the North American Great Lakes Basin (Phosphorus Management Strategies Task Force, 1980), is provided in Table 9.4. A listing of methods for controlling runoff from urban areas with separate and combined sewer systems is provided in Tables 9.5 and 9.6, respectively.

Agricultural non-point sources. In attempting to reduce the potential impacts of agricultural activities on the eutrophication of lakes and reservoirs, two major control goals are:

1. Application of natural and mineral fertilizers in a manner that inhibits their transport and entrance into waterbodies (including maximizing the uptake of applied fertilizer by crops); and

2. Prevention of soil erosion to the maximum degree.

There are normally many more variables affecting agricultural non-point source nutrient runoff than for non-point urban sources. As noted earlier, the two major variables regarding the types and quantitites of nutrient loads from urban sources are the extent of impermeable surface areas (mostly street surface) and the presence or absence of street sweeping. In contrast, the major variables affecting agricultural nutrient runoff include soil type, intensity of land usage, extent of fertilizer application, type and density of crops grown, specific land management practices, etc. (see Chapter 7).

Table 9.4 Possible remedial measures for urban non-point sources, based on experiences in the North American Great Lakes Basin (modified from Phosphorus Management Strategies Task Force, 1980)

---

1. *SOURCE CONTROL*: Measures for reducing pollution in urban stormwater have been directed primarily toward control at the source, control in the collector system, and storage with treatment. Water quality problems identified with urban stormwater frequently have only been associated with combined sanitary and storm systems. Stormwater alone has been considered to be of a relatively low control priority. Control at the source may be accomplished with a variety of approaches, including:

Street sweeping – the efficiency of street sweepers in reducing the total pollutant load is dependent upon a number of factors, including sweeping frequency, condition of pavement, type of equipment used (vacuum or mechanical sweepers), frequency of rainfall, and public attitudes. Reliable predictions of the effectiveness of these measures are generally not available. However, estimates indicate that phosphorus removal levels may range between 4–44 per cent for vacuum sweepers, and between 5–16 per cent for brush sweepers, depending on the frequency of operation.

Catchment basin cleaning – now performed only about once or twice per year. A more regular programme, combined with street sweeping, may improve quality by 25–50 per cent.

2. *URBAN RUNOFF CONTROL*: A reduction in quantity, and an increase in the runoff period of urban stormwater also will reduce the total loading of pollutants from urban areas. This can be accomplished either with a reduction in the total flow ultimately reaching a sewer system and/or a redirection of flow overland, thereby increasing the time available for natural infiltration before introduction to the artificial conveyance system. Available measures include: (a) Roof top ponding; (b) Temporary check dams; (c) Detention/retention ponds; (d) Infiltration basins; (e) Drainage swales; and (f) Porous pavements.

3. *STORMWATER TREATMENT*: Measures are available to treat urban stormwater runoff. However, the available options are generally very expensive and would likely be cost-effective only where severe local water quality problems occurred and where more than one pollutant parameter was of concern. Available measures include: (a) Physical–chemical systems; (b) Swirl concentrators; (c) Stationary screens; (d) Treatment lagoons; (e) Air flotation; (f) Trickling filters; (g) Contact stabilization; and (h) Horizontal and/or vertical shaft rotary screens.

4. *URBAN CONSTRUCTION*: The most important source of sediment in urban areas is from developing urban lands, where yields may be 1000 times greater than from a comparable agricultural land. There are a variety of proven measures available for dealing with this problem, including: (a) Sedimentation ponds; (b) Organic mulch; (c) Temporary mulch and seed; (d) Chemical soil stabilizers; (e) Staged land clearing; (f) Retaining walls; (g) Hydro seeding; and (h) Matting and netting for slope stabilization.

Agricultural nutrient control measures include the proper construction and operation of suitable manure depots, as well as development and adoption of guidelines for the proper application of manures and fertilizer. Ideally, fertilizers (whether manures or commercial fertilizers) should be applied at the start of the vegetation period or, in the cold temperate climates, after the snow has melted. Fertilizer application may be done in several steps in some instances, in accordance with specific plant needs.

Control of eutrophication

Table 9.5 Runoff control methods for urban areas with separate storm sewer systems (from United States Environmental Protection Agency, 1980)

Street sweeping – includes sweeping of parking lots, to remove accumulation of dust, dirt, and debris.

Catch basin cleaning – removal of accumulations on a regular basis to maintain ability to intercept solids.

Eliminate cross-connecting – illegal connections of sanitary sewage can significantly increase strength of storm sewage.

Permeable sewers and catch basins – in areas with soil of sufficiently high permeability, runoff flow and loads can be attenuated.

Detention basins – various types. Provides for sedimentation, percolation, or increased evaporation to reduce flows and loads. Flow reductions may help reduce subsequent erosion.

Wetlands treatment – routing storm flows through an area of vegetation in a controlled manner, to remove nutrients, metals, and solids.

Public education and ordinances – can reduce runoff loads by reducing litter accumulations, controlling fertilizer applications, eliminating dumping of oil or other objectionable materials.

Table 9.6 Control methods for urban areas with combined storm sewer overflow systems (CSO) (from United States Environmental Protection Agency, 1980)

Storage basins – intercept and retain CSOs for return to sewers during dry periods, or equalize flows for more efficient treatment prior to discharge.

Sedimentation – provides gravity settling of suspended solids. Chemical coagulation may be employed to enhance removal efficiency.

Dissolved air flotation – separates solids by flotation and surface skimming. Uses less area than sedimentation. Chemical coagulation may be used.

Swirl concentrator – developed for CSO applications. Compact device. Separates settleable solids and floatable material.

Helical bend separator – developed for CSO applications. Provides solids separation. Installed in line. Low head loss.

Screens and microscreens – remove particles by straining. Wide range of apertures from several inch opening to very fine mesh (23 μ). Finer screening devices are mechanical.

Disinfection – with chlorine or other agents. Required facilities include chemical feeders, mixing, and contact chambers.

Sewer flushing – removes solids accumulations, which occur at critical points in sewer system in advance of scour during storm runoff.

Street flushing – transports street surface contaminate to sewer during dry weather, when they will reach treatment plants without overflow.

The applied fertilizer should always be worked into the soil. The amount of fertilizer used should not exceed the actual need of the crop grown, as determined by appropriate soil tests. Applications of phosphate sufficient to provide several years fertilizer needs at one time should not be practiced.

It is noted that it may be difficult to combat eutrophication on a large scale by reducing the quantities of mineral fertilizers applied to agricultural lands. To do so effectively, the amount of phosphate in the uppermost layer of soil probably would have to be reduced to such a degree that a reduction in the crop yield could occur. Moreover, legislative restrictions on the use of fertilizers can be avoided relatively easily. Thus, it is often more sensible to attempt to demonstrate to farmers that the associated costs of excessive fertilizer applications often can be higher than the profit one might expect with the resulting increase in crop yield.

Measures to decrease erosion, and associated nutrient loss, include vegetative buffer strips, contour cultivation, cross slope tillage and strip cropping. In developing countries, the reforestation of steeply sloping land areas, and the prevention of overgrazing by stock, are very important control measures. The use of forests located on slopes for grazing, and marshes for pastures in the immediate vicinity of receiving waters, should be prohibited.

An auxiliary means of reducing rural or agricultural pollution is by legislative means to establish 'protected zones', as mentioned previously. The normal purpose of such zones is to protect a lake or reservoir from the harmful influences of civilization. However, they also can be useful for the control of eutrophication, for example, via restrictions on development and building, and prohibition of fertilizer use in the vicinity of receiving waters. It also is possible to ban phosphate detergents in the drainage basin, if great amounts of phosphates emerge from septic tanks, milk rooms, etc.

A listing of possible remedial measures for agricultural non-point sources of nutrients, as identified in the North American Great Lakes Basin, is provided in Table 9.7. The control methods for reducing nutrient losses from agricultural lands are highlighted in Table 9.8, while measures for reducing erosion from croplands are identified in Table 9.9.

A summary of the overall effectiveness of a large number of control measures for different non-point pollution sources is presented in Table 9.10. This summary table is based on the results of the previously cited study of non-point source pollution in the North American Great Lakes Basin (PLUARG, 1978a; Johnson et al., 1978). A more detailed description of these non-point source control methods, including their costs and relative effectiveness, is provided by Monaghan (1977) and Skimin et al. (1978).

ASSESSMENT OF COSTS OF ACHIEVING PHOSPHORUS CONTROL GOALS

Practical experience suggests that reducing nutrient inputs to a lake or reservoir by treatment of point source nutrient loads (especially municipal wastewater treatment plant effluents) is a cost-effective approach in many cases. This is due in part to the fact that the necessary technology for this approach is well documented, and past experiences have shown this

Table 9.7 Possible remedial measures for agricultural non-point sources, based on experiences in the North American Great Lakes basin (modified from Phosphorus Management Strategies Task Force, 1980)

1. *CONSERVATION TILLAGE* – includes no till, disk plant and chisel plow:

   No till – with the exception of a narrow strip penetrated by a fluted coulter (5–8 cm), there is no disturbance of residue cover or soil profile. Seeding, fertilizing and herbicide application usually are carried out in one operation. Up to 90 per cent reductions in soil loss over conventional tillage have been attributed to this technique, with no sacrifice in crop yields on some soils in the Great Lakes Basin. Sediment phosphorus reductions between 10–90 per cent have been reported when evaluated on a watershed basis.

   Disk plant – may consist of one or two passes over the field, followed by the planting operation. System loosens entire soil surface to 10 cm, breaks up crop residue and mixes it with the top surface soil. Soil loss reductions between 70–90 per cent have been reported on some soils.

   Chisel plow – results in soil breakup to a depth of 20 cm, but does not turn over soil surface layer. Thus, large portion of crop residue is maintained on surface. Chisel plowing normally is followed by at least one disking. Sediment loss reductions between 30–90 per cent have been reported in the Great Lakes Basin.

2. *VEGETATIVE BUFFER STRIPS* – proven effective in reducing overland flow rate, and trapping sediment in suspension, when maintained adjacent to water courses. A 20-cm wide strip may remove up to 50 per cent of the sediment load. Level of utility is affected by width, vigor and height of vegetation and slope. In addition to buffer strip development and maintenance costs, the remaining long-term operating cost is the loss of production agricultural land.

3. *CONTOUR CULTIVATION* – requires that cultivation be done parallel to natural contours of the land. Surface runoff which normally flows down the fall line is now diverted laterally and more gently to the base. Average soil losses can be reduced by 50 per cent on moderate (3–7 per cent) slopes.

4. *CROSS-SLOPE TILLAGE* – in contrast to contour cultivation, this approach requires only that cultivation proceed at right angles to the slope direction. Probably more easily adopted than contouring, although sediment reductions may only be half as much.

5. *STRIP CROPPING* – by alternating strips of close-grown crops (hay and grains) with row crops (corn) across the slope, the slope length is effectively broken. This increases the absorption, and reduces the velocity, of overland flow. It also is effective in reducing wind erosion. Strip cropping may reduce sediment yields up to 85 per cent. This technique encourages crop rotating by farmer.

6. *FERTILIZER APPLICATION* – poor timing and placement of fertilizer can increase the levels of soluble inorganic phosphorus leaving a watershed. Although levels are usually low (<9 per cent), this form of phosphorus is the most readily available to algae. Incorporation of applied fertilizer into the soil profile during the time of maximum crop need will assist in reducing the loss of fertilizer nutrients due to surface runoff.

7. *SUBSURFACE (TILE) DRAINAGE* – in areas of poorly drained soils, subsurface drain construction can reduce runoff and subsequent soil loss. On these soils, installation of subsurface drains also may improve crop yields, when conservation tillage practices are followed.

8. *MANAGEMENT OF LIVESTOCK MANURE* – although this source may be less important in relation to total phosphorus loads, considerable improvements are possible in existing operations. Factors such as the location of feeding operations relative to water courses, size and designs of manure storage facilities, frequency of winter spreading, and distance of spreading from streams and open ditches can affect the delivery of manure-derived phosphorus to streams.

Table 9.8 Methods for reducing nutrient losses from agricultural lands (From United States Environmental Protection Agency, 1980)

Control practice	Practice highlights
ELIMINATING EXCESSIVE FERTILIZATION:	– May cut nitrate leaching appreciably; reduces fertilizer costs; has no effect on yield.
LEACHING CONTROL:	
Timing nitrogen application	– Reduces nitrate leaching; increases nitrogen use efficiency; ideal timing may be less convenient.
Using crop rotations	– Substantially reduces nutrient inputs; not compatible with many farm enterprises; reduces erosion and pesticide use.
Using animal wastes for fertilizer	– Economic gain for some farm enterprises; slow spreading problems.
Plowing under green legume crops	– Reduces use of nitrogen fertilizer; not always feasible.
Using winter cover crops	– Uses nitrate and reduces percolation; not applicable in some regions; reduces winter erosion.
Controlling fertilizer release or transformation	– May decrease nitrate leaching; usually not economically feasible; needs additional research and formation development.
CONTROL OF NUTRIENTS IN RUNOFF:	
Incorporating surface applications	– Decreases nutrients in runoff; no yield effects; not always possible; adds cost in some cases.
Controlling surface applications	– Useful when incorporation is not feasible.
Using legumes in pastures and haylands	– Replaces nitrogen fertilizer; limited applicability; difficult to manage.
CONTROL OF NUTRIENT LOSS BY EROSION:	
Timing fertilizer plow-down	– Reduces erosion and nutrient loss; may be less convenient.

approach to be successful in achieving eutrophication control goals in many situations. Nevertheless, there are situations in which the often high initial costs of building (and subsequent operation and maintenance) of a municipal wastewater treatment plant are not justified by the degree of expected water quality improvement. In other cases, this approach simply may not be economically feasible. In such cases, the use of in-lake eutrophication methods may provide a workable alternative approach, as discussed in

Table 9.9 Measures for control of cropland erosion (modified from United States Environmental Protection Agency, 1980)

No.	Erosion control practice and highlights
E1	*No-till planting in prior crop residues*: Most effective in dormant grass or small grain; highly effective in crop residues; minimizes spring sediment surges and provides year-round control; reduces labor, machine, and fuel requirements; delays soil warming and drying; requires more pesticides and nitrogen; limits fertilizer and pesticide placement options; some climatic and soil restrictions.
E2	*Conservation tillage*: Includes a variety of no-plow systems that retain some surface residues; more widely adaptable but somewhat less effective than E1; advantages and disadvantages generally same as E1, but to less degree.
E3	*Sod-based rotations*: Good meadows lose virtually no soil and reduce erosion from succeeding crops; total soil loss greatly reduced, but losses unequally distributed over rotation cycle; aid in control of some diseases and pests; more options for fertilizer potential transport of water soluble phosphorus; some climatic restrictions.
E4	*Meadowless rotations*: Aid in disease and pest control; may provide more continuous soil protection than one-crop systems; much less effective than E3.
E5	*Winter cover crops*: Reduce winter erosion where corn stover has been removed and after low-residue crops; provide good base for slot-planting next crop; usually no advantage over heavy cover of chopped stalks of straw; may reduce leaching of nitrate; water use by winter cover may reduce yield of cash crop.
E6	*Improved soil fertility*: Can substantially reduce erosion hazards as well as increase crop yields.
E7	*Timing of field operations*: Fall plow facilitates more timely planting in wet springs, but generally increases winter and early spring erosion hazards; optimum timing of spring operations can reduce erosion and increase yields.
E8	*Plow-plant systems*: Rough, cloddy surface increases infiltration and reduces erosion; much less effective than E1 and E2 when long rain periods occur; seedling strands may be poor when moisture conditions are less than optimum. Mulch effect is lost by plowing.
E9	*Contouring*: Can reduce average soil loss by 50 per cent on moderate slopes, but less on steep slopes; loses effectiveness if rows break over; must be supported by terraces on long slopes; soil, climatic, and topographic limitations; not compatible with use of large farming equipment on many topographies. Does not affect fertilizer and pesticide rates.
E10	*Graded rows*: Similar to contouring, but less susceptible to row break-overs.

(...continued)

Available techniques for treating eutrophication

Table 9.9 (*continued*)

E11	*Contour strip cropping*:	Row crop and hay in alternative 20 to 40 m strips reduce soil loss to about 50 per cent of that with the same rotation contoured only; Fall-seeded grain in lieu of meadow only about half as effective; alternating corn and spring grain not effective; area must be suitable for cross-slope farming and establishing rotation meadows; favorable and unfavorable features similar to E3 and E9.
E12	*Terraces*:	Support contouring and agronomic practices by reducing effective slope length and runoff concentration; reduce erosion and conserve soil moisture; facilitate more intensive cropping; conventional gradient terraces often incompatible with use of large equipment, but new designs have alleviated this problem; substantial initial cost and some maintenance costs.
E13	*Grassed outlets*:	Facilitate drainage of graded rows and terrace channels with minimal erosion; involve establishment and maintenance costs and may interfere with use of large implements.
E14	*Ridge planting*:	Earlier warming and drying of row zone; reduces erosion by concentrating runoff flow in mulch-covered furrows; most effective when rows are across slope.
E15	*Contour listing*:	Minimizes row breakover; can reduce annual soil loss by 50 per cent; loses effectiveness with post-emergence corn cultivation; same disadvantages as E9.
E16	*Change in land use*:	Sometimes the only solution. Well-managed permanent grass or woodland effective where other control practices are inadequate; lost acreage can be compensated for by more intensive use of less erodible land.
E17	*Other practices*:	Contour furrows, diversions, subsurface drainage, land forming, closer row spacing, etc.

previous sections. Such possibilities will have to be examined on a case-by-case basis.

Relative costs of point and non-point source phosphorus control measures

As noted in Chapter 3, there are situations in which the use of non-point source control methods can be less expensive than very stringent point source control methods. An example was provided for the North American Great Lakes Basin (PLUARG, 1978a). In this effort, the United States and Canadian federal governments developed 'target' phosphorus loads to be achieved for each of the five Great Lakes, based on certain desired water quality conditions for each of the lakes. The target loads, the existing (1975) loads, and the necessary phosphorus load reductions to achieve the target loads, are summarized in Table 9.11.

To achieve these phosphorus target loads, PLUARG considered several remedial measures, including:

Table 9.10 Matrix of remedial measures for non-point source pollutants (from PLUARG, 1978a)

Remedial Techniques	Land Use										
	Urban	Agriculture	Recreation	Forest	Extractive	Transportation	Liquid Waste Disposal	Deepwell Disposal	Solid Waste Disposal	Lakeshore & Riverbank Erosion	Shoreline Landfilling
1 Chemical Soil Stabilizers	Sn	Sn				Sn				S	
2 Roof Top Ponding	sn										
3 Dutch Drain (Gravel filled ditches with option drainage pipe in base)	sn										
4 Porous Asphalt Paving	sc					sc					
5 Precast Concrete Lattice Blocks and Bricks	sn c		Sn			sn c					Sn
6 Seepage Basin or Recharge Basin (Single Use)	sn c										
7 Recharge-Detention Storage Basins (Multi-Use)	sn c	sn									
8 Seepage Pits or Dry Wells	sn c										
9 Pits, Gravity Shafts, Trenches and Tile Fields	sn c										
10 Recharge of Excess Runoff by a Pressure Injection Well	sn c										
11 Conservation Construction Practices	S			S	S	S			S		
12 Temporary Mulching and Seeding of Stripped Areas	S			S	S	S					

Significantly Effective in
Reducing Magnitude of Pollutant

C—chemicals
N—nutrients
P—pesticides
S—sediments

Moderately Effective in
Reducing Magnitude of Pollutant

c—chemicals
n—nutrients
p—pesticides
s—sediments

(continued)

Available techniques for treating eutrophication

Table 9.10 (continued)

	Remedial Techniques	Urban	Agriculture	Recreation	Forest	Extractive	Transportation	Liquid Waste Disposal	Deepwell Disposal	Solid Waste Disposal	Lakeshore & Riverbank Erosion	Shoreline Landfilling
13	Conservation Cultivation Practices on Steep Slopes	S	S				S					
14	Temporary Diversions on Steeply Sloping Sites & Temporary Chutes	S	S		S	S	S				S	
15	Temporary Check Dams on Small Swales and Watercourses	S	S			S	S					
16	Seeded Areas Protected with Organic Mulch	S	S		S	S	S				S	
17	Seeding Areas protected by Netting or Matting	S			S	S	S				S	
18	Single Family Aerobic Treatment Systems	Nc	Nc	Nc								
19	Contour Listing		S									
20	Disposal of Treated Sewage Effluent by Spray Irrigation	sN c	sN c	sN c				sN c	sN c			
21	Surface Water Diversion	S	S		S		S					
22	Terraces (Diversion Terraces)	S	S		S		S					
23	No-Tillage Cultivation (Shoot Planting, Zero Tillage)		S									
24	Pesticide Application Methods	P	P	P	P		P					
25	Alternative to Chemical Pesticides	P	P	P	P		P					

Significantly Effective in
Reducing Magnitude of Pollutant

C—chemicals
N—nutrients
P—pesticides
S—sediments

Moderately Effective in
Reducing Magnitude of Pollutant

c—chemicals
n—nutrients
p—pesticides
s—sediments

(continued)

Control of eutrophication

Table 9.10 (continued)

Remedial Techniques	Urban	Agriculture	Recreation	Forest	Extractive	Transportation	Liquid Waste Disposal	Deepwell Disposal	Solid Waste Disposal	Lakeshore & Riverbank Erosion	Shoreline Landfilling
26 Slow Release Fertilizers		Nc		Nc							
27 Placement of Fertilizer		Nc		Nc							
28 Timing of Fertilizer Application		Nc		Nc							
29 Roughening of the Land Surface		S									
30 Promotion of Soil Clods or Aggregates		S									
31 Stripcropping		S		S							
32 Miscellaneous Tillage Alternatives		S									
33 Conservation Tillage		S									
34 Sod-Based Crop Rotation		S									
35 Improved Soil Fertility		S									
36 Winter cover crops		S									
37 Timing of Field Operations		S									
38 Contouring or Contour Cultivation		S									
39 Grassed Outlets	S	S				S				S	
40 Direct Dosing of Alum to a Septic Tank	N	N	N								
41 Swirl Concentrator for Runoff Treatment	Sn										
42 Retention Basins for the Treatment of Wet-Weather Sewage Flows	Sn	Sn									

Significantly Effective in
Reducing Magnitude of Pollutant

C — chemicals
N — nutrients
P — pesticides
S — sediments

Moderately Effective in
Reducing Magnitude of Pollutant

c — chemicals
n — nutrients
p — pesticides
s — sediments

(continued)

Available techniques for treating eutrophication

Table 9.10 (continued)

Remedial Techniques	Urban	Agriculture	Recreation	Forest	Extractive	Transportation	Liquid Waste Disposal	Deepwell Disposal	Solid Waste Disposal	Lakeshore & Riverbank Erosion	Shoreline Landfilling
43 Stationary Screens	Sn										
44 Horizontal Shaft Rotary Screen	Sn										
45 Vertical Shaft Rotary Fine Screen	Sn										
46 Treatment Lagoons *	sN		sN								
47 Rotating Biological Contactors *	N	N	N								
48 Trickling Filters *	N	N	N								
49 Contact Stabilization	N										
50 Air flotation	Sn										
51 Physical-Chemical Systems	sN										
52 Reverse Osmosis of Mine Tailings Effluent					C						
53 Chemical Adsorption onto Clays in Experimental Environment		P		P							
54 Surface Water Division		Sn c		Sn c	Sn c				Sn c		
55 Reducing Ground or Mine Water Influx					nC				nC		
56 Underdrains for Mineral Stockpiles or Tailings					nC						
57 Evaporation Ponds					nc						
58 Street Cleaning	Snc					Snc					

Significantly Effective in Reducing Magnitude of Pollutant

C—chemicals
N—nutrients
P—pesticides
S—sediments

Moderately Effective in Reducing Magnitude of Pollutant

c—chemicals
n—nutrients
p—pesticides
s—sediments

(continued)

Control of eutrophication

Table 9.10 (continued)

Remedial Techniques	Urban	Agriculture	Recreation	Forest	Extractive	Transportation	Liquid Waste Disposal	Deepwell Disposal	Solid Waste Disposal	Lakeshore & Riverbank Erosion	Shoreline Landfilling
59 Interception of Aquifers		nc	nc		nc				nc		
60 Neutralization of Mine Acid Waste					c						
61 Stream Neutralization					nc						
62 Improved Methods of Sludge Disposal on Land	nc	nc					nc				
63 Annual Storage and Land Application of Livestock Wastes		N									
64 Sewer Flushing	Sn c										
65 Combined Sewer Overflow Regulators	SN C										
66 Overburden Segregation	Sn				Sn						
67 Mineral Barriers or Low Wall Barriers					Sn c						
68 Longwall Strip Mining					Sn c						
69 Modified Block Cut or Pit Storage					Sn c						
70 Head-of-Hollow-Fill					Sn c						

Significantly Effective in
Reducing Magnitude of Pollutant
C—chemicals
N—nutrients
P—pesticides
S—sediments

Moderately Effective in
Reducing Magnitude of Pollutant
c—chemicals
n—nutrients
p—pesticides
s—sediments

(continued)

Available techniques for treating eutrophication

Table 9.10 (continued)

Remedial Techniques	Urban	Agriculture	Recreation	Forest	Extractive	Transportation	Liquid Waste Disposal	Deepwell Disposal	Solid Waste Disposal	Lakeshore & Riverbank Erosion	Shoreline Landfilling
71 Box Cut Mining					Sn / c						
72 Area Mining					Sn / c						
73 Auger Mining					sn / c						
74 Reducing Surface Water Infiltration					nc						
75 Road Planning & Design				S		S					
76 Blocking					C						
77 Check Dams	S					S				S	
78 Retaining Walls for Road Construction for Sleeper Slopes				S		S					
79 Revegetation-Reforestation Cut Areas and Bare Slopes	S			S	S	S					
80 Vegetative Buffer Strips	S	S			S	S				S	
81 Sediment Basin	S	S			S	S	Sn		S		
82 Rip Rap Bank Protection	S					S				S	S
83 Protectin of Culvert Outlet, Chute Outlets, etc.	S	S		S						S	
84 Dolos (Offset asymmetric tetrapods)						S					S

Significantly Effective in Reducing Magnitude of Pollutant

C—chemicals
N—nutrients
P—pesticides
S—sediments

Moderately Effective in Reducing Magnitude of Pollutant

c—chemicals
n—nutrients
p—pesticides
s—sediments

(continued)

Control of eutrophication

Table 9.10 (continued)

	Remedial Techniques	Land Use											
		Urban	Agriculture	Recreation	Forest	Extractive	Transportation	Liquid Waste Disposal	Deepwell Disposal	Solid Waste Disposal	Lakeshore & Riverbank Erosion	Shoreline Landfilling
85	Engineering Design & Management For Shoreline Landfilling											Sn c
86	Revegetation of Mines Tailings: Stabilization					S						
87	Slope Lowering of Spoil and Tailings Stockpiles					S						
88	Package Sewage Treatment Plants (Multi-Family Use)	sN c	sN c	sN c								
89	Waste Exchange for Resource Recovery									C		
90	Head Gradient Control									sn C		
91	Biological Treatment									SN C		
92	Streambank Protection with Vegetation	S	S								S	
93	Grass Channels or Waterways	Sn	Sn				Sn				S	
94	Permanent Diversions	Sn	Sn		Sn		Sn				Sn	
95	Bank Protection by Jetties, Deflectors	S	S				S				S	S

Significantly Effective in
Reducing Magnitude of Pollutant

C —chemicals
N —nutrients
P —pesticides
S —sediments

Moderately Effective in
Reducing Magnitude of Pollutant

c —chemicals
n —nutrients
p —pesticides
s —sediments

(continued)

Available techniques for treating eutrophication

Table 9.10 (continued)

	Land Use	Urban	Agriculture	Recreation	Forest	Extractive	Transportation	Liquid Waste Disposal	Deepwell Disposal	Solid Waste Disposal	Lakeshore & Riverbank Erosion	Shoreline Landfilling
	Remedial Techniques											
96	Reduction and Elimination of Highway Deicing Salts	C					C					
97	Septic Tank/Tile Bed Sewage Disposal	N	N	N								
98	Miscellaneous Methods to Reduce Storm Runoff	Sn					Sn					
99	Exclusion of Livestock From Watercourses		Sn									
100	Land Smoothing		S									
101	Gabion Baskets	S	S								S	S
102	Miscellaneous Erosion Control Fabrics and Materials	S	S								S	S
103	Miscellaneous Individual Wastewater Treatment Systems	N	N	N								
104	Clivus Multrum	N	N	N								
105	Controlling Feedlot Runoff		N							C		
106	Landfill Liners											
107	Hydroseeding	S			S	S	S				S	
108	Catch Basin Cleaning	Snc					Snc					
109	Plant Materials For Bank and Slope Stabilization	S	S	S	S	S	S				S	S

Significantly Effective in Reducing Magnitude of Pollutant
C—chemicals
N—nutrients
P—pesticides
S—sediments

Moderately Effective in Reducing Magnitude of Pollutant
c—chemicals
n—nutrients
p—pesticides
s—sediments

Table 9.11 Target loads and recommended reductions in annual phosphorus loads (t) for the lower Great Lakes (modified from Johnson et al., 1978)

Basin	Estimated 1975 load[a]	Target load[b]	Recommended reduction[c]
Lake Erie	13 400	11 000	2 400
Lake Ontario	9 400	7 000	2 400

[a] The 1975 load after an effluent phosphorus concentration of 1 mg/l has been achieved at all municipal treatment plants discharging >3800 m^3/day.
[b] Based on the 1978 Great Lakes Water Quality Agreement.
[c] Calculated as the difference between the 1975 estimated load and the target load.

1. Reduction of phosphorus levels in effluents of municipal wastewater treatment plant (discharging >3800 m^3/day) to 1, 0.5 and 0.3 mg/l;
2. Urban non-point measures: Level 1 (reduction at the source, street sweeping); and Level 2 (Level 1 measures, plus detention and sedimentation of urban stormwater); and
3. Agricultural non-point measures: Level 1 (good land 'stewardship', including proper incorporation of fertilizers and manures into soil, no excessive fertilizer use, and general conservation plowing techniques); Level 2 (Level 1 measures, plus conservation tillage, buffer strips, strip cropping, better drainage construction, and winter cover crops in areas where row crops are grown on fine-textured soils); and Level 3 (Level 2 measures, plus increased crop cover, spring plowing, establishment and management of pasture, critical area protection, gradient terracing, and grassed waterways in areas where row crops are grown on fine-textured soils).

The estimated costs (1975 US dollars) for Level 2 and Level 3 agricultural measures in the North American Great Lakes Basin are as follows:

Level 2 – conservation tillage: $3.80–$23.80/ha;
   – contour strip cropping: $13.75/ha;
   – cover crops: $10/ha;

Level 3 – improved crop cover/conservation tillage: $12.50/ha;
   – spring plowing: $2.50/ha;
   – establish/manage pastures: $125/ha;
   – critical area protection – $125/ha;
   – improved drainage: $15–$125/ha;
   – gradient terracing: $2.50–$70/ha;
   – grassed waterways: $2.50–$3.75.

It was thought that Level 1 agricultural measures would result in a net benefit to the farmer. Therefore, PLUARG determined that Level 1 agricultural measures would have a 'minimal cost' and result in a ten per cent reduction in the rural phosphorus load. Level 2 measures would have an average cost of $15/ha and result in a 25 per cent reduction in the rural phosphorus load, while Level 3 measures would have an average cost of

$55/ha and result in a 40 per cent reduction in the rural load.

As an example of a process for selecting the most cost-effective mix of point and non-point source phosphorus control programmes, these estimated costs are applied to Lake Erie. Based on Table 9.11, an annual phosphorus load reduction of 2400 t is necessary to achieve the phosphorus target load for Lake Erie (assuming that a 1 mg/l effluent phosphorus limitation has been achieved for all municipal wastewater treatment plants discharging >3800 m^3/day). A listing of the general types of remedial measures available for Lake Erie, and their estimated achievable phosphorus load reductions and associated costs, is provided in Table 9.12. The cost-efficiency ($1000/t phosphorus removed) is also provided.

There are, of course, several combinations of control measures which would allow the necessary 2400 t phosphorus load reduction for Lake Erie to be achieved. However, Table 9.12 shows that this reduction cannot be achieved using only point source controls. A reduction in effluent phosphorus concentrations even down to the stringent level of 0.3 mg/l would only allow an overall load reduction of 1950 t (i.e. 1305 + 645), which is less than the necessary 2400 t reduction.

Table 9.12 indicates that several combinations of control measures are possible. For example, a municipal wastewater effluent phosphorus reduction from 0.5 to 0.3 mg/l would result in an annual reduction of 645 t, compared to the Rural Level 2 reduction of 350 t. However, the cost/t reduction for the former is nearly $100 000, compared to only $64 300 for the Rural Level 2 measures. The Rural Level 3 measures would provide

Table 9.12 Relative costs and effectiveness of phosphorus reduction options for Lake Erie (modified from PLUARG, 1978a)

Remedial measure[a]	Estimated incremental phosphorus reduction (t/yr)	Estimated incremental annual costs[b] (million $)	Thousand dollars/t phosphorus reduction
I. Urban point sources Reduction of phosphorus in wastewater treatment plants effluents:			
(a) 1 mg/l to 0.5 mg/l	1 305	10.5	8.0
(b) 0.5 mg/l to 0.3 mg/l	645	61.0	95.5
II. Rural non-point sources[a]			
Level 1	450	'minimal'	0
Level 2	350	800	64.3
Level 3	305	53	174.0
III. Urban non-point sources[a]			
Level 1	445	36.5	82.0
Level 2	615	96.5	156.9

[a] See text for description of non-point source remedial measures.
[b] All costs based on 1975 US dollars.

almost as great a reduction as the Urban Level 1 measures. However, the cost/t phosphorus reduction is $174 000 for the former, compared to $82 000 for the latter. Based on these types of assessments of the incremental costs of various phosphorus control options, PLUARG determined the least expensive combination of point and non-point source control measures for the North American Great Lakes Basin, as summarized in Table 9.13. This process is discussed in more detail by PLUARG (19798a), Johnson et al. (1978), Johnson & Berg (1979), and Rast (1981b).

It is emphasized that the above example assumed that costs were the sole consideration in developing the phosphorus control programme. Futhermore, the approach was directed solely to phosphorus reduction at the source (as contrasted with use of in-lake methods). However, as mentioned in Chapter 3, other factors often must be considered in specific situations (e.g. need for even higher quality water, aesthetic considerations, public acceptability of control measures, relative costs of labor, materials and expertise in different countries, etc.). Such factors can strongly influence the selection of phosphorus control programmes in different locations and socioeconomic situations.

Table 9.13 Example of least expensive combination of phosphorus control options for Lake Erie[a] (modified from PLUARG; Johnson et al., 1978)

Remedial measures	Phosphorus removal (t)	Annual cost[b]
1. Rural Level 1 measures	450	'minimal'
2. 0.5 mg/l phosphorus limitation for municipal wastewater treatment plant effluent	1 305	$10.5 million
3 Rural Level 2 measures	350	$22.5 million
4 Urban Level 1 measures	445	$36.5 million
TOTALS:	2 550	$69.5 million

[a] Assumes an effluent phosphorus concentration of 1 mg/l has already been achieved for municipal wastewater treatment plants discharging $>3800 m^3/day$
[b] Based on 1975 U.S. dollars

Use of optimization models in selection of phosphorus control options for lakes and reservoirs

As illustrated above, in many cases one may have to choose between several eutrophication control options in the development and implementation of an effective eutrophication control programme. Futhermore, one usually has to consider the simultaneous impacts of various factors (including relative costs and effectiveness) associated with a given control programme. These considerations can make it difficult to select the 'best' phosphorus control strategy to use in a given situation, particularly when an overriding desire is to keep costs to a minimum.

To assist with such management decisions, a category of mathematical models known as 'optimization models' has been developed (see Chapter 6 for a discussion of watershed and waterbody models). Optimization models have also been called 'prescriptive' models by Biswas (1981). For the purpose of this document, such models would normally be used to maximize (or minimize) specific conditions or criteria, when assessing eutrophication management alternatives. Examples of such criteria would include such concerns as the optimum use of available resources, determination of minimum costs for achieving a given goal, etc.

A feature of optimization models is that they allow one to attempt to determine the optimal or 'best' solution to a management problem from among several possibilities, given multiple constraints or factors whose individual effects must be considered simultaneously. Loucks (1976) has suggested that even if optimization models do not allow one to find the best solution, at least they allow one to eliminate the worst solutions from further consideration the worst solutions. Examples of constraints considered in such models include technological limitations, limited funding, necessity to maintain certain water quality standards, etc.

As with watershed and waterbody models (Chapter 6), optimization models can be static (reflecting long-term, average conditions) or dynamic (reflecting the temporal dynamics) in nature. In terms of eutrophication management, only a few optimization models have been used to date.

One example of the use of optimization models was that conducted during the Lake Balaton study by researchers at the International Institute for Applied Systems Analysis (Laxenburg, Austria). This effort included models for determining tributary phosphorus loads to the lake and for predicting the response of the lake to different phosphorus loads. Management alternatives, such as the secondary and tertiary treatment of municipal wastewaters, the construction of shallow pre-reservoirs, the management of urban runoff, and the management of agricultural activities (e.g. fertilizer and erosion control), were assessed. Two basically different formulations were considered with these models. These formulations were maintenance of optimal water quality under specific budgetary limitations, and selection of the 'best' compromise solution between two conflicting criteria or goals (e.g. economic realities and desired water quality). Discussion of the use of these models is provided by Bogardi et al., (1983), Somlyódy & Wets (1985), and Somlyódy & van Straten (1986).

The dynamic optimization models are similar to the general dynamic models previously discussed (Chapter 6). They allow one to attempt to assess the time sequence, duration and intensity of possible management alternatives. An example would be selection of the optimal combination of control measures for achieving a desired trophic status (or water quality condition) at the minimal cost. The philosophy of this type of model is to consider all the available management options, their relative costs, and the potential ecological impacts of the management options on the waterbody. In an example discussed by Kalčeva et al. (1982) and Straškraba (1982), five simple management alternatives were considered, including a decreased influent phosphorus concentration, biomanipulation by regulation of fish stocks, mixing of the eplimnion, use of dyes to decrease light penetration into the waterbody, and manipulation of eplimnetic discharges from the reservoir. The reader is referred to these references for further discussion of the use of dynamic optimization models.

The International Institute of Applied Systems Analysis (Laxenburg, Austria) has published a number of reports dealing with the use of optimization models for lake and reservoir water quality, concentrating on Lake Balaton. For example Duckstein et al. (1982) have presented a discussion of a multi-objective modeling approach for controlling the nutrient loading of a small watershed to Lake Balaton.

It also is mentioned that the use of an ecological model (SALMO) for assessing management alternatives for a multiple-use reservoir is discussed in Chapter 11.

CHAPTER 10

THE POSSIBILITIES FOR THE REUSE OF NUTRIENTS

On a worldwide scale, there is a growing demand for food. In this regard, nitrogen and phosphorus compounds can be considered extremely valuable resources, because of their necessity for cultivation of crops. Thus, unwanted growths of algae or macrophytes in productive lakes and reservoirs represent the loss of materials which otherwise could be used to stimulate agricultural production. On the other hand, productive waterbodies also can be used for the production of fish as a food source. In many tropical areas, for example, particularly southern and southeastern Asia, more than 50 per cent of the protein requirements of the population are met from inland fisheries sources. Annual fish yields of 5–6 t/ha can be achieved with minimal nutrient inputs (Pantulu, 1982). In these situations, high productivity of a waterbody may be the principal management goal.

In contrast to the previous chapters, this chapter focuses on the positive properties of eutrophication in relation to the production of foodstuffs, as well as other positive uses of the products of fertile lakes and reservoirs. Whether reducing or stimulating the symptoms of eutrophication, the basic principles are the same. Thus, the topics covered in this chapter can be viewed as an alternative to simply 'controlling' eutrophication. The opportunistic harvest of biota, or the reuse of nutrients from eutrophic waterbodies, constitutes a legitimate and desirable use of the water resource in many cases. In some situations, these may even be the primary use of a waterbody.

USE OF ALGAE AND MACROPHYTES FOR NUTRIENT REMOVAL

Phytoplankton

Algae (phytoplankton) are very effective at removing nutrients from the water column, and algal growth is actually one of the most effective forms of wastewater treatment available (Gordon et al., 1982). Since phytoplankton can convert waste nutrients into potentially valuable resources, the application of algal culture techniques for treatment of municipal wastewaters holds considerable promise. In addition, as discussed below, microscopic algae can exhibit very high growth rates (up to 50 t dry weight/ha.yr). Furthermore, phytoplankton are currently being considered as human and animal foodstuffs, and as natural sources for biogas generation (Shelef & Soeder, 1980).

The use of microalgae for such purposes, however, is problematical at present. For example, many algal species lack the desired chemical makeup or nutritional value, or worse, are toxic or nutritionally useless (Sykora et al., 1980). Consequently, commercial algal culturalists try to maintain unialgal cultures of those species whose chemistry and food value are desirable (Azov et al., 1980). Algologists attempt to optimize production of desired algal species primarily by maintaining optimal nutrient and pH levels for the growth of the desired algae. Therefore, such algal culture on a large scale requires trained personnel, inputs of nutrients, and small ponds in which the water chemistry can be easily manipulated. This task can be difficult in ponds receiving sewage of varying quality. The culture of algae would seem to be feasible primarily where nutrient sources can be identified, where adequate resources are available for pond construction and purchase of land, and where experienced personnel are available (Azov et al., 1980).

The method of harvesting cultured algae can also be a problem. Algae can be conveniently precipitated with various chemicals (e.g. alum), but the alum–algae precipitate has a much reduced value (Oswald & Golueke, 1968). Centrifugation of the algae shows promise where the value of the algae justifies the expense. Drying of the algae is feasible in those regions with sufficient water and a hot climate, although the dried algae is often contaminated by soils, which lowers its market value. Algae also may bioaccumulate toxins and heavy metals which can render the algae unfit for human or animal consumption (Gordon et al., 1982). Such contaminated algae could be used for the generation of biogas or for organic chemical extraction. However, these techniques merely postpone the problem of nutrient disposal, and the algae are not as valuable as uncontaminated algae.

Nevertheless, the harvest of algae from lakes and reservoirs is reasonable in certain situations. In waterbodies with concentrated blooms of net phytoplankton (algal species with diameters > 50 μm) it appears economically feasible to strain the algae from the natural system (Oswald, 1976). If the algal species has a high monetary value, the costs of harvest are recoverable. As an example, some of the world's supply of *Spirulina* currently comes from the harvest of natural algal populations. Unfortunately, naturally occurring algal blooms composed of species with a high economic value do not occur frequently. Thus, at present, the primary use of algal harvest techniques can be considered only a mechanism for improving the aesthetic aspects of algal blooms.

Aquaculture of phytoplankton, therefore, remains largely experimental. In a practical sense, reliance on a possibly varying nutrient source, and possible algal contamination with pathogens and heavy metals, hampers attempts to culture and market phytoplankton. Furthermore, the mechanics of harvesting natural algae are limited at present to certain large forms that may occur on an irregular basis. Natural algal assemblages rarely consist primarily of species with sufficient market value to warrant their harvest. Thus, although phytoplankton show great promise as scavengers of nutrients in nutrient-rich waters and wastewaters, and as potential products, the necessary technology and markets are developed only in a few situations.

Harvesting of macrophytes and filamentous algae

The cultivation of macrophytes for the purpose of cleansing sewage-laden waters has been suggested previously, primarily because such plants usually have prodigious growth rates (hence, large nutrient requirements) and because they are relatively easy to culture. In lakes and reservoirs with extensive littoral areas, the removal of nutrients from the littoral zone as a result of intensive macrophyte harvesting programmes can be very high. The harvesting of rooted macrophytes, however, may have little effect on the water quality of such eutrophic waterbodies, since the plants can derive the majority of their nutrients from the sediments, rather than from the water column (Carignan & Kalff, 1980). A large portion of sediment-bound nutrients is irrevocably buried in the sediments and would not otherwise be available to the waterbody, except for macrophytes. Moreover, decomposition of such plants, if they are not removed from the waterbody, can function as a nutrient 'pump' from the sediment back into the water column. In these cases, the aquaculture of these plants in natural systems may actually increase the nutrient content and, hence, the degree of eutrophication of the lake or reservoir.

These problems do not apply to treatment of wastewaters, however, when non-rooted macrophytes are used to treat the water. Normally such treatment occurs outside the lake or reservoir. McNabb (1976) recommends that macrophyte beds used in treatment of sewage waters be planted in soils of low nutrient content, so that the water is the only source of plant nutrients.

The harvesting of macrophytes and macroscopic algae has been used in several lakes and, in the short run, can alleviate the problems associated with the excessive growth of such plants (Dunst et al., 1974; Petersen et al., 1974). The plants are easily harvested and some (e.g. the water hyacinth *Eichhornia*) grow as fast or faster than any other plants under cultivation

(Reedy & Tucker, 1983). Thus, their culture and harvest could represent one workable method for the removal of nutrients from lakes and reservoirs.

It is mentioned, however, that the cultivation of macrophytes can lead to weed beds, which are not aesthetically pleasing, and which can be sources of disease vectors and other pests. Furthermore, floating macrophytes such as duckweed (*Lemna*) and water hyacinth do not normally release oxygen into the water column (Pokorny & Rejmankova, 1983). They can also shade the submerged plants and algae that do release oxygen. For these reasons, cultivation of floating macrophytes at present appears to be restricted to limited-use systems, and is somewhat incompatible with fisheries aquaculture.

Use of macrophyte biomass as animal food

Because of the increasing costs of attempting to control macrophyte growths, harvesting the macrophyte biomass is becoming more and more attractive as an alternative management option. On a worldwide scale, both food shortages and aquatic weed problems often exist simultaneously in the same locality. The use of such plants as food sources, therefore, could represent conversion of a weed problem into a valuable crop (Mitchell, 1974; Jayaraman, 1981; Vass & Zutshi, 1983). The types of harvestable plants from floating-leafed to submerged and/or emergent (such as rushes and reeds). Submersed vegetation is often long and stringy and cannot be easily handled. Both floating and submerged vegetation normally are bulky and very high in moisture (in most cases, 90 per cent moisture content or more). The transport of such vegetation, therefore, to distant places can be prohibitively expensive, unless onshore dehydration is possible at reasonable cost. One person can harvest about 1500 kg of emergent or floating-leafed plants per day from moderately dense stands of most species. For larger-scale harvesting efforts, various mechanical devices have been developed (Breck *et al.*, 1979).

In many tropical and sub-tropical countries, vast areas of shallow lakes, waterways and irrigation channels are infested with water weeds, such as *Eichhornia crassipes* (water hyacinth), *Pistia stratiotes* (water lettuce), and *Salvinia rotundifolia*, *Salvinia molesta* and *Salvinia auriculata* (floating-leafed ferns). The first approach to possible economic usage of such plants usually is its conversion to animal feed, particularly for ruminants. The nutritional value of *Eichhornia* and *Hydrilla* compare favorably to that of bermuda grass (*Cynodon dactylon*).

There are considerable variations in the chemical and biochemical composition of different species of aquatic plants, and even of different stages of maturity for the same species. Therefore, generalizations regarding their nutritive value as a livestock feed are not possible. In shallow lakes on the Peruvian altiplano, for example, submerged macrophytes are used directly as food by grazing cattle (Löffler, 1982). In China, several aquatic plants (e.g. *Eichhornia, Pistia, Wolffia, Lemna* and *Azolla*) are harvested from canals or ditches, and are even cultivated and distributed directly to ponds as a feed for grass carp. The quantity of food required to produce 1 kg of grass carp varied among the plants as follows: *Wolffia*–37 kg, *Lemna*–41 kg, and *Vallisneria*–101 kg (FAO, 1983).

In warm climates, the use of harvested macrophytes as animal feed may be hampered by rapid microbial decomposition of the macrophytes. Thus, prompt and proper processing of the harvested macrophytes is imperative. Such biomass may be chopped, pressed, and ensilaged or dried (Bagnall *et al.*, as cited in Breck *et al*, 1979). The simplest type of dehydration is to allow the plants to dry in thin layers in the sun. To decrease microbial decay, it is important that the plants be turned over at regular intervals. Only species with a very low moisture content (80 to 85 per cent) can be used directly as food without undergoing dehydration. Costs for drying can be high if energy sources other than solar radiation must be used. Such expenses can be eliminated, however, by converting the plant material to silage. To produce an acceptable silage, moisture must be reduced, and an additional source of carbohydrates (such as citrus pulp or chopped paddy straw) must be added. Weise & Jorga (1981) discuss attempts to use submerged macrophytes as cattle feed in ensilaged form, after drying on cropped grassland. However, beef cattle often refuse to eat feeds prepared from aquatic weeds (Varshney & Rzóska, 1976). Edwards (1980) and Appler & Jauncey (1983) also discuss the potential use of macrophytes as animal fodder.

When given in feeding trials as a food supplement for pigs, fresh duckweed (*Lemna minor* and *Spirodela polyrhiza*) produced an additional 20 per cent growth, due to the large content of high quality proteins and carotenoids (Sperling, 1962). Pelleting of dried water hyacinth and *Hydrilla* improved the acceptance of such macrophytes as food by animals in feeding trials performed in the United States. In these trials, diets containing up to 30 per cent aquatic plants were possible.

Thus far, only a small, although important, number of aquatic plants are used as human food sources. Examples are rice (*Oryza sativa*), water chestnut (*Trapa natans*), *Ipomoea aquatica* and watercress (*Nasturtium officinale*).

Use of plant biomass as raw materials for commercial purposes

Aquatic macrophytes also have actual or potential uses as sources of fiber for paper making, and as raw materials for chemical products. Viewed from this perspective, high areal yields of waterweeds in eutrophic lakes and reservoirs represent a potential source of raw materials, the value of which increases with shortages of fossil fuels and timber, particularly in warm climates. Reeds, such as *Phragmites communis*, represent a valuable raw material for cellulose processing, and the demand can often substantially exceed the local supply. However, the use of floating-leafed plants, such as *Eichhornia*, as a pulp source is hampered by its relatively low fiber content. To satisfy the cellulose demand of a medium-sized pulp factory, transport of such plants over long distances may be necessary, which can substantially increase their costs. However, more efficient technologies are expected to be developed for mechanical, chemical and biochemical processing of aquatic plants, particularly as other sources of raw materials become more expensive.

Another use of macrophyte biomass is as a carbon source for the production of biogas. A farm project which converts organic wastes into biogas, fertilizer and animal feed, has been described by Wolverton & McDonald (1976). J. C. Goldman (1979) and Benemann (1981) also discuss the use of macrophytes for the production of biogas.

In waters with an abundant nutrient supply and a high rate of primary production, the incorporation of nitrogen and phosphorus by macrophytes can be very high, attaining levels of $0.34\,g\ N/m^2 day$ and $0.034\,g\ P/m^2 day$ for *Eichhornia crassipes* (Boyd, 1982). Thus, as noted earlier, aquatic plants can be useful in removing nutrients from eutrophic waters, if such plants are subsequently harvested. However, if such plants are not harvested, they act as 'pumps' for nutrients from the sediments into the water column as the plants decompose. Macrophytes also can be used in some cases as green manure or as compost for agriculture or aquaculture (see E.C.S. Little, 1968; Reedy & Tucker, 1983). Although most aquatic weeds are not an ideal composting material because of their high water content, compost from aquatic plants has been found to be useful as an organic fertilizer for fish ponds in southeast Asia. *Eichhornia* compost is especially good for such uses because of its high nutrient content (Mitra & Banerjee, 1976, as cited by Varshney & Rzóska, 1976).

In summation, although macrophytes have rapid growth rates, and although methods for harvesting macrophytes are relatively well established, their market value can be low. Thus, culturing and harvesting macrophytes (as a means of removing excess nutrients from the water column) presently shows limited promise as a general mechanism for ameliorating the consequences of eutrophication (see Seidel, 1976). Edwards (1980) and FAO (1979) provides a review of the economic potential of various macrophytes, and exceptions to the above observations.

AQUACULTURE FOR THE PRODUCTION OF FISH AND OTHER AQUATIC FOOD ORGANISMS

The use of aquaculture and the maintenance of fisheries as a method of nutrient reuse has a long, well-developed history. The aquaculture systems and fisheries of the developing nations have traditionally relied on nutrient inputs from animal and vegetable wastes. The integrated use of aquaculture to process animal wastes is a well-known agricultural process (e.g. see Allen & Hepher, 1978; Roger et al., 1979). It must be stressed that the vast majority of such cases represent methods of enhancing fish production, rather than the amelioration of the symptoms of eutrophication. Fortunately, however, the results can be very complementary.

Aquaculture for the control of nutrients

Shallow waterbodies, either natural or man-made, can be used for the cultivation of finfish, crayfish and freshwater prawns. Thus, a nutrient-rich waterbody can yield a high value, protein-rich product. Generally speaking, there is a tendency for an increase in fish yield in lakes and reservoirs as the level of primary production increases (although the types and quantity of the fish may change, as discussed in a following section). Eutrophic or polytrophic conditions in a lake or reservoir are often indicative of a high level of fish production. Bulon & Winberg (1981) reported that the yield of fish from drainable ponds amounts to approximately one per cent of the primary production of the waterbody (without feeding of fish), while the yield from non-drainable ponds is about 0.1 per cent of the primary production. These differing values are due mainly to complete harvesting of fish in the former case, and incomplete harvesting in the latter case. Such yields, however, can vary in specific cases. Oglesby (1977), for example, reported fish yields (carbon 'transfer efficiencies') ranging between 0.7–1.5 per cent for small, shallow, tropical/sub-tropical lakes. Liang et al. (1981) suggest that the carbon transfer efficiency in intensively managed ponds in such locations as China, Malaysia and Israel would likely exceed one per cent. They also reported that a group of eighteen shallow, eutrophic ponds and lakes located in China exhibited carbon transfer efficiencies ranging between 1–11 per cent.

Current attempts to manage nutrient levels with the use of fish primarily involve aquaculture, using point sources of nutrients. While there are problems associated with the taste and social acceptance of fish raised in various types of wastewaters, it is undeniable that the use of such wastes in extensive aquacultural systems can enhance considerably the production of fish (Allen & Hepher, 1978; Roger et al., 1979). Furthermore, while there is dispute as to its effectiveness for treating wastewaters (see Schroeder, 1975; S. Henderson, 1983; Nash & Brown, 1979), aquaculture can effectively remove much of the excess nutrient content of wastewaters (Gordon et al., 1982). Fish also have a developed and easily exploited market potential and they are relatively easy to harvest. Thus, this method of nutrient reuse involving aquatic organisms is currently the only one in relatively widespread use.

However, as noted in Chapter 9, fish can also excrete a large portion of the nutrients they consume back into the waterbody. Therefore, they are not nearly as effective as plants in removing nutrients from the water column. Thus, in contrast to plant culture, animal aquaculture cannot be seen as a reliable tertiary treatment process for wastewaters (Gordon et al., 1982; Nash & Brown, 1979). Moreover, maximum fish production rates are only about ten per cent of those of algae and macrophytes, usually being less than ten t wet weight/ha·yr.

Fisheries can remove a substantial quantity of nutrients from a eutrophic system (Barthelmes & Kliebs, 1978; Dunst et al., 1974). For example, the extensively managed fisheries of North Lake, China, result in a removal of an estimated 40 kg/ha·yr of phosphorus in fish flesh from this productive lake (Liang et al., 1981). However, there are few examples of fish harvesting actually reducing the degree of eutrophication of a lake (Sreenivasan, 1980). There are two possible reasons for this fact. First, fisheries tend to deplete those fish with high economic value. Thus, many eutrophic lake systems tend to have large concentrations of lower quality, rough fish. These latter fish normally are not fished at rates adequate to substantially reduce the nutrient content of the entire system. Furthermore, in many of the developing countries (where eutrophication is not viewed with as much disdain as in industrialized countries) heavily-fished lakes will often be fertilized intentionally, so that the yields of future harvests will not be reduced (Liang et al., 1981). Nevertheless, fish stocks can contain up to half the total phosphorus content of an aquatic system (Kitchell et al., 1979). Thus, intensive fishing efforts would seem to be a plausible mechanism for attempting to reduce the eutrophication of a lake or reservoir, while making a profit at the same time. One would expect this mechanism to function best in those areas with well-developed markets for the less valued fish species.

One can attempt to ameliorate the effects of eutrophication by managing fish or animal stocks. A salient example is the use of phytophagous fish to control macrophyte growths. Herbivorous fish can be quite successful in reducing the extent and range of such weeds. Further, many of these herbivorous fish (e.g. grass carp, tilapia) have relatively high market values. Manipulation of excessive phytoplankton growth with the use of fish, however, has been decidedly less successful. One approach maintains a high density of piscivorous fish in the waterbody, which should work to keep the zooplanktivorous fish at very low densities (Hrbaček, 1966; Shapiro, 1979; Shapiro et al., 1975). The reduced numbers of zooplanktivorous fish usually allow zooplankton to thrive. The zooplankton, in turn, should keep the algal density at low levels. Initial testing of this approach, however, has produced mixed results (e.g. Shapiro & Wright, 1984, compared with Benndorf et al., 1984b; also see Oskam, 1978).

There are several factors which probably limit the use of fish management techniques for controlling the effects of eutrophication. First, the larger predaceous fish generally are the most prized fish, and most susceptible to commercial and sport fishing pressures. Thus, application of this approach may be successful only in areas where fishing pressure is kept artificially low. Furthermore, it is not completely clear whether or not zooplankton can actually control algal growth in eutrophic lake systems, since many phytoplankton can avoid grazing by zooplankton. In addition, fishless eutrophic systems with large zooplankton populations (e.g. sewage treatment ponds and some fish ponds) also may have extremely large phytoplankton populations. Thus, zooplankton may control algal biomass by sequestering limited nutrients, not by continual grazing of the algae. This suggests zooplankton may be able to control algal biomass only in less eutrophic lake systems, where the residual nutrients not incorporated into the zooplankton and fish biomass are insufficient to produce blooms of inedible algae. However, in sewage treatment ponds without fish, or in carp ponds with low population densities, growths of *Daphnia* or *Moina* grazing on phytoplankton can produce a 'clean-water' state (Uhlmann, 1958; Dinges, 1973). However, large growths of rotifers in a waterbody do not appear to be able to sustain a clean water state over a long period.

Another fish management proposal involves the use of filter-feeding fish to control phytoplankton populations. Although the results have been inconsistent, most experience with this technique suggests it has not produced reductions in phytoplankton biomass. In fact, several attempts to use this approach have even resulted in increases in phytoplankton biomass when algae-eating fish were added (Opuszynski, 1979; Dunseth, 1977; D.W. Smith, 1985; Cooper et al., 1985). Since fish consume only larger phytoplankton, it is unlikely that fish alone can effectively control phytoplankton biomass. Because the fish also consume zooplankton, which prey upon the smaller species of phytoplankton, the introduction of filter-feeding fish can remove both the competitors and predators of the smaller species of algae, thereby leaving these smaller species to flourish to excessive levels. Thus, using fish alone to attempt to control algae may only change the size structure of the phytoplankton population, rather than the total biomass (D.W. Smith, 1985; Cooper et al., 1985).

A new method, which uses both zooplankton and filter-feeding fish as biological control agents, has recently been tested (D.W. Smith, 1985; Cooper et al., 1985). The method maintains the coexistence of filter-feeding fish and large numbers of zooplankton in the waterbody, by excluding the fish from part of the water column. This exclusion provides the zooplankton a refuge from fish predation. Initial testing, done in large tanks, resulted in a 95 per cent reduction in algal biomass, compared to control tanks with filter-feeding fish (silver carp) and no zooplankton refuge. While this method shows promise, two factors limit its current application. First, the method requires some sort of fish barrier, which means it can be applied only in small lakes, ponds, or cages. Furthermore, this technique relies on an unknown, optimal mixing rate across the fish barrier.

It should be noted that short-term improvements in the eutrophication status of a waterbody, via manipulation of fish populations, often are inconsistent with strategies aimed at maximizing fish yield for profit and/or nutrient retrieval. For example, heavy stocking of grass carp can result in quick removal of bothersome macrophytes, after which the excessive nutrients will persist in forms unusable to the fish (e.g. either in phytoplankton biomass or as raw nutrients; see Leslie *et al.*, 1983). Thus, the basic problem would simply endure in another form, and fish growth would be considerably slowed.

Fish culture, therefore, has certain strong overall advantages, the foremost of which is a developed market for human and animal consumption. In addition, methods for harvesting fish are well developed and effective. However, animals also excrete a sizable per cent of the nutrients they consume. Thus, compared to plants, fish are relatively ineffective at removing nutrients from the water column (Gordon *et al.*, 1982). Consequently, fish have less potential as agents of water treatment than either macrophytes or phytoplankton. Nevertheless, fish biomass can comprise a large portion of the nutrients in an aquatic system in an easily marketable, easily harvestable form. For these reasons, aquaculture and fisheries represent the most appropriate strategies at present for the reuse of excess nutrients in lakes and reservoirs.

Estimation of fish yield

Various investigators have reported positive relationships between the fish yield or the standing crop of a lake or reservoir and: (a) *chemical/physical factors* associated with eutrophication (Jenkins, 1967; Hrbaček, 1969; Jenkins & Morais, 1971); (b) *mean depth* (Rawson, 1955; Srisuwantach, 1978); (c) *bottom fauna standing crop* (Matuszek, 1978); (d) *phytoplankton biomass* (Oglesby, 1977; Almazan & Boyd, 1978; J.R. Jones & Hoyer, 1982), (e) *phytoplankton productivity* (McConnell, 1963; Wolny & Grygierek, 1970; Melack, 1976; McConnell *et al.*, 1977; Toews & Griffith, 1979; Srisuwantach & Soungchomphan, 1981; Liang *et al.*, 1981; Bulon & Winberg, 1981); (f) *total phosphorus concentration* (Hanson & Leggett, 1982); and (g) *total phosphorus load* (Lee & Jones, 1981a,b).

To examine a few of these examples, Hanson & Leggett (1982) assessed the relationship between the total phosphorus concentration (*TP*) and fish yield (FY) in temperate lakes, and developed the following relationship (Figure 10.1):

$$\text{FY (kg/ha} \cdot \text{yr)} = 0.072\, TP\,(\mu g/l) + 0.92 \qquad (10.1)$$
$$(n = 21;\ r^2 = 0.84).$$

The possibilities for the reuse of nutrients

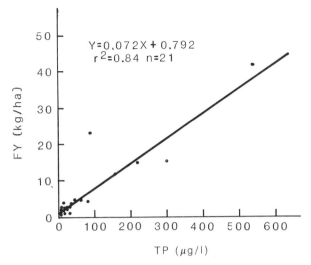

Figure 10.1 Relationship between total phosphorus (TP) and fish yield (FY) (from Hanson & Leggett, 1982)

Considering mean depth ($\bar{z}$, expressed in m) as an additional variable, these authors obtained the following relationship:

$$FY = 0.071\,TP + 0.165\,\bar{z} - 1.164 \tag{10.2}$$
$$(r^2 = 0.96)$$

Both equations appear to give a better estimation of the potential fish yields of a waterbody than the often used Morphoedaphic Index (MEI), which is based on total dissolved solids or conductivity divided by mean depth (Ryder et al., 1974; Oglesby, 1977; Matuszek, 1978). Nevertheless, a rough estimate of potential fish yields in both temperate and tropical lakes and reservoirs also can be obtained with the MEI (Wijeyaratne & Costa, 1981; Srisuwantach & Soungchomphan, 1981).

Lee and Jones (1981a,b; R.A. Jones and Lee, 1982) developed a logarithmic relationship (Figure 10.2) between the 'flushing corrected average annual phosphorus inflow concentration' (OECD, 1982; see Equation 6.10 for definition of terms) and fish yield (FY; expressed as g dry weight/$m^2 \cdot$ yr), as follows:

$$FY = 0.7\log\left[(L_p/q_s)/(1 + \sqrt{t_w})\right] - 1.86$$
$$(r^2 = 0.86). \tag{10.3}$$

The authors point out that this relationship does not give any information about the types or quality of fish to be expected.

Oglesby (1977) previously derived a logarithmic relationship between algal biomass (expressed as mean summer chlorophyll concentration, Chls)

223

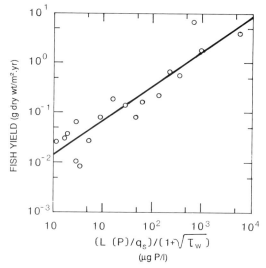

Figure 10.2 Relationship between flushing-corrected, annual phosphorus load and fish yield (from Lee & Jones, 1981a)

and fish yield (FY; g dry weight/m² · yr), as follows:

$$\log FY = 1.17 \log Chls\,(\mu g/l) - 1.92$$
$$(n = 19;\ r^2 = 0.84) \qquad (10.4)$$

More recently, J.R. Jones & Hoyer (1982) developed a relationship between the potential sportfish yield and the mean summer phytoplankton standing crop (expressed as chlorophyll *a*) for lakes and reservoirs in the midwestern United States, as shown in Figure 10.3.

Bulon & Winberg (1981) (also see Breck (1981)) reported a nearly linear positive relationship between phytoplankton primary productivity (*PP*; kcal/m².yr) for all types of lakes, including temperate and tropical impoundments, and fish yield (FY). They developed the following equation:

$$FY = (2.24 \pm 1.09) \cdot 10^{-3} \cdot PP^{0.950 \pm 0.118} \qquad (10.5)$$

or, simplified:

$$FY = (1.8 \pm 0.99) \cdot 10^{-3} \cdot PP. \qquad (10.6)$$

Equation 10.6 was based on 42 waterbodies, exhibiting phytoplankton production ranging from 170–14,000 kg/m² · yr and fish yields ranging from 0.4–26 kg/m² · yr (the latter corresponding roughly to 4–260 kg/ha · yr). As a general observation, the full range of fish yield–phosphorus or fish yield–primary productivity relationships appear to exhibit a logistic (sigmoid) curve.

In temperate lakes, fish production in carp ponds is about one t/ha. This

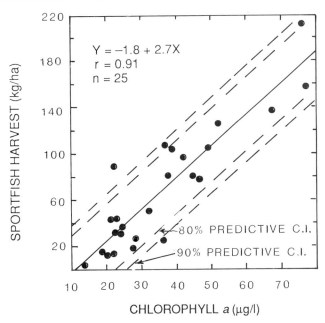

Figure 10.3 Relationship between angler sportfish harvest and mean summer chlorophyll *a* concentration for waterbodies in Missouri and Iowa, United States, (C.I. = confidence intervals; from J.R. Jones & Hoyer, 1982)

is not much less than the magnitude of animal protein which can be produced on terrestrial agricultural areas (Spet, 1972). As already mentioned, the fish yield also can be substantially higher. Fish production in Laguna Lake, Philippines, has been estimated at 1.3–1.5 t/ha · yr (personal communication, T.C. Rey, Laguna Lake Development Authority, Manila, 1984). Among 18 suburban lakes and ponds near Wuhan, China, seven had an annual fish yield of more than seven t/ha (Liang *et al.*, 1981). In three of these cases, the yield ranged between 11.0 and 14.6 t/ha. There was a strong correlation between gross photosynthesis and fish yields. The efficiency of carbon transfer from photoplankton primary productivity to fish yield is greater than one per cent in the most productive ponds. No fish food was added to the ponds, but the waters were enriched with domestic sewage and, in some cases, with inorganic fertilizer.

It is also pointed out that the quantity of nutrients incorporated into the fish flesh can be insignificant, related to the total pool of nitrogen and phosphorus in a lake. With respect to the nutrient load from the surroundings, however, it may become important with higher yields of fish. Thus, an annual fish yield of 5–6 t can result in a depletion of 2.5–3 g P/m^2 · yr in a waterbody.

Considerations for the management of fish ponds and small waterbodies

Shallow waterbodies with high nutrient loads tend to have fish populations dominated by carp (cyprinids) or cichlids, which are less affected by the detrimental effects of eutrophication than more prized fish species. However, high levels of primary productivity in temperate region lakes and reservoirs can have negative impacts on cold water fisheries, especially in thermally-stratified waterbodies. Nevertheless, if dissolved oxygen is still present in this habitat, cold water fish can survive in a microhabitat water layer just below the thermocline. For optimum cold water fisheries, therefore, a balance is necessary between algal growth sufficient to serve as a fish food source, but not so high as to cause detrimental hypolimnetic oxygen depletion (Lee & Jones, 1981b). Hypolimnetic aeration can be used as a suitable remedial measure in some cases (see Chapter 9).

It is clear that management of eutrophication for optimum fisheries in shallow waters normally requires high nutrient loads. Fish kills can occur, however, as a consequence of an excessive nocturnal oxygen depletion. Fish kills also may result from an excessively high pH level, in combination with an increased concentration of ammonium, resulting in ammonia intoxication of fish (Schäperclaus, 1979). Such effects also depend on seasonal variations in the physiological condition of fish. Consequently, such negative impacts do not automatically occur in waterbodies with high levels of algal biomass.

In about 25 per cent of Canadian prairie lakes, most of which are shallow, the survival of stocked trout is adversely affected by oxygen depletion resulting from mid-summer algal die-offs (Ayles & Barica, 1977). Carp are less sensitive than trout to such conditions. Carp ponds in middle Europe, subjected to combined phosphorus and nitrogen fertilization (instead of only phosphorus fertilization), have been shown to selectively grow small green algae (e.g. *Scenedesmus*), which are less likely than blue-green algae to cause hypolimnetic oxygen depletion problems. In contrast, phosphorus fertilization in the absence of nitrogen fertilization favors the growth of N_2-fixing blue-green algae (Barthelmes, 1981, 1983; Belsare 1984).

Carp are known to stir up bottom sediments, thereby causing the water to become turbid and inhibit light penetration into the littoral zone. As a result, a waterbody initially choked by dense growths of weeds can be converted into a waterbody dominated by dense phytoplankton growths.

In shallow waterbodies which develop an ice cover lasting for several months, winterkills of fish may occur if the snow cover is sufficient to reduce light penetration into the waterbody below the critical level necessary for photosynthetic oxygenation of the water. In artificial ponds, this phenomenon normally can be controlled by the mechanical removal of the snow cover or by artificial aeration of the waterbody (Boyd, 1982).

In eutrophic or polytrophic lakes and reservoirs, management for optimum fisheries depends to a large degree on the mean depth and wind fetch of the waterbody. Thermally stratified, eutrophic lakes with an areal phosphorus load of $0.5 \text{ g/m}^2 \cdot \text{yr}$ or more may display detrimental hypolimnetic oxygen depletion and accumulation of hydrogen sulfide. In such cases, planktogenic matter accumulating on bottom muds is not available for fish, or even for most of the bottom invertebrates (which are themselves important food sources for fish). Under these conditions, the less desirable species and size classes of fish often predominate. On the other hand, shallow lakes and artificial ponds may yield an optimum quantity of well-developed (i.e. non-stunted) fish, even with an areal phosphorus load as high as $5 \text{ g/m}^2 \cdot \text{yr}$ or more.

From the perspective of water quality management, integrated pond systems which combine the use of point sources of nutrients (e.g. domestic sewage; animal and human excrements) with the production of fish protein (see Figure 10.4) may be particularly valuable. In fact, in China and some other countries, fish culture is integrated with other animal and plant production. Animal wastes are used as manures to fertilize fish ponds. In warm climates, artificial ponds rich in nutrients may also be used effectively for the production of freshwater prawns. A listing of reference books on aquaculture is presented by Coche (1982).

Harvesting fish from highly productive waterbodies in tropical areas normally is based upon stocking the waterbodies with planktivorous species such as silver carp (*Hypophthalmichthys molitrix*) and big-head carp (*Aristichthys nobilis*). These species feed primarily on zooplankton, but are able to utilize photoplankton and detritus as food sources. Among the fish used for multi-species stocking of ponds in tropical regions, *Catla catla* and the Indian carps (*Cirrhina mrigala* and *Labeo rohita*) are good herbivorous plankton feeders (Gupta & Roy, 1975). In ponds rich in nutrients and stocked with phytoplankton-feeding fish, the phytoplankton growth normally is sufficient that additional food for the fish is unnecessary.

Under conditions of low water transparency, the phytoplankton can outcompete the phytobenthos. Therefore, dense stocking with grass carp (*Ctenopharngodon idella*), is useful only if the growth of macrophytes is to be controlled. In tropical polyculture ponds, a few specimens of this grass carp frequently are stocked together either with the phytoplankton-feeding fish species mentioned above, with tilapia (*Sarotherodon piceus*), or with some black carp (*Mylopharyngodon piceus*). Molluscs play the most important role in the diet of this latter fish. Snails (vectors of bilharzia disease in warm climates) can be controlled in this manner.

In contrast to fish culture in warm climates, temperate zone ponds for common carp (*Cyprinus carpio*) normally are operated with addition of fish food. Thus, zoobenthos, zooplankton, phytoplankton, detritus or macrophytes are of lesser importance in such ponds, even though they are produced and utilized at high rates.

It is emphasized again that care should always be taken when introducing exotic fish into new habitats, since valuable endemic fish stocks may become greatly reduced in number, or even eliminated, as a result.

Control of eutrophication

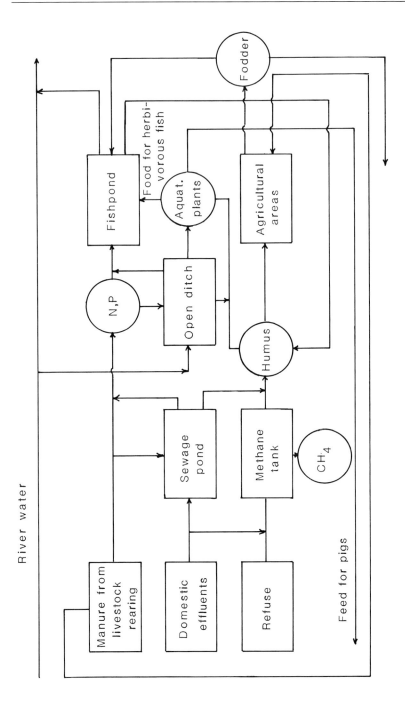

Figure 10.4 Schematic representation of different possibilities for the integrated management of nutrient-rich waterbodies in tropical settings; (China and Indochina) (adapted from various sources)

OTHER NUTRIENT REUSE POSSIBILITIES

The sludge from fish ponds is used as a fertilizer for agricultural areas in China. In fact, more than 25 per cent of all fertilizers used in Chinese agriculture are said to come from dredging waterbodies, mainly fish ponds (ADCP, 1979).

Dredging is considered an effective lake and reservoir restoration method (see Chapter 9). Since dredging costs normally are very high, the restoration gains may be increased substantially if there also is a concomitant need to improve the land by enriching sandy soils. For example, large fruit plantations near the Havel River achieved a 20–30 per cent increase in crop production by the application of lake sediment to the land, at a rate of $300 \, m^3/ha$ (Kalbe, 1976). However, muds with a calcium carbonate content of more than 10 per cent, and an organic matter content of less than 15 per cent, do not appear to be suitable for this special purpose.

If land and water ecosystems are considered simultaneously, agriculture, aquaculture and lake management appear to be closely interrelated in many cases. For irrigation use, for example, waters rich in dissolved nitrogen and phosphorus compounds may be highly preferred. Thus, even hypolimnetic waters can be a valuable resource in such cases. As pointed out by Klapper (1980), the season characterized by the highest demand for irrigation waters normally occurs simultaneously with the period of thermal stratification and increased hypolimnetic accumulation of dissolved nutrients in eutrophic lakes and reservoirs. By means of deep-release with an Olszewski tube (see Chapter 9), waters with high nutrient concentrations can be diverted from a lake or reservoir and used for such purposes as spray or furrow irrigation of vegetables, fruit plantations and other crops, or for pastures.

A nearly complete reuse of nutrients may become possible with use of treated domestic wastewater effluents for irrigation. The introduction of nitrogen and phosphorus into waterbodies also can be largely prevented in this manner. The Werribee farm near Melbourne, Australia, for example, uses up to $570\,000 \, m^3/day$ of sewage (supporting 20 000 cattle and 50 000 sheep with no supplementary food), on the basis of alternating pasture irrigation and grass filtration (MMBW, 1976). During the rainy season, however, overflows from saturated irrigation lands, or from overloaded sewage reticulation systems do occur. Nevertheless, this method appears to be well-suited for areas with warm and dry climates. Under the climatic conditions of Zimbabwe, for example, about 90 per cent of the nutrients originating from treated sewage and storm waters from the capital of Harare can be prevented from entering the water courses in this manner (Rowe, 1982; McKendrick, 1982).

A close integration of fish culture, agriculture and livestock farming is practiced in China (see Figure 10.4). Since the main cultivated fish species are either herbivorous or plankton-feeders, domestic effluents can be used effectively. Furthermore, floating-leafed plants can be used as feed for pigs, with up to 150 pigs being fed from one hectare of water surface (FAO, 1983).

EUTROPHICATION AND ACIDIFICATION OF LAKES

Recent work suggests that the increased productivity inherent in the eutrophication process also may be useful in counteracting the negative impacts of acidification of a lake or reservoir. For example, Kerekes *et al.* (1984) examined the chemical and biological characteristics of two lakes in Nova Scotia (Canada). Both lakes were acidified due to acid inputs. However, one lake (Little Springfield Lake) was oligotrophic, while the other (Drain Lake) received inputs of domestic sewage and was eutrophic. Kerekes *et al.* (1984) found that, in contrast to Little Springfield Lake, Drain Lake exhibited greatly increased biological activity at many trophic levels. As a result, Drain Lake also contained some species (i.e. fish and amphibians) normally not seen in highly-acidic lakes.

Based on studies in three Canadian Shield lakes, Kelly *et al.* (1982) reported that bacterial and/or algal chemical and biological reactions may play a significant role in neutralizing acid deposition impacts on waterbodies. A model based on their study results suggests that artificial eutrophication of the hypolimnia of two lakes in the Experimental Lakes Area (Canada) potentially could produce enough 'persistent' alkalinity to neutralize acid precipitation inputs to the lakes. According to the model, a non-eutrophic lake could not produce such results. Kilham (1982b) also reported on the interactions between acid inputs and biologically mediated production of in-lake alkalinity in a lake in Michigan, United States.

Schindler (1985) and Schindler *et al.* (1985) provided an enlightening discussion of the effects of aquatic organisms (primarily bacteria and phytoplankton) on the geochemical cycling of elements in lakes. Based on whole-lake studies in the Experimental Lakes Area, Schindler concluded that interactions and couplings between biological organisms and geochemical cycling of nutrients in waterbodies are quite resistant to many environmental stresses. This suggests that biological communities can adapt to changing aquatic stresses (including lake acidification), at least at lower trophic levels. Schindler also outlined the various biologically mediated reactions that can work to counteract the increasing acidification of acid-sensitive lakes.

Although still of a preliminary nature, such results suggest that further research on the interactions between lake/reservoir eutrophication and acidification is warranted. As a minimum, the above-cited results suggest that even under conditions of acidity, a waterbody can exhibit a trophic response to nutrients similar to that seen in non-acidic lakes. Even more important, these initial results suggest that artificial fertilization *of suitable waterbodies* may be a useful approach for attempting to reverse the negative impacts that normally occur as a result of increasing lake acidification.

CHAPTER 11

SELECTION OF EFFECTIVE STRATEGIES FOR THE MANAGEMENT OF EUTROPHICATION

GENERAL CONSIDERATIONS

No single approach or control measure will successfully treat all cases. As noted earlier in the description of the eutrophication process and the various factors which can affect it (see Chapters 4 and 5), the extent of present scientific knowledge is still not sufficient to produce a completely foolproof eutrophication control programme. Nevertheless, our present knowledge is sufficient to develop a generalized approach which, if used in conjunction with an adequate monitoring programme and continuing scrutiny of the measured data, will usually work in the majority of cases likely to be encountered. Indeed, the results of the OECD international eutrophication study (OECD, 1982) illustrate that one can attempt to assess the response of a waterbody to its nutrient input by studying the statistical behavior of a large group of similar waterbodies. This was shown in the individual OECD regional projects (Rast and Lee, 1978; Ryding, 1980; Fricker, 1980a; Clasen & Bernhardt, 1980; Janus and Vollenweider, 1981) and with the composite data base derived from all the regional projects (OECD, 1982). Of course, one always should remain aware of the uncertainty and potential error associated with such data bases, in order to use them effectively for predictive purposes. The use of statistically analyzed data bases and derived quantitative relationships allows the development of a reasonable, generalized approach for attempting to assess and control eutrophication of lakes and reservoirs.

The most feasible control option in a given situation will vary from location to location, depending on the circumstances. As noted earlier, it is generally believed that the control of external nutrient (especially phosphorus) inputs represents the most effective, long-term strategy for attempting to control eutrophication of natural lakes and reservoirs. Nevertheless, it is important to be realistic in selecting specific control measures, both in terms of how much reduction in the external phosphorus loads can be expected and how much such control measures will be likely to cost. An unrealistic management plan can undermine popular support for phosphorus control efforts if it is observed that a given plan will not achieve the desired phosphorus control goals, or that it is inappropriate from the point of view of cost-effectiveness.

It must be recognized that, in a given situation, the observed differences in the characteristics of natural lakes and reservoirs (see Table 4.5) may affect the selection of control measures based on reduction of the external phosphorus load. For example, compared to a natural lake, the relatively larger size of a reservoir drainage basin may require that one concentrate initially on the major phosphorus sources located nearest to the reservoir rather than the ones located more distant from the reservoir. This is because the eutrophication impact of distant phosphorus sources may be lessened by the 'effective transmission' of phosphorus (see Chapter 7). Furthermore, the larger basin size of a reservoir may cross more administrative boundaries, thereby complicating implementation of necessary control measures.

It must be recognized that it may not be possible to achieve the desired water quality and trophic conditions in all cases, even after implementation of the most feasible phosphorus control efforts. If so, additional control efforts (often considerably more expensive) will be necessary to achieve desired in-lake conditions. Alternatively, one can accept the in-lake conditions resulting from the 'achievable' phosphorus control efforts as the best that can be obtained under the circumstances. This decision should be made by those who are most familiar with the specific circumstances.

WATER QUALITY AS RELATED TO DESIRED WATER USE

As noted previously, 'good' or 'bad' water quality is often defined on the basis of several in-lake parameters (see Tables 4.3-4.4). Furthermore, one can relate the trophic status of a lake or reservoir to specific boundary levels for several of these parameters (see Tables 4.1–4.2). Because of such relationships, it is also possible to relate desired water uses to the optimal (or minimally acceptable) water quality for such uses. Therefore, a logical approach for establishing an effective eutrophication control programme is to determine the necessary water quality and/or trophic conditions for a desired water use (or uses), and design the programme to achieve these necessary conditions.

It is important to remember that one cannot always define the trophic status or intended water use of a lake or reservoir in an unequivocal manner. Further, the same waterbody can exhibit conditions indicative of one trophic state based on one water quality parameter, and a different trophic state based on a second parameter (see Table 4.2). Nevertheless, it is usually possible to identify ideal or acceptable water quality for given water uses. For example, a lake or reservoir used as a drinking water supply ideally would have water of such good quality that it can be treated easily, using standard inexpensive methods, to yield a water suitable for human consumption. The content of phytoplankton and their metabolic products in waterbodies used for such purposes should be as low as possible to facilitate this goal. Furthermore, water used for swimming and similar recreational pursuits should be free of nuisance blooms of planktonic organisms, which can cause such physical symptoms as allergic skin reactions and conjunctivitis. Excessive macrophyte growths in shallow, nearshore areas also can affect recreational water uses in a negative way.

If a waterbody has one primary use, selection and implementation of the control measures for achieving the necessary water quality can be based on this single use. In many cases, however, there may be multiple, competing uses for the same waterbody. In these cases, determination of the desired water quality can be based on the single use of highest priority. This use may require the highest standards of water quality in some situations, while less stringent water quality may be sufficient in other cases. It is cautioned, however, that using water quality standards less stringent than those required for the most sensitive water use may produce water quality conditions unsuitable for the most sensitive use over the long term. Thus, decisions on a primary water use of a waterbody used for multiple purposes are best made on the basis of specific knowledge of the lake or reservoir in question.

The above descriptions point out only a few types of water use impairment related to eutrophication. To reiterate, the 'usability' of a waterbody is dependent on the water quality which, in turn, is directly influenced by its trophic state. The properties of a lake or reservoir that are largely influenced by its trophic state include the following (also see Table 4.4):

1. Water transparency;
2. Extent of macrophyte growth;
3. Concentration and composition of algae;
4. Supply of fish food organisms;
5. Dissolved oxygen concentrations;
6. pH;
7. Concentration of divalent metal ions;
8. Concentration of organically-derived detrital materials; and
9. Concentration of plant nutrients.

Water transparency can be influenced significantly by the presence of large quantities of inorganic sediments (as well as algal biomass) in the water column. This is especially true for reservoir systems (see Table 4.5). In such cases, high concentrations of suspended solids do not correlate directly to the trophic condition of a waterbody. A more-turbid waterbody may have a lower chlorophyll concentration than a less turbid waterbody.

This is due both to diminished light levels for algal photosynthesis and to the possible adsorption of biologically available phosphorus by suspended sediment particles. Nevertheless, inorganic turbidity can still significantly hinder some water uses because of the necessity to remove the materials causing the turbidity prior to the water use (e.g. drinking water). Thus, both biotic (e.g. chlorophyll) and abiotic (e.g. suspended sediments) conditions may have to be considered for some waterbodies. In either case, the control measures for decreasing water turbidity would be similar.

Based on practical experience with temperate zone lakes and reservoirs, a summary of intended water uses and their optimal ('required') and minimally acceptable ('still tolerable') trophic state for such uses is provided in Table 11.1. Determination of the specific trophic conditions of a lake or reservoir is discussed in detail in Chapter 8 and Annex 1.

A SIMPLE APPROACH FOR DEFINING AND SELECTING AN EFFECTIVE EUTROPHICATION CONTROL PROGRAMME

General overview

A logical sequence of decisions to be made by a water manager is outlined in Figure 11.1. The items considered in Figure 11.1 are similar to the steps

Table 11.1 Intended lake and reservoir water uses as related to trophic conditions (modified from Bernhardt, 1981)

Desired utilization	Trophic state	
	Required	Still tolerable
Drinking water production	oligotrophic	mesotrophic
Bathing purposes	mesotrophic	slightly eutrophic
Low-water improvement		
with long distance supply line	—	mesotrophic
without long distance supply line	—	slightly eutrophic
Fish culture		
salmonid waterbodies	oligotrophic	mesotrophic
cyprinid waterbodies	—	eutrophic
Providing process water	mesotrophic	slightly eutrophic
Cooling water production	—	eutrophic
Water sports		
(without bathing)	mesotrophic	eutrophic
Landscaping in recreation		
areas	—	slightly eutrophic[1]
Irrigation		
(by means of channels)	—	strongly eutrophic
Energy production	—	strongly eutrophic[2,3]

[1]Within the scope of landscaping, a eutrophic state caused by the natural aging process, can even be desirable.
[2]Without consideration of the eventual water quality requirements for the receiving canal.
[3]Not valid for river power plants, which may be impaired by macrophyte and algal growths.

Selection of effective strategies for the management of eutrophication

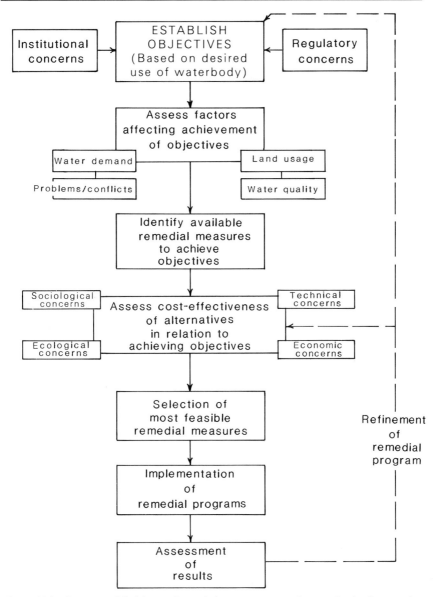

Figure 11.1 Sequence of decisions to be made by water manager in regard to implementation of remedial programmes

outlined in Chapter 3 in regard to formulation of eutrophication control policies and programmes. It is pointed out here that the final selection of an appropriate control strategy should be a 'multi-judgement', based on the relevant social, technical, economical and ecological aspects. It is also very important to set up a responsive monitoring programme both for defining the necessary pre-treatment condition of the waterbody and for properly evaluating the final outcome of the remedial measures enacted.

One point previously made in Chapter 3 is reiterated here; namely, that one is advised to start with a simple approach, and then add more detail and complexity as further knowledge and experience are gained. In this way, one can build on one's successes and generally reinforce one's goals.

A simplified, practical approach for selecting appropriate eutrophication control measures is outlined below in Figure 11.2. A 'decision-tree' approach is taken, with the answers to key questions dictating the direction to be taken. While the non-technical decisionmaker also may gain insight into effective eutrophication control efforts by reviewing this overall approach, the individual components of Figure 11.2 are directed primarily at the technical concerns.

This approach relies primarily on control of the external phosphorus and nitrogen loads to a lake or reservoir. The rationale for this approach was discussed earlier in Chapter 4. The eutrophication models presented in Figure 11.2 focus on the nutrient status of a waterbody. This focus appears to be appropriate for temperate and tropical lakes and reservoirs, and for sub-arctic lakes based on initial evaluation (McCoy, 1983; V.H. Smith et al., 1984). J.A. Thornton (1979; 1980; J.A. Thornton & Walmsley, 1982), for example, has applied the statistical phosphorus loading models of the type developed by Vollenweider (1976a) and concluded that they generally were adequate for assessing African lakes, although the boundary phosphorus concentrations denoting the transition between mesotrophic and eutrophic waterbodies may be too low to accurately describe tropical lake conditions (see Table 4.6). The similar model developed by Dillon & Rigler (1974) also appeared to work well for 31 southern African lakes (J.A. Thornton & Walmsley, 1982). Walmsley & Thornton (1984) and J.A. Thornton et al. (1986) also have evaluated the applicability of the type of models developed in the OECD (1982) eutrophication study and concluded they were reliable predictors for southern African impoundments.

Such studies support the use of the nutrient control philosophy inherent in Figure 11.2 as a rational approach for management of eutrophication of lakes and reservoirs on a global scale.

If it is determined that the eutrophication problem is sufficiently severe to justify the implementation of a eutrophication control programme, one should focus first on the nutrient status of the waterbody. The rationale for nutrient control was discussed earlier (see Chapter 4); namely, that nutrient reduction (especially phosphorus) appears to be the most effective approach at present for the long-term control of eutrophication. Thus, examination of the nutrient and phytoplankton dynamics of a waterbody can give valuable information on the nutrient status of the waterbody and the logical nutrient(s) to be controlled. One also may wish to assess whether or not nitrogen is a

Selection of effective strategies for the management of eutrophication

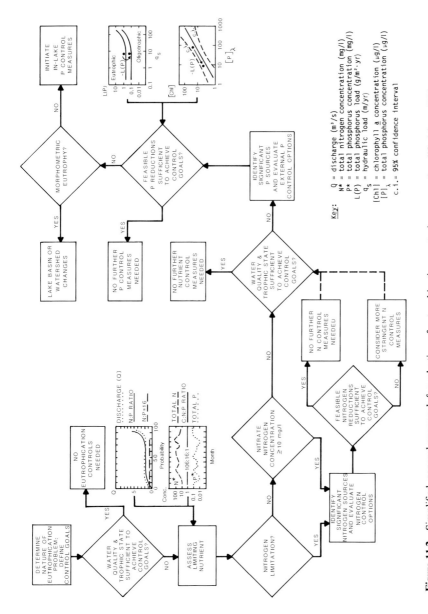

Figure 11.2 Simplified, general approach for selection of eutrophication control measures (modified from Uhlmann, 1984)

nutrient and/or health problem. If so, after identifying the significant nitrogen sources in the drainage basin and the available options for their control, one may consider initiation of some degree of nitrogen control. Nevertheless, except for its health implications for drinking water supply reservoirs (i.e. the nitrate–nitrogen concentration should be less than 10 mg/l to be considered safe for human consumption), it is better to focus primarily on phosphorus control. As pointed out earlier by Golterman (1975), it is not so important that phosphorus is the specific algal limiting factor in a waterbody at a given time, but rather that it can be made limiting by reducing the external load to the waterbody or by in-lake control measures, since phosphorus is 'the only essential nutrient that can easily be made to limit algal growth.' This conclusion is supported both by technical and economic considerations (e.g. Vollenweider, 1968; Lee, 1971; Schindler, 1977; Rast & Lee, 1978; PLUARG, 1978a; Ryding, 1980; OECD, 1982).

Thus, if nitrogen control alone cannot produce the desired improvement in water quality, one should identify the significant sources of phosphorus in the drainage basin, especially the biologically available phosphorus sources (see Chapter 4). One then can evaluate the available control options. Chapter 9 discusses various phosphorus control options and their relative costs and feasibility in given situations. One then can evaluate whether or not the level of external phosphorus reduction achievable with a given control measure will meet the desired eutrophication control goals. It is pointed out that one may be willing to spend large quantities of funds and go to considerable efforts in a given situation. Such decisions, of course, will have to be made in relation to the importance of achieving the control goals.

Simple eutrophication models, such as those developed in the OECD (1982) eutrophication study, can be very useful as initial methods for evaluating the potential water quality and/or trophic state improvements to be expected from a given phosphorus control programme (see Chapter 6 for discussion of available eutrophication models). Rast *et al.* (1983) have shown the simple OECD models to be useful and accurate for a wide range of temperate waterbodies, while J.A. Thornton & Walmsley (1982) have shown them also to be useful in assessing tropical waterbodies. More recently, Walker (1981; 1982; 1985) has assessed the applicability of simple, empirical models to a large group of reservoirs in the United States. In using such approaches, the degree of expected improvement in eutrophication-related water quality can be compared to the associated costs (both economic and social) of achieving this improvement. This also will allow one to assess the comparative cost-effectiveness of alternative phosphorus control options.

Specific steps to follow

There is a logical sequence of events outlined in Figure 11.2. Each of these individual steps is discussed further below:

Selection of effective strategies for the management of eutrophication

1. Assess eutrophication problem/define eutrophication goals

One must first determine the nature of the eutrophication problem and decide on the goals of a control programme. The eutrophication problem in a given situation may be excessive growths of algae and/or macrophytes, decreased water transparency, hypolimnetic oxygen depletion and related fish kills, nutrient regeneration or water quality deterioration due to the regeneration of reduced chemicals, taste and odor problems in drinking water supply reservoirs, etc., or a combination of these types of problems (see Chapter 4 for discussion of eutrophication related water quality deterioration). In this manner, one can relate the major water use (or uses) of the lake or reservoir to the necessary water quality to achieve the use, based on the categories presented in Table 11.1. Obviously, if the existing trophic state of a waterbody is compatible with the water use, no action is necessary in regard to phosphorus loading conditions. If not, both point and non-point phosphorus control measures may be necessary.

An important yardstick for assessing the efficiency of external phosphorus control measures is the reply to the question of whether or not an economically feasible reduction in the phosphorus content of an effluent results in a change towards an oligo- or mesotrophic state. Once the problem is clearly understood and defined in this manner, one can determine whether or not the problem is severe enough (or will become severe enough) to consider the control programme.

2. Assess limiting nutrient

If a eutrophication control programme is necessary to achieve the desired water quality goals for a lake or reservoir, one must then begin assessing the logical measures to take in a given situation. The strategy outlined in Figure 11.2 to assess the situation is based on simple and readily accessible criteria normally provided by a routine sampling programme for chemical and biological data. Since an effective, long-term control measure is usually to control the external nutrient load (see Chapter 4), the next step is to determine the likely nutrient to be controlled.

The trophic state of the waterbody must be considered to make a realistic estimate of the role of nitrogen and phosphorus as potential algal growth-limiting nutrients. The absolute concentrations of the biologically available nutrients (see Chapter 4) are especially important in this assessment. If the biologically available nitrogen and phosphorus concentrations decrease below approximately $20\,\mu g$ N/l or $5-10\,\mu g$ P/l, respectively, during an algal bloom peak, that nutrient is likely the limiting one. If both nutrients decrease below this value, both may be limiting.

The simple stoichiometric atomic ratio between $C:N:P$ of $106:16:1$ in plankton cells (which corresponds to a mass ratio of approximately $40:7:1$) also has proved to be useful in deciding whether nitrogen and/or phosphorus is the nutrient most limiting to algal growth. Under the assumption that the ratio in algal cells reflects the relative proportion needed by algae for growth and reproduction, measurement of the quantities of these nutrients in the water column can be used to determine which nutrient is not present in the needed proportions (i.e. which nutrient is limiting).

As indicated earlier in Chapter 4, carbon is seldom an algal growth-limiting nutrient in most lakes and reservoirs. Thus, the N:P mass ratio is usually sufficient, and can be calculated as the ratio of the concentration of inorganic nitrogen to dissolved reactive phosphorus (expressed in common units) collected in the water column during a growing season algal bloom peak. An N:P mass ratio below about 7:1 suggests nitrogen is the primary algal growth-limiting nutrient, while a ratio above this value suggests potential phosphorus limitation. With such information, control measures can be directed to the major sources of the biologically available forms of the limiting nutrient in the drainage basin.

The N:P ratio of the dissolved inorganic compounds in the tributaries of lakes and reservoirs also usually reflects the major land uses in the watershed, since both the concentrations and loads of nitrogen and phosphorus are governed by different mechanisms, and normally can be attributed to different sources. The relative significance of point and non-point loading sources, therefore, can be indicated by the N:P loading ratio. In domestic effluents, for example, the N:P ratio (5:1) is much lower than in the runoff from agricultural areas (30–50:1).

The often observed marked seasonality regarding the role of nitrogen and phosphorus as primary growth-limiting elements can frustrate attempts to get a conclusive picture. Therefore, a duration curve or observation series for a whole year of the relationship between the amounts of nitrogen and phosphorus can be very helpful in estimating the relative impact of these nutrients on the aquatic environment for the period of special concern (see Figure 4.2).

3. Assess need for control of nitrogen

Even if nitrogen is not the limiting nutrient, it may be necessary to take measures to control nitrogen if the critical concentration for drinking water supply health criterion is exceeded. Since drinking water supply is a principal use of lakes and reservoirs, excess nitrate levels require a high priority in the context of the management of lakes and reservoirs. Control measures should be implemented as far as possible from the water treatment plant, and as close as possible to the nitrate sources. Obviously, the successful application of preventive measures presupposes that the principal sources in the drainage basin have been correctly identified.

Among non-point sources, the relative importance of nitrogen can be greater than that of phosphorus. This is because the adsorption of phosphate on soil particles of fertilized croplands is much more important than for nitrate. Land use changes, as well as treatment of inflowing waters, are much more efficient for phosphorus elimination than for dissolved nitrogen compounds. It should be noted that measures which result in a high infiltration rate can be very efficient in reducing excess nitrate levels from the soils (although this can greatly increase the levels in ground waters). Furthermore, in temperate climates, the efficiency of some of the nitrogen control measures described in Chapter 9 can decline substantially during the cold season.

4. Assess alternative phosphorus control options

Before adequate control programmes can be identified and implemented, it often is necessary to use generally applicable predictive tools. Although the concept of trophic states may be difficult for decisionmakers to appreciate, it is possible to estimate the expected in-lake phosphorus concentration resulting from a given control programme, based on knowledge of the phosphorus loading, water retention and mean depth of a lake. By employing known relationships between phosphorus and common eutrophication variables found in multi-lake studies, algal biomass (in terms of chlorophyll) and water transparency can be predicted (e.g. OECD, 1982). The transparency of the water is an easily understood parameter, and it is informative when explaining lake water quality data to laymen. Ryding (1983) has presented a nomogram for transforming phosphorus values to chlorophyll and transparency values (Figure 11.3), based on data from 30 Swedish lakes. Since this nomogram simply represents the correlation between the in-lake, summer average total phosphorus and chlorophyll concentrations (top of graph) and between chlorophyll concentration and Secchi depth (bottom of graph), one can also develop a site-specific nomogram based on local data, if adequate.

Alternatively, one may choose to use a more detailed approach, such as computer scenario analysis, to assess eutrophication control options. This approach is discussed in a following section.

As an aid in the management policymaking process, there is often a use for mathematical models, with which future trends in water quality for a given waterbody under a changing pattern of nutrient input can be assessed. This assessment is not only in terms of the 'average' values for a specific time period. Extreme situations (e.g. worst possible conditions) are also of special interest to both lake/reservoir managers and the public. However, such situations can be hard to define when conducting routine monitoring programmes. Consequently, relationships between the average and maximum values of chlorophyll (OECD, 1982) can be very useful for making predictions for water management. An estimation of the annual peak chlorophyll value could be derived by multiplying the annual average

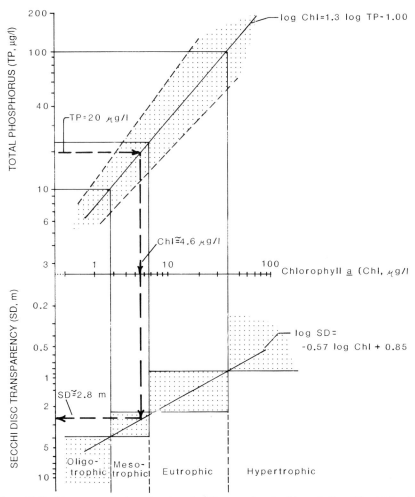

Figure 11.3 Nomogram for transforming phosphorus values to chlorophyll and Secchi depth values (based on Swedish lake data; from Ryding, 1983)

value by a factor of three or the average summer value by a factor of 1.5–2.0 (also see R.A. Jones *et al.*, 1979). These relationships have been shown to be valid for several eutrophication-related variables in Swedish lakes, both for standing and flowing waters (Table 11.2). Such simple models provide sufficient information for many situations; nevertheless, one may also wish to consider the use of a more temporally or spatially detailed dynamic model. These more-detailed models are also useful in some situations, as discussed earlier in Chapter 6.

With regard to standing waters, the main public interest is usually focused on the water quality during the summer period, the period of maximal use.

Table 11.2 Linear correlations between maximum (y) and mean (x) values for common water quality variables in Swedish lakes (from Ryding, 1981a)

Variable	Lake waters[a]		Flowing waters[b]	
	Correlation coefficient	Equation	Correlation coefficient	Equation
Discharge			+0.99	$y = 2.73x + 1.06$
Transparency	+0.93	$y = 0.73x - 0.04$		
Absorbance –suspended matter	+0.96	$y = 1.59x - 8.75$		
Absorbance –color	+0.96	$y = 1.64x - 79.8$		
pH	+0.92	$y = 0.93x + 1.26$		
Conductivity	+0.99	$y = 1.06x + 1.11$		
Suspended matter	+0.98	$y = 1.36x + 2.63$	+0.92	$y = 1.35x + 2.56$
NH_4-N	+0.84	$y = 1.92x + 0.14$	+0.93	$y = 2.52x + 0.06$
NO_2-N	+0.86	$y = 2.91x + 0.00$		
NO_3-N	+0.82	$y = 1.70x + 0.10$	+0.95	$y = 2.04x + 0.17$
Organic-N	+0.97	$y = 1.53x + 0.00$	+0.88	$y = 1.91x - 0.15$
Total-N	+0.96	$y = 1.42x + 0.11$	+0.93	$y = 1.65x + 0.27$
PO_4-P	+0.99	$y = 1.53x + 0.02$	+0.96	$y = 2.68x - 0.01$
Residual-P	+0.98	$y = 1.55x + 0.00$	+0.96	$y = 2.32x - 0.01$
Total-P	+0.98	$y = 1.39x + 0.02$	+0.97	$y = 2.43x - 0.03$
Organic matter	+0.97	$y = 1.33x + 0.13$	+0.89	$y = 1.40x + 0.30$
Chlorophyll a	+0.96	$y = 1.77x + 7.62$	+0.94	$y = 2.60x + 0.84$

[a] Average values, June–September, n = 2300.
[b] Annual average values, n = 3640.

In flowing waters, recreational use may be prolonged throughout the entire year.

5. Assess need for further (in-lake) control measures

If the expected improvement in water quality and/or trophic conditions from external phosphorus control measures will not be sufficient (based on model predictions or post-treatment monitoring) to achieve the eutrophication control goals, one can also consider supplemental in-lake control measures. The predicted water quality improvement, for example, following a phosphorus load reduction of 75–90 per cent may still represent eutrophic conditions in some cases, especially for shallow waterbodies. Shallow waterbodies can be especially sensitive because their water mass is more susceptible to mixing by wind action, their algae biomass is more frequently present in the euphotic zone, etc.

In such cases, one may consider such options as alterations in the lake basin morphometry (e.g. dredging) or initiation of in-lake nutrient control measures. As pointed out in Chapter 3, the latter measures are usually only temporarily effective. Such measures can, however, be very useful when the primary method of external nutrient control alone is either inadequate for achieving the goals, or too expensive to be implemented in a given situation. In-lake controls (see Chapter 9 and Annex 2) include such measures as nutrient inactivation, hypolimnetic aeration, harvesting of macrophytes, application of algicides, etc. Biological controls (e.g. enhancement of certain food chain pathways by introduction or replacement of specific food chain organisms) may also be considered, although the long-term, ecological effects of this approach are largely unknown at present.

Since some of the internal protection measures are closely interrelated, the decision process at this point must clearly recall the particular goal of the control programme. In many cases, excessive and/or sustained growths of phytoplankton (particularly blue-green algae) are the primary control target. In this context, one can attempt to give attention to the factors which will increase or decrease the competitive advantage of planktonic blue-green algae over other types of algae.

Recent work suggests that one can attempt to manipulate the ratios of nitrogen and phosphorus in a waterbody to improve the 'quality' of the algal populations to be expected in the waterbody. As noted in Chapter 4, for example, V.H. Smith (1983) reported that blue-green algae generally were not seen in waterbodies with in-lake total nitrogen: total phosphorus ratios (TN:TP) greater than about 30:1. This observation is supplemented by the work of others who have shown, in a range of waterbodies, that growth of N_2-fixing blue-green algae was generally inhibited if nitrogen was added to a waterbody or if the N:P loading ratio was increased (e.g. Schindler, 1977; Lindmark, 1979; Leonardson & Ripl, 1980; Barcia et al., 1980; Flett et al., 1980).

Consequently, V.H. Smith (1983; 1985) suggests one approach the control of eutrophication by calculating the 'desirable' loading rates of nitrogen and phosphorus to a waterbody (based on minimizing the in-lake total phosphorus concentration) and then adjusting the nitrogen input to provide an in-lake TN:TP ratio > 30:1. The addition of a nutrient to a waterbody, of course, contradicts most conventional eutrophication management thought, and this approach will require more study and experience before its benefits can be accurately assessed.

6. Assess effectiveness of control programme:

In most of the cases studied so far, economic optimization with respect to water quality is concerned primarily with control measures in three major areas:

1. Nutrient source control in the watershed (external control);
2. Temporal detention in the waterbody (internal control); and
3. Treatment plants (off-line control), in the case of water used as a water supply.

The ultimate benefit that can be realized will usually be substantially higher if optimization is related to all three control categories. This integrated approach is useful for control of both nitrogen and phosphorus. As noted earlier, it is preferable over the long term to reduce or eliminate the sources of the substances (e.g. phosphorus) causing eutrophication, rather than temporarily ameliorating the symptoms of eutrophication.

Finally, if neither the external or internal control measures are sufficient to achieve the eutrophication control goals, one may simply have to accept the achievable water quality as the best that can be attained under the circumstances. However, while all eutrophication control goals may not be achieved in a given situation, practical experience suggests that the basic condition of the aquatic ecosystem will usually be improved over the long term with the implementation of eutrophication control programmes.

What if the waterbody does not respond as expected?

It must be recognized that there may be differences in the internal structure or the morphometric/hydrologic environment of a waterbody, which may result in unexpected differences in the water quality responses of otherwise similar lakes and reservoirs. This concern also applies to differences between eutrophication model predictions and the observed responses of waterbodies to control measures. Other factors which potentially can cause complications in predicting the responses of a lake or reservoir to eutrophication control measures include the effects of toxic substances on phytoplankton, algal predation by zooplankton, nutrient limitation by elements other than phosphorus, and the light attenuation capacity of the water column.

Consequently, as a general rule, one should attempt to identify, to the maximum extent possible, the factors that may be responsible for uncertainties in the predictive capability of existing eutrophication models.

Some of the most important elements of uncertainty to be considered in the management of eutrophied lakes and reservoirs are identified and discussed below.

1. Assess response time of lake or reservoir ('lag period')

If a nutrient behaved chemically as a conservative substance (i.e. the substance would not undergo any biological or chemical transformations in a waterbody), one could normally calculate the time necessary for a lake or reservoir to respond to a nutrient control programme solely on the basis of the flushing rate of the waterbody. This 'response time' constitutes the so-called 'lag period'. For example, if the input of a nutrient to a waterbody were altered (e.g. initiation of a nutrient control programme), the time necessary for the waterbody to reach a new steady state condition (i.e. to 'respond' to the control programme) would depend on the hydraulic residence time. In the case of a conservative substance, the response time would equal approximately three times the hydraulic residence time (Rainey, 1967; Vollenweider, 1969; Sonzogni et al., 1976).

However, aquatic plant nutrients such as phosphorus are non-conservative substances in aquatic systems. They undergo various chemical transformations in the water column, are assimilated by phytoplankton, sink to the bottom sediments, etc. In such cases, one would use the residence time of the nutrient to calculate the response time (or lag time) of a lake or reservoir to altered nutrient inputs (increases or decreases), as contrasted with the hydraulic residence time.

Sonzogni et al. (1976) provide a detailed derivation of the phosphorus residence time calculation. Rast and Lee (1978) also reported a simple approach for calculating the phosphorus residence time, R(P), as follows:

$$R(P) = [P]_l / [P]_{in} \qquad (11.1)$$

where: $[P]_l$ = annual mean total phosphorus mass in waterbody (e.g. kg P); and

$[P]_{in}$ = annual total phosphorus input to waterbody (e.g. kg P/yr).

The basis for this equation is that the annual average mass of total phosphorus in the waterbody will differ from the annual input as a direct function of the combined impact of various in-lake transformations and losses, and the flushing rate of the nutrient from the waterbody. A similar expression can be developed for the nitrogen residence time (for nitrogen-limited waterbodies). However, in the latter case, the relationship between the nitrogen levels and nitrogen residence time would necessarily be more complex and uncertain, since one must also consider a gaseous phase in the aqueous chemistry of nitrogen, due to nitrogen fixation and denitrification reactions (Torrey & Lee, 1976).

Rast & Lee (1978) have shown that, in most cases, the phosphorus residence time is shorter than the hydraulic residence time (usually by at least several-fold) because of the environmental aqueous chemistry of phosphorus. Thus, the new steady state condition in a lake or reservoir subjected to a phosphorus control programme will require a time period ('lag time') equal to approximately three times the phosphorus residence time. The exceptions appear to be relatively unproductive, oligotrophic lakes and reservoirs, which exhibit response times approaching their hydraulic residence times. Overall, eutrophic waterbodies appear to exhibit the shortest phosphorus residence times. Thus, such waterbodies should normally exhibit a 'response' in the shortest time period, following initiation of a control programme. However, the magnitude of the response (especially the publicly perceptible signs) must be compared to the initial degraded condition of the waterbody, when assessing the extent of water quality improvement resulting from a control programme.

2. Reassess limiting nutrient

Due to its relatively low solubility and low concentrations in natural waters, phosphorus limits maximum attainable algal biomass in many inland waters. Thus, eutrophication problems are often caused by an excessive input of phosphorus. However, in heavily polluted lakes (where phosphorus can be in excess of algal needs), other nutrients or factors can play a decisive role in limiting maximum algal growth. Yet, virtually all lake/reservoir models derived so far for managerial purposes (e.g. for predicting trophic changes resulting from alterations in loading conditions) normally focus only on phosphorus as the limiting nutrient. It is very important therefore, to determine if phosphorus is actually the growth-limiting nutrient. Guidelines for evaluating nitrogen and/or phosphorus as limiting nutrients were discussed previously in Chapter 4.

3. Reassess phosphorus–chlorophyll relationships

The strong relationships derived between in-lake phosphorus and chlorophyll concentrations (e.g. OECD, 1982) provide useful tools for water management. For example, several OECD investigations have related measures of phosphorus and chlorophyll at various seasons. The slopes of the derived regression lines are often greater than one, indicating the algal biomass increases at a faster rate than the phosphorus concentration. In fact, the phosphorus:chlorophyll ratio is often greater at higher concentrations of phosphorus than at low concentrations (i.e. a greater portion of the phosphorus is incorporated into the algal populations in lakes with higher phosphorus concentrations, a phenomenon commonly referred to as algal 'luxury consumption').

A large scatter is often observed when plotting the phosphorus–chlorophyll relationship for large data sets. This scatter may be due to:

1. The effects of other limiting nutrients;
2. Different requirements for phosphorus by different algal compositions;
3. The effects of zooplankton grazing of phytoplankton;
4. The utilization of phosphorus by other organisms of the aquatic food chain;
5. The complexity of the phosphorus cycle;
6. Sampling or analytical errors;
7. Waterbodies in different climatic zones; and/or
8. Different quantities of chlorophyll per algal cell, depending on the species communities and light and mixing regime.

Phosphorus–chlorophyll relationships provide a means of estimating the algal densities in lakes and reservoirs, and allow estimates to be made of the expected effects of altered phosphorus levels on algal biomass It must be stressed, however, that the data set analyzed with a predictive model should be from the same climatic region as that from which the predictive model is derived.

4. Reassess nutrient loading

In addition to analysis of water for its nutrient content, measuring the water discharge will always give the most accurate nutrient loading values for a given waterbody. In the absence of directly measured nutrient loads, a traditional approach is to estimate the annual nutrient load from a drainage basin using unit area loads for specific types of land use in the drainage basin (see Chapter 7). These estimates, however, are valid only for 'normal' conditions during an average hydrologic year.

As previously discussed, the presence of large quantities of inorganic suspended matter (especially in reservoirs) can significantly affect the phosphorus–chlorophyll relationships (i.e. less chlorophyll is produced per unit of total phosphorus in a turbid waterbody).

Another problem with obtaining an accurate estimate of the annual nutrient load concerns the hydrological regime in the lake basin itself. For lakes with a short water residence time (i.e. a rapid water flushing rate), a certain portion of the nutrients entering the lake at the beginning of the year may be flushed through the lake before the growing season commences. On the other hand, nutrients entering a lake after the growth period, even if in large quantities, are of no consequence in the biological response (i.e. algal growth of the growth period just completed. Therefore, consideration of the actual hydrological conditions (i.e. the amount of nutrients which theoretically influence the water quality during the growth period) is most important in regard to assessing the ecologically relevant nutrient input. This topic was discussed previously in Chapter 7, in regard to estimating the nutrient load.

A comparatively high concentration of nitrogen and phosphorus in the water column, compared to the measured external load of these elements, indicates the existence of substantial internal sources (e.g. ground water seepage, nitrogen fixation or phosphorus release from sediments). Comparing a given data set to the generally observed relationship between the loading and in-lake concentrations of nitrogen and phosphorus can indicate the occurrence of such conditions. Predictive lake models should not be applied to lakes affected by internal sources of nutrients, unless the internal load can be estimated and added to the total load value, in order to obtain a realistic total load estimate.

As pointed out in Chapter 7, various physical, chemical and biological processes can alter nutrient concentrations during the period of downstream travel of the nutrient from the upper part of a drainage basin to a downstream receiving lake or reservoir. Water quality can exhibit considerable variation along a river stretch, due to such processes, as well as successive inputs of additional nutrients from other sources along the tributary. Consequently, nutrient inputs in the upper parts of a drainage basin can have a less significant impact on water quality in a downstream lake or reservoir than nutrient sources closer to the waterbody. This reality necessitates a selective approach to eutrophication control, incorporating external nutrient (phosphorus) control as the basic component of an effective water management programme, augmented by additional measures as needed. An example of such an integrated approach is provided by Ryding (1986).

5. Reassess sampling programme

Due to potentially large and intermittent variations in water quality often observed in many eutrophied waterbodies, an efficient and economically feasible sampling programme is not always a routine matter (see Chapter 8). It is logical, therefore, to suggest that samples be taken as frequently as possible to properly describe actual in-lake conditions.

Average annual values of any relevant parameter cannot be obtained if the sampling effort is distributed irregularly over the year. The difficulties of forecasting hydrological and meteorological conditions from one year to another, especially because flow-flood periods and ice-breaking periods are difficult to predict reliably over the long term, emphasizes the need for evenly distributed sampling intervals over the annual cycle as a baseline sampling/monitoring programme. Spring flood and major rainfall events should also be sampled. Furthermore, large year-to-year variations can make predictions based on only one sampling year rather suspect. A sampling period of at least three years normally should be regarded as the minimum effort for the most accurate assessment of a waterbody.

Another potential problem with a sampling programme is the 'patchiness' phenomena, often noted for planktonic organisms. Because plankton may not be uniformly distributed throughout a waterbody, several samples distributed over the lake surface area, especially for large lakes, are necessary to overcome the problem. Furthermore, if planktonic organisms are irregularly distributed vertically, sampling at many depths may also be necessary.

6. Consider geographic factors

Different physical factors may have a significant impact on the ultimate bioproductivity of a waterbody (e.g. mean air temperature, rainfall, snow cover, overland flow, soil erosion, etc.). Furthermore, most stratified waterbodies in tropical regions have hypolimnion temperatures $>20°C$, which can substantially promote microbial conversion of nitrate to molecular nitrogen.

7. Consider morphometric factors

The shape of a lake basin can substantially influence its response to changes in nutrient loading. Generally it is difficult to define the 'average conditions' for waterbodies with substantial longitudinal gradients in water quality (see Chapter 8). In reservoirs, the main station used for comparison is normally near the water outlet. In lakes consisting of several sub-basins, or in hypertrophic lakes exhibiting a 'patchiness' phenomenon, it may be difficult to obtain a characterization of epilimnetic water quality as a whole. The morphometric conditions of a lake basin are also of vital importance in terms of restricted mean depths leading to homothermal conditions, and increased contact between sediments and water with substantial release of nutrients from bottom sediments. Furthermore, reservoirs subject to periodic drawdown can appear to have several widely differing morphometries during an annual cycle.

8. Consider hydrodynamic factors

The mixing regime comprises such elements as the wind fetch in different seasons, the ratio of hypolimnion to epilimnion volume, the ratio of mixing depth to mean depth and the ratio of littoral area to epilimnion. Resuspension of sediments by wind action not only increases the mobilization of phosphorus, but also decreases the N:P ratio by denitrification, causing elements other than phosphorus (e.g. nitrogen) to act as a growth-limiting nutrient. Uncertainties about the potential effects of control measures often increase with water residence time, but decrease with depth. Thus far, no critical phosphorus loading levels have been identified for oligomictic or polymictic tropical lakes. The phosphorus loading capacities in these two cases will likely be quite different from those of temperate waterbodies.

9. Consider geochemical and chemical factors

The relationship between nutrient load and trophic reaction is different for lakes with hard waters on one hand, and for drainage basins with igneous rocks on the other. Because this is of practical importance in water quality management, the hardness of the water should be included in the rating scales for selecting trophic categories (see Table A.6 in Annex 1). Furthermore, in hard-water lakes, the biota is not as sensitive to the presence of toxic elements as in soft-water lakes.

10. Consider biological factors

The significant role of the biological community structure in the nutrient response of a waterbody should not be overlooked. In this regard, the selective feeding of fish on invertebrates (zooplankton and bottom fauna) is an important process (see Chapter 10). In the absence of fish, 'overgrazing' by zooplankton can reduce phytoplankton concentrations to very low levels, even if the dissolved phosphorus concentration is extremely high.

DETAILED SCENARIO ANALYSIS AS AN AID IN SELECTING EUTROPHICATION CONTROL MEASURES

The strengths and limitations of dynamic mathematical models as assessment and predictive tools for the eutrophication process were discussed previously in Chapter 6. Prominent limitations were the increased data requirements (compared to simpler empirical models) and the difficulty of applying such models to different waterbodies without considerable revision of the types of necessary model variables and/or values of these variables.

One of the strengths of dynamic models, however, is their typically greater ability to provide insight into the internal processes controlling eutrophication (e.g. why or how a relationship or process occurs). They also allow multiple changes to be considered simultaneously. Manipulation of the types and/or values of the state variables in the model can provide potentially valuable insight into what factors exert primary control in the eutrophication process (Rast, 1981a).

Dynamic models can also be a useful tool in situations where development of a eutrophication control strategy depends on controlling internal ecological processes. One can alter the values and/or relationships describing the ecological processes of interest, and use the model to simulate the effects of such changes on water quality and other relevant variables. Alternatively, one can compare the simulated effects of nutrient reduction programmes against those involving control of internal ecological processes. Based on the results of such comparisons, as well as the measured limnological conditions, one can attempt to select optimal control strategies in a given situation.

One example of the use of a dynamic model to aid in the decisionmaking process regarding eutrophication control alternatives is provided here. This example uses an analysis of the simulated effects of different control scenarios to aid in selection of an optimal eutrophication control strategy for the multi-purpose Bleiloch Reservoir in the southern region of the German Democratic Republic (Benndorf & Recknagel, 1982).

Bleiloch Reservoir was experiencing hypolimnetic oxygen depletion and the production of hydrogen sulfide, as a result of biologically treated industrial effluents discharged to the waterbody during the period of thermal stratification. To combat the water quality deterioration, construction of an advanced physico-chemical treatment plant at a large paper mill discharging to the reservoir was being considered. Such a plant would reduce the quantities of oxygen-demanding materials entering the reservoir, thereby reducing the oxygen depletion and hydrogen sulfide production problems. However, the plant would also have the potential, of reducing or eliminating brown-colored recalcitrant sewage materials in the water. These compounds, which greatly reduced the transparency of the reservoir water, were believed to be an inhibiting factor for algal growth as a result of light limitation. The basic question being considered, therefore, was whether or not the benefit of reducing the pollutant load from the pulp mill would be offset by an increased level of algal biomass and associated symptoms of eutrophication.

In this example, a general ecological model, called SALMO, was used to assess the possible impacts of the anticipated control measure (Benndorf & Recknagel, 1982). This model provides an example of the scenario analysis procedure. With this procedure, each specific scenario or set of conditions (e.g. specific boundary conditions, constraints, nutrient loads, and one or more internal control variables) are assessed by means of a model simulation run. The various conditions are then changed to reflect a new scenario being considered, and another simulation run is made using the new conditions. After simulation of all the desired or specified scenarios, the predicted results of all the simulation runs can be compared to each other. Based on these results, an optimal eutrophication control measure (or combination of measures) can be selected for the waterbody.

In the case of Bleiloch Reservoir, three control variables were explicitly considered in the models, as follows:

1. Reduction of light extinction (i.e. increase in water transparency) as a result of building the treatment plant;
2. Reduction of light extinction, plus artificial destratification (mixing) of the waterbody during the summer stagnation period;
3. Reduction of light extinction, plus an increase in phytoplankton grazing by zooplankton (resulting in a reduced algal biomass); this would be done by stocking the reservoir with predaceous fish, which feed on zooplankton-eating fish; and
4. All three variables considered simultaneously.

The details of the model structure and parameters, input variables, boundary conditions, etc., are provided by Benndorf & Recknagel (1982) and Recknagel & Benndorf (1982).

The results of this scenario analysis are presented in Figure 11.4. The y-axis indicates the predicted level of bioproduction resulting from each of the scenarios being considered, while the x-axis represents time (months). The horizontal lines in each of the graphs indicate the critical boundary level for acceptable biological production. Thus, the hatched areas above the lines indicate excessive (unacceptable) biological production.

Examination of Figure 11.4 shows that a combination of the advanced treatment plant and artificial mixing (scenario 2) provides the optimal single management strategy for Bleiloch Reservoir under the conditions

Selection of effective strategies for the management of eutrophication

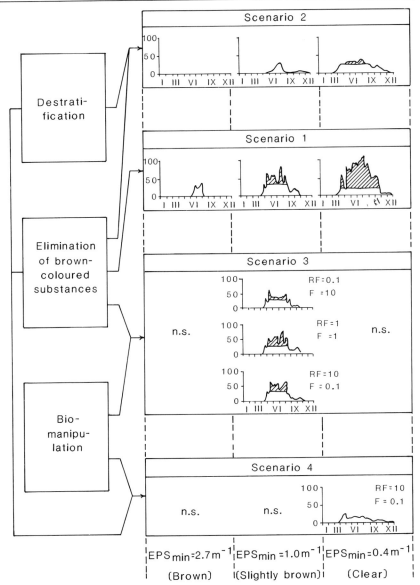

Explanation of terms: y-axis = bioproduction (g wet weight/m² . day); x-axis = time (months); RF = predacious fish; F = zooplankton-eating fish (relative units); EPS_{min} = light extinctioin coefficient for water without suspended particles; ns = not simulated; hatched areas = bioproduction exceeding critical levels; see text for discussion of other terms.

Figure 11.4 Example of scenario analysis for Bleiloch Reservoir, using the dynamic model SALMO (from Benndorf & Recknagel, 1982)

considered. Destratification alone, and destratification plus treatment of most of the pulp wastes (i.e. $EPS_{min} = 1.0/m$; a slight brown color remains), does not result in biological production above the critical level. The treatment plant alone (i.e. $EPS_{min} = 0.4$) would result in a small excess biological production during only a portion of the year. Neither scenario 1 nor 3 are acceptable, due to the excessive production predicted by the model. Scenario 4 will also allow the desired conditions to be achieved, but requires all three control options to be applied. Thus, scenario 2 appears to be the most appropriate control strategy, based on the conditions considered, the control goals and the predictive capability of the model utilized.

This application provides an example of the utility of the general type of ecological model (SALMO) considered above. While the limitations of such models must be considered (see Chapter 6), this approach also allows one to attempt to simulate conditions not readily modeled with the simpler load–response models discussed in Chapter 6. Such applications are useful under the conditions described, and support the further development of dynamic models and refinement of their use in assessing alternative eutrophication management options.

POST-TREATMENT MONITORING

In order to obtain sufficient information for a judicious selection of eutrophication control measures, extensive studies of the chemical and biological conditions of the waterbody of concern and its tributaries, usually are required. Upon completion of such studies, (i.e. after control measures have been planned and carried out), one may conclude that further studies are not necessary. Such a conclusion is false. Even after eutrophication control programmes have been initiated (e.g. reducing the nutrient influx), post-treatment studies should be continued for at least several more years. This should be done to compare the condition of the waterbody before and after the start of eutrophication control measures, and to ascertain whether or not the results expected from model calculations have actually been achieved. Only then can one be certain whether or not (or to what degree) the corrective action taken was correct, and whether or not the monetary investment was a financially responsible one.

Should it occur that, in spite of very careful planning and use of all available knowledge, the observed results obtained fall short of the expected results, post-treatment measurements still can be used to improve the model predictions in question. This will also work to decrease the uncertainty of model predictions for future planning purposes.

The period of time necessary for carrying out such post-treatment measurements is dependent on the individual case. The longer the lake is expected to take to recover, the longer the period of time such measurements will have to be taken. Even after a prompt recovery of a given lake, it may still be necessary to continue monitoring studies for several years, in order to be sure that the situation has been correctly assessed. This is especially necessary for cases in which large annual differences in water quality can occur before eutrophication control measures are initiated. Examples are lakes with relatively short water retention times, and lakes which are situated in areas with very changeable weather conditions.

Post-treatment monitoring and evaluation also provide valuable information to others concerned with similar eutrophication management problems, and help guide future efforts (e.g. building the information and experience base for improved lake and reservoir management technology).

ANNEX 1

CLASSIFICATION OF WATERBODIES IN RELATION TO THEIR DESIRED USES

This annex presents two approaches for classifying inland waters related to their desired usage. The first approach describes a simple empirical system, based on the trophic classification scheme developed from the OECD (1982) international eutrophication study. The second approach describes a more detailed classification system, based on a Technical Standard used in the German Democratic Republic (Technical Standard, 1982). This Technical Standard, in effect since 1974, has been used successfully for lakes and reservoirs located primarily in temperate zones, as well as in some non-temperate situations.

Lake McIlwaine, a tropical African, man-made lake in the Republic of Zimbabwe, is used as the study waterbody in both approaches presented in this annex. This lake is of interest because it has multiple competing uses (e.g. drinking water supply, recreation, irrigation, fisheries), and because it provides an example of the response of a tropical region waterbody to a control measure (nutrient diversion) normally used in temperate zone lakes and reservoirs (see footnote, Table A.8).

A SIMPLE LAKE/RESERVOIR CLASSIFICATION SYSTEM

The data for this assessment and classification were taken from J.A. Thornton & Nduku (1982), who presented a detailed discussion of the history of cultural eutrophication of Lake McIlwaine, and the subsequent changes in lake water quality and trophic status following initiation of a nutrient control programme.

The lake classification approach used here was developed as a component of the OECD Cooperative Programme for Monitoring of Inland Waters, an international eutrophication study (OECD, 1982). This study involved over a hundred waterbodies of various trophic states and water quality in a number of countries around the world, though primarily in the temperate zone. Relevant load – response relationships were developed from the large data base, for use as simple eutrophication assessment and management tools. The reader is referred to OECD (1982) for information and data on the overall study, and to the individual reports of the four regional projects (Rast & Lee, 1978; Ryding, 1980; Fricker, 1980a; Clasen & Bernhardt, 1980; Janus and Vollenweider, 1981) for details regarding the individual studies.

The relevant models used in this classification scheme were identified previously in Chapter 4, particularly the fixed and open trophic boundary classification schemes (Tables 4.1 and 4.2). These tables present approximate boundary values, based on the OECD study results, for several commonly-used trophic and water quality indicators.

The questions to be addressed are, what is the general trophic condition of Lake McIlwaine (based on its post-treatment characteristics), and what water uses would be hindered or prohibited on the basis of the lake's trophic condition? The relevant data for answering these questions, as summarized by J.A. Thornton & Nduku (1982), are presented in Table A.1

According to the OECD fixed trophic boundary system (Table 4.1), Lake McIlwaine presently (i.e. following its eutrophication control treatment) can be considered eutrophic on the basis of all the parameters, except total inorganic nitrogen (Table A.1). Total soluble phosphorus was used in place of total phosphorus, since the latter data were not available. Thus, the mean total phosphorus concentration should be higher than the $40 \mu g/l$ value listed in Table A.1. Since there were no values given for the maximum (peak) chlorophyll, this value was estimated on the basis of the derived OECD (1982) relationship between mean and maximum chlorophyll (Figure A.1). Assuming the line of best fit strictly applied to Lake McIlwaine, the mean chlorophyll value of $9 \mu g/l$ in Table A.1 translates into a peak value of $30 \mu g/l$, which is also in the eutrophic range, based on the boundary conditions in Table 4.1 (Note: this value was subsequently substantiated by personal communication with J.A. Thornton). Based on the 80 per cent

Table A.1 Relevant water quality parameters for Lake McIlwaine, circa 1980 (from J.A. Thornton & Nduku, 1982)

Parameter	Value
Total soluble phosphorus concentration	$40 \mu g/l$
Mean inorganic nitrogen concentration[1]	$110 \mu g/l$
Mean chlorophyll concentration	$9 \mu g/l$
Maximum (peak) chlorophyll concentration[2]	$30 \mu g/l$
Mean Secchi depth	1.8 m

[1] Sum of ammonia, nitrate and nitrite nitrogen.
[2] Projected value from Figure A.1, based on mean chlorophyll concentration of $9 \mu g/l$.

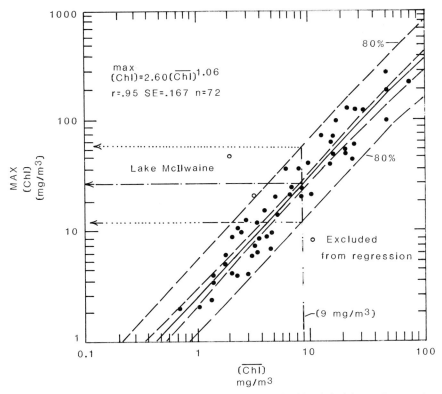

Figure A.1 Relationship between mean and peak chlorophyll levels in lakes and reservoirs, showing peak value for Lake McIlwaine (modified from OECD, 1982)

confidence intervals in Figure A.1, the projected peak chlorophyll levels lie in a range of 12–64 µg/l, which essentially spans the eutrophic–mesotrophic boundary values in Table 4.1.

It is noted that all the values in Table A.1 (which describe its 'present' condition), except for mean Secchi depth, approximate the eutrophic–mesotrophic boundary values presented in Table 4.1, in contrast to its previous eutrophic condition. This suggests a positive response of Lake McIlwaine to its nutrient control programme.

If one uses the open boundary system (Table 4.2), virtually the identical conclusions are reached, particularly when the ±1 standard deviation range for each parameter is considered. The overlapping nature of the trophic parameters at the boundary conditions is clearly seen.

One can most easily visualize this open and overlapping trophic boundary system with a visual representation of the trophic boundaries suggested in Table 4.2, as illustrated in Figures A.2–A.5. Based on the values presented in Table A.1, as well as the projected peak chlorophyll value of 30 µg/l, the

Table A.2 Trophic classification of Lake McIlwaine, based on simplified classification system

Parameter	Major Trophic Probability
Mean total phosphorus*: (*total soluble phosphorus)	55% mesotrophic; 38% eutrophic
Mean chlorophyll:	50% eutrophic; 40% mesotrophic
Projected peak chlorophyll:	50% eutrophic; 34% mesotrophic; 16% hypertrophic
Mean Secchi depth:	50% eutrophic; 38% hypertrophic; 10% mesotrophic

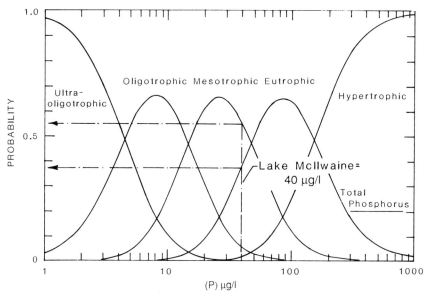

Figure A.2 Probablity distribution of trophic categories based on total phosphorus (modified from OECD, 1982)

trophic state possibilities for Lake McIlwaine are summarized in Table A.2.

Based on these four water parameters one would conclude that Lake McIlwaine is most likely still in a eutrophic condition, but exhibiting a shift in its trophic character to a mesotrophic condition. If one compares these trophic possibilities to the 'still tolerable' (i.e. still usable) trophic conditions for the indicated water uses shown in Table 11.1, one would conclude that Lake McIlwaine waters were impaired in regard to use as a drinking water supply, for low-water production and for the production of salmonidae fish. In fact, this is a reasonable indication of its current conditions.

If one uses the more stringent 'required' category (i.e. optimal trophic condition) in Table 11.1, then Lake McIlwaine is impaired (without supplemental water treatment) for virtually all uses except irrigation, landscaping and energy production.

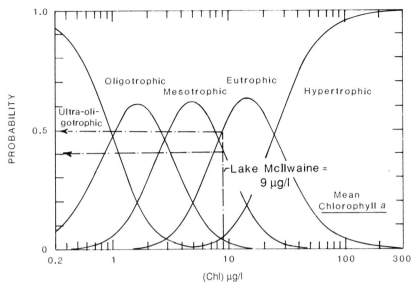

Figure A.3 Probablity distribution of trophic categories based on mean chlorophyll (Modified from OECD, 1982)

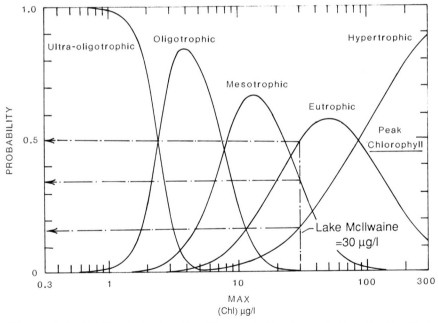

Figure A.4 Probability distribution of trophic categories based on peak chlorophyll (modified from OECD, 1982)

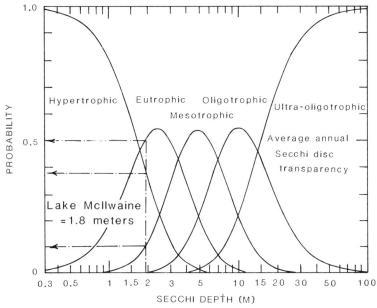

Figure A.5 Probability distribution of trophic categories based on mean Secchi depth (modified from OECD, 1982)

A DETAILED CLASSIFICATION SYSTEM

The minimum data requirements for assessing the state of a waterbody were discussed previously (see Table 8.1). Additional criteria useful for a more detailed assessment of lake/reservoir water quality are summarized in Table A.3. Based on these types of data (both for simple and detailed classification schemes), classification of a waterbody should incorporate the following items:

1. The waterbody should be considered as a part of its catchment area;
2. The interaction between lake morphology and water quality should be considered;
3. Water quality should be described as an integrated summary of a waterbody's physical properties, chemical composition and biological components;
4. The classification should be sensitive to water quality changes due to nutrient loads and the efficiency of restorative measures;
5. The classification should include information about the suitability of the waterbody for various water uses. If possible, the classification should also provide relevant information about the potential impacts of these uses on water quality; and
6. The classification scheme should supply usable and understandable information to the decisionmakers responsible for water management.

Annex 1

Table A.3 Additional information necessary for a detailed classification of water quality related to water use[a] (modified from Technical Standard, 1982)

Parameter	units
1. Drainage basin and external nutrient load:	
Area of drainage basin	km²
Seasonal variation of monthly air temperature	°C
Seasonal variation of monthly rainfall	mm
Seasonal variation of volume flow of main tributaries (at least monthly means)	m³/day
Portion of forest, crop area, meadows and pastures, and built-up area	%
Total population in drainage basin	—
Amount of domestic effluents	m³/day; kg P and N/day; kg BOD_5/day
Type of sewage and waste treatment facilities	—
Numbers of types of livestock	—
Amount of industrial wastes	m³/day; kg P and N/day; kg BOD_5/day
Areal phosphorus loss from agricultural and forest areas	kg/ha·yr
Areal nitrogen loss from agricultural and forest areas	kg/ha·yr
Areal phosphorus load to lake	g/m²·yr
Areal nitrogen load to lake	g/m²·yr
2. Morphometric conditions:	
Length and width of waterbody	km
Shore length	km
Shore development (L = shoreline length; A = lake surface area)	$L/(2\sqrt{\pi \cdot A})$
Sediment bottom area	per cent of total lake surface area (A)
Nature of sediments	—
3. Basic chemistry of lake water:	
Total dissolved solids	mg/l
Electrical conductivity	µS/cm
Alkalinity	meq/l
Acidity	meq/l
Major ions (Ca^{2+}, Mg^{2+}, Na^+, K^+, Cl^-, SO_4^{2-}, CO_3^{2-}, and HCO_3^- from alkalinity)	mg/l; meq/l

Color or equivalent criteria (e.g. dissolved organic carbon or chemical oxygen demand in the case of brown waters; humic compounds, lignosulfuric acids)

4. Hydrodynamic conditions:	
Thermal stratification (vertical profiles along longitudinal axis, including deepest points)	—
Epilimnion:hypolimnion ratio	—

(*continued*)

Table A.3 (*continued*)

5. *In-lake conditions*:	
Free carbon dioxide	mg/l
Total suspended solids	mg/l
Vertical extinction coefficient	1/m (400–700 nm)
Bio-volume of phytoplankton and zooplankton	mm^3/l
Dissolved iron; dissolved manganese	mg/l
Diurnal variation in oxygen saturation	per cent
Hydrogen sulfide in hypolimnetic waters	mg/l
6. *Hygienically-relevant conditions*:	
Dissolved heavy metals (cadmium, chromium^{3+}, chromium^{6+}, cobalt, copper, lead, mercury, nickel, zinc)	mg/l; µg/l
Other elements (aluminum, arsenic, beryllium, fluoride)	mg/l; µg/l
Organic compounds (water vapor volatile phenols, detergents, organochloric compounds, polycyclic aromatic hydrocarbons)	µg/l; ng/l; pg/l
Microbial pollution (coliforms, enterococci)	number/100 ml

[a]Also see Table 8.1 for criteria deemed essential for lake and reservoir classification.

The detailed approach presented in this section (Technical Standard, 1982) is similar to a scheme used by Kudelska *et al.* (1981). The class limits of the six part scale for chemical and biological criteria are almost identical in the two proposals. Both approaches have been tested in lakes of the Baltic Basin. They also appear to be applicable for waterbodies located outside of the temperate zone.

Overall, this detailed classification scheme evaluates water quality on the basis of several classes of criteria, as follows:

1. Hydrographic and territorial criteria;
2. Trophic criteria; and
3. Salt content, special and hygienically relevant criteria.

These three criteria classes are further sub-divided into three groups of characteristics. Classification of a lake or reservoir is based on several criteria from each group of the characteristics being evaluated. Therefore, a detailed evaluation of these three classes of criteria results in a survey of possible utilizations. The multitude of possible characteristics suggests this classification scheme may be useful for a broad range of limnological conditions.

Hydrographic and territorial criteria

These criteria (Tables 8.1 and A.3) can be subdivided into three groups of characteristics:

1. Waterbody morphology;
2. Hydrographic relations between catchment area and waterbody; and
3. Anthropogenic nutrient load.

Groups 1 and 2 describe the natural status of a waterbody. Group 3 represents the anthropogenic nutrient load. For an unpolluted lake, the hydrographic and territorial criteria (Table A.4) represent an expectation corresponding to that based on trophic criteria (see Table A.5). Comparison of the expected value for the 'natural state' criteria with that based on the trophic criteria allows one to assess the chances of success of external and internal control measures.

Table A.4 Classification based on hydrography

Criterion	Quality class					
	1	2	3a	3b	4	5
1. *Morphometry*:						
Type of stratification:	–stratified (holomictic or meromictic)[a]		–no stable stratification– (polymictic)			
– Mean depth, $\bar{z}$ (m)	$\geqslant 15$	$\geqslant 10$	< 10	2 to 10	1 to 2	< 1
– Maximum depth, $\bar{z}_{max}$ (m)	$\geqslant 30$	$\geqslant 20$	< 20	– not applicable –		
– Hypolimnion volume/ epilimnion volume (mean value of stagnation period)	$\geqslant 1.5$	$\geqslant 1.0$	< 1.0	– not applicable –		
– Mean water retention time (yr) (= waterbody volume/ annual inflow volume)	$\geqslant 10$	$\geqslant 1$	$\geqslant 0.2$	$\geqslant 0.1$	< 0.1	< 0.1
2. *Drainage basin (DB)*:						
– Volume quotient $(km^2/10^6 \cdot m^3)$ (= area of DB/water volume)[b]	$\leqslant 3$	$\leqslant 5$	$\leqslant 10$	> 10	not applicable	
– Area quotient (= area of DB/area of waterbody)	$\leqslant 30$	$\leqslant 60$	$\leqslant 300$	> 300	not applicable	
– Per cent of DB in forest use	$\geqslant 80$	$\geqslant 50$	$\geqslant 20$	$\geqslant 10$	< 10	
3. *Load*[c]:						
– L $(PE/10^6 \cdot m^3)$ (= population equivalent, (PE)/water volume)	$\leqslant 50$	$\leqslant 500$	$\leqslant 2500$	$\leqslant 5000$	> 5000	
– P-Load (g orthophosphate $P/m^2 \cdot yr$) (To be determined for strong through-flow lakes and dams)	The upper limits shall be based on the mean depth and mean water retention time of the waterbody, based on P-load model (Vollenweiden 1976a). For seston-rich inlets, use 50 per cent of the total phosphorus load, rather than the orthophosphate load.					
N-Load[d] (g inorganic $N/m^2 \cdot yr$)	< 5	< 10	< 15	– not applicable –		

[a] For meromictic waterbodies, the non-circulating deepwater zone (= monimolimnion) should be subtracted from the z_{max} and z values.
[b] For dams and reservoirs, the values refer to the mean water elevation.
[c] For calculating loading criteria (L) from direct use, see Table 7.9. For calculating P-load based on waste treatment facilities in catchment area, see Table 7.2. For classification purposes, multiply the population equivalents from direct use by 10 (because of their great influence on water quality), and add the sum to the load from catchment area.
[d] Considered only if nitrogen is a phytoplankton growth-limiting nutrient.

Table A.5 Classification based on trophic criteria

Criterion	Quality class					
	1	2	3a	3b	4	5
1. *Oxygen balance*:						
— Mean variation of O_2-saturation at surface, in per cent (summer, calm weather, diurnal cycle; see Chapter 7)	90–120	80–150	60–200		20–300	0–500
— O_2-content in hypolimnion[a], in mg/l (at end of stagnation period in warm season)	⩾6	⩾1	anaerobic		– not applicable –	
2. *Nutrient budget*[b]						
(A) At the beginning of spring overturn period in temperate climates; vernal overturn (mean values for Z_{mix})						
— Orthophosphate[d] (mg P/l)	⩽0.005	⩽0.01	⩽0.03		⩽0.05	>0.05
	⩽0.005	⩽0.015	⩽0.2		⩽1.2	>1.2
— Total phosphorus[d] (mg P/l)	⩽0.015	⩽0.025	⩽0.04		⩽0.06	>0.06
	⩽0.015	<0.045	⩽0.3		⩽1.5	>1.5
— Dissolved inorganic nitrogen[e] (mg N/l)	⩽0.3	⩽0.5	⩽1.0		⩽1.5	>1.5
(B) During stagnation period in summer (mean value for epilimnion)[f]						
— Orthophosphate (mg P/l)	0–0.002	0–0.005	0–0.1		>0.1	>0.5
— Total phosphorus (mg P/l)	⩽0.015	⩽0.04	0.04–0.3		>0.3	>0.5
— Dissolved inorganic nitrogen (mg N/l)	⩽0.01	⩽0.03	⩽0.1		>0.1	>0.5
3. *Bioproduction*:						
(A) Phytoplankton primary production (gross production):						
— Annual primary production (g C/m² · yr)	⩽120	120–150	250–400		400–500	>500
— Ratio of primary production in the 1-m layer of maximum production to the total primary production (mg C/m³ · day)/(mg C/m² · day)	⩽15	15–30	30–75		75–90	>90

(continued)

Among the morphometric criteria, the stratification properties of a waterbody are of prime importance. Waterbodies in which trophogenic and tropholytic processes are separated by thermal stratification exhibit a lower rate of nutrient cycling and productivity than shallow lakes (Patalas, 1968; Richardson, 1975). In contrast, polymictic, shallow lakes generally exhibit a trend toward the massive growths of phytoplankton. A condition of high water transparency may occur, however, as a result of an intense zooplankton grazing pressure on the phytoplankton (Uhlmann, 1958).

Polymixis comprises a broad range of water quality conditions. The

Annex 1

Table A.5 (continued)

Criterion	Quality class					
	1	2	3a	3b	4	5
3. *Bioproduction*: (continued)						
(b) Phytoplankton biomass in the epilimnion:						
— Phytoplankton volume (counting and measurement)	⩽1.5	⩽5	5–10	10–20	20–30	>30
— Chlorophyll a^{665} (April-Sept. average)	⩽3	<10	10–20	20–40	40–60	>60
(c) Transparency:						
— Secchi depth (m)	⩾6	⩾4	⩾1		⩾0.5	<0.5
— Vertical extinction coefficient at 400–700 nm wavelength (m^{-1}) (April-Sept. average)	⩽0.5	⩽0.6	⩽1.3		>1.3	>2.5
(d) Zooplankton biomass in epilimnion^g:						
— Zooplankton (g dry weight/m^3 (April-Sept. average)	<0.1	<0.3	<0.8		>0.8	0->0.8
4. *Trophic state*:	oligo-trophic	meso-trophic	eu-trophic stratified	eu-trophic unstratified	poly-trophic	hyper-trophic

^{a}O$_2$ measurements taken at 5 m above bottom.
^{b}For water residence times >1 year; for residence time <1 year, use annual areal load (g/m^2.yr).
^{c}Z$_{mix}$ = total water column involved in turnover.
^{d}The lower values refer to low alkalinity waters (<70 mg CaO/l); the higher values refer to high alkalinity waters (>70 mg CaO/l).
^{e}Considered only if nitrogen is the algal growth-limiting nutrient.
^{f}Included only if no data are available at the beginning of the overturn period, or if no loading data are available.
^{g}To be used as additional information if measurements are taken at biweekly intervals.

internal phosphorus load is usually increased as the frequency of wind mixing of the water down to the bottom of the waterbody increases. Further, the mean and maximum depth also give insight into the quality of a waterbody, since deeper waterbodies generally exhibit better water quality than shallow ones.

Morphometric factors affecting water quality are not considered further here. However, they may yield relevant information about the potential fisheries productivity of a waterbody (see Chapter 10), based on the following factors:

1. The percentage of the lake bottom in the epilimnetic zone;

2. The shape of shoreline (shore length divided by the circumference of a circle with equal area A; i.e. $U = \text{shore length}/2\sqrt{\pi \cdot A}$);
3. The embankment angle as a measure of littoral development.

Group II of the hydrographic and territorial characteristics describes the drainage basin, which is related both to the volume of the lake (volume quotient) and to the lake area (area quotient). Some authors have related lake area plus catchment area to lake volume (Schindler, 1971a; also see Chapter 5). An increasing erosion and nutrient loss from the catchment area generally occurs in the following order: forest, permanent grassland, forage crop cultivation, root and tuber crop cultivation. Point sources are considered concurrently with non-point sources.

Two classifications are possible on the basis of the anthropogenic load characteristics. The first classification requires an inventory of all the direct uses of the waterbody (see Table 7.9) and of the major land uses in the catchment area. The phosphorus and nitrogen population equivalents, related to the water volume (see Table 7.2), become the basis for the classification. The second possibility is that if sufficient analytical data are available, it is possible to calculate the annual areal load and the ratio of mean depth to mean retention time directly (see Figure 6.1). In this section, the Vollenweider (1976a) loading diagram has been modified to consider orthophosphate, rather than total phosphorus.

Trophic criteria

The trophic criteria (Table A.5) are concerned with the balance of materials in standing waters, and can be subdivided into the following groups of characteristics:

1. Oxygen balance;
2. Nutrient balance; and
3. Bioproduction

The diurnal variation of oxygen saturation in the epilimnion is especially important in regard to fisheries.

Dissolved nutrients are a prerequisite for primary production to occur in a waterbody. Thus, information on the dissolved nutrient concentrations existing in a lake or reservoir prior to periods of massive phytoplankton growths (in the early spring in the temperate zone) can give relatively reliable information about the extent of primary production and algal biomass to be expected. In waterbodies rich in carbonate, a part of the dissolved phosphorus in the water column can be precipitated during the course of biogenic decalcification. Consequently, the level of primary production in the water column would remain below the expected value. Because of this phenomenon, the classification based on the phosphate concentration should be performed separately for high and low alkalinity waters. It is noted, however, that the epilimnetic dissolved nutrient concentration, measured during the summer stagnation, often is so variable that classification based on this parameter may not be very reliable.

Dissolved solids, special and hygienically relevant criteria

These criteria comprise groups of additional characteristics which are significant from the perspective of water uses, as follows:

1. Dissolved solids;
2. Special criteria; and
3. Hygienically relevant criteria (Table A.6).

Obviously, the criteria in this grouping are heterogeneous in nature.

Dissolved solids are important mainly with regard to industrial uses of the water (e.g. boiler feeding waters and irrigation). The group of special criteria mainly comprise unusual water conditions (e.g. acidic artificial lakes in mining areas, which are also rich in iron and humic substances; formerly these lakes would be described as 'dystrophic'). Overall, these are mainly criteria which disturb the trophic relations between nutrient supply and bioproduction 'calibrated' for neutral clear-water lakes.

Lake classification in regard to possible water uses

Based on the previously cited criteria, to classify standing waterbodies in regard to particular types of utilization, one must examine the summary determinations of the quality classes calculated on the basis of Tables A.4–A.6. Potential water utilizations for various trophic conditions are listed in Table A.7 (also see Table 11.1).

For proper management of lakes and reservoirs, management objectives have to be stipulated, and utilization priorities must be taken into account when assessing multiple uses. If water quality regulations exist, these must also be considered when evaluating particular water uses. If the major water uses require a higher water quality than exists in a given waterbody, suitable treatment measures will be necessary to improve the quality. In special cases, the water utilization class may be determined with a single criterion (e.g. nitrate, with regard to drinking water).

How to classify a waterbody

The minimum data requirements for assessing the eutrophication status of a waterbody were discussed previously in Chapter 8. As in the previous section, Lake McIlwaine, Zimbabwe, is used to demonstrate the application of the GDR Technical Standard (Technical Standard, 1982) for the classification of lakes. Relevant data are provided in Table A.1, and additional data are given by J.A. Thornton (1982).

Table A.6 Salinity, special and hygienically relevant criteria

Criterion	1	2	3a	3b	4	5
1. Salinity:						
— Calcium (mg Ca^{2+}/l)	<60	<100	<150		<250	>250
— Magnesium (mg Mg^{2+}/l)	<25	<50	<100		<150	>150
— Sodium (mg Na^+/l)	<30	<70	<150		<300	>300
— Chloride (mg Cl^-/l)	<50	<100	<250		<500	>500
— Sulfate (mg SO_4^{2-}/l)	<100	<150	<350		<500	>500
— Total hardness (CaO)	≤100	≤150	≤300		≤500	>500
— Temporary hardness (CaO)	≤70	≤120	≤250		—	—
— Total salinity (mg/l)	350	750	1500		2500	2500
2. Special Criteria:						
— Iron (mg total Fe/l)	≤0.1	≤0.5	≤1.0		≤3.0	>3.0
— Manganese[1] (mg Mn/l)	≤0.02	≤0.1	≤0.2		≤0.5	>0.5
— pH value						
– in epilimnion of neutral waters:	6.5–8	7–8.5	7–9	7–9.5	6.5–10	6–11
– in minerogenic acidotrophic waters:	>6	5–6	4–5		3–4	<3
— Ammonium in epilimnion (mg NH_4^+/l)	ND	ND	≤0.1		≤1.0	>1.0
— Hydrogen sulfide in epilimnion (mg S^{2-}/l)	ND	ND	ND		ND	>0.01
— Hydrogen sulfide in hypolimnion (mg S^{2-}/l)	ND	ND	>0.01			
— Humic compounds, as chemical oxygen demand (humic acid) (mg O_2/l)	<3	<5	<10		<20	>20
— Humic standard	oligo-humous	oligo-humous	oligo-humous		meso-humous	poly-humous
3. Hygienically relevant Criteria:						
— Nitrate (mg NO_3^-/l):						
– maximum concentration:	<15	<30	<40		>40	
– mean annual concentration:	<10	<20	<30		>30	
— Fluoride (mg F^-/l)	<1.0	<1.0	<1.2		<5.0	>5.0
— Phenols, water vapor volatile (mg Phe/l)	ND	ND	≤0.005		≤0.5	>0.5
— Detergents (mg Ten/l)	ND	≤0.1	≤0.2		≤2.0	>2.0
— Dissolved heavy metals:						
– Copper (mg Cu/l)	ND	ND	≤0.05		≤1.0	>1.0
– Chromium III (mg Cr^{3+}/l)	ND	≤0.1	≤0.2		≤1.0	>1.0
– Chromium VI (mg Cr^{6+}/l)	ND	≤0.01	≤0.02		≤0.1	>0.1
– Lead (mg Pb/l)	ND	≤0.03	≤0.05		≤0.5	>0.5
– Arsenic (mg As/l)	ND	≤0.01	≤0.05		≤0.2	>1.2

(*continued*)

Table A.6 (*continued*)

	Quality class					
Criterion	1	2	3a	3b	4	5
– Zinc (mg Zn/l)	ND	ND	$\leqslant 0.01$		$\leqslant 0.1$	> 0.1
– Cadmium (mg Cd/l)	ND	ND	$\leqslant 0.001$		$\leqslant 0.01$	> 0.01
– Cobalt (mg Co/l)	ND	$\leqslant 0.01$	$\leqslant 0.1$		$\leqslant 1.0$	> 1.0
– Nickel (mg Ni/l)	ND	$\leqslant 0.05$	$\leqslant 0.2$		$\leqslant 1.0$	> 1.0
– Mercury (mg Hg/l)	ND	ND	$\leqslant 0.001$		$\leqslant 0.01$	> 0.01
— Polycyclic aromatic hydrocarbons (mg PAK/l)	ND	ND	$\leqslant 0.0001$		$\leqslant 0.001$	> 0.001
— Organophosphoric hydrocarbons (mg POK/l)	ND	ND	$\leqslant 0.001$		$\leqslant 0.01$	> 0.01
Bacteriological criteria:	not detectable in:				colonies/ml:	
– Coliforms	10 ml	1 ml	$\leqslant 100$		$\leqslant 1000$	> 1000
– Enterococci	100 ml	10 ml	$\leqslant 10$		$\leqslant 100$	> 100

[a]In stratified waters, use the mean hypolimnetic value; otherwise, use the mean value of the whole water column.

Class determination: The numerical values of the quality classes can be determined on the basis of Tables A.4 to A.6, as follows:

1. The arithmetic mean of the individual class values for each of the three sub-groups is determined (see Table A.8). Using the Quality Class based on hydrographic and territorial criteria in Table A.4 as an example (i.e. Characteristic 1: Morphometry), Lake McIlwaine is scored as follows:

   Mean depth = 9.4 m class = 3
   Maximum depth = 27.4 m class = 2
   Hypolimnion volume/epilimnion volume = 0.75 class = 3
   Mean water retention time = 0.8 yr. class = 3
   Sum of (A)+(B)+(C)+(D) = 11/4 = 2.8

2. The mean of the mean values of characteristics 1 to 3 in each Quality Class is determined, and rounded to one decimal place to obtain the class value of the complex of characteristics. Using the hydrographic and territorial criteria from step (1), Lake McIlwaine is scored as follows:

   Characteristic 1: Morphometry mean = 2.8
   Characteristic 2: Drainage basin mean = 3.0
   Characteristic 3: Load mean = 3.0
   Sum of 1+2+3 = 8.8/3 = 2.9 (rounded up to 3)

Taking the mean in two partial steps facilitates decision making with regard to methods of quality improvement, and insures uniform consideration of the subgroups for class formation.

Table A.7 Classification of possible uses

Type of use	Quality Class					
	1	2	3a	3b	4	5
Drinking water (see explanation at bottom of table)	simple treatment	normal treatment	comprehensive treatment; treatment partly complicated	complicated treatment; treatment partly impaired	unusable	unusable
Industrial water	simple treatment	normal treatment	comprehensive treatment; treatment complicated at times	complicated treatment; treatment partly impaired	unusable with reservations for particular purposes	unusable
Cooling water	perfect	perfect	perfect	can be used	usable with reservations	usable with reservations or unusable
Irrigation – Organic load	perfect	perfect	perfect	perfect	usable	usable with reservations
– Mineral load	perfect	perfect	usable	usable	usable with reservations	unusable
– Hygienic load	perfect	perfect	usable	usable	usable with reservations	unusable
Recreation	perfect	perfect	perfect	usable	usable with reservations	unusable

(continued)

3. The value which represents the 'total class' of water quality of the waterbody is calculated as the arithmetic mean of the values of the complexes, rounded to the next whole number. The mean value before rounding should also be indicated.

Table A.7 (continued)

Type of use	Quality Class					
	1	2	3a	3b	4	5
Bathing	inadmissible when used as drinking water, otherwise perfect	inadmissible when used as drinking water, otherwise perfect	inadmissible when used as drinking water, otherwise perfect	inadmissible when used as drinking water, otherwise perfect	to be queried	unusable
Fisheries	on a natural feeding basis	on a natural feeding basis	when used as drinking water or for bathing on a natural feeding basis	when used as drinking water or for bathing on a natural feeding basis	for intensive use with artificial feeding; usable with reservations	intensive use with artificial feeding at risk (O_2, NH_4^+, pH)
Waterfowl	inadmissible	inadmissible	inadmissible	inadmissible	admissible	admissible
Navigation and boating with internal combustion engines	inadmissible	inadmissible	inadmissible when used as drinkwater; usable with reservations when used for bathing	inadmissible when used as drinkwater; usable with reservations when used for bathing	admissible	admissible

Examples of drinking water treatment of raw water taken from dams and lakes with increasing trophic state:

Simple: Rapid filtration, chlorination.
Normal: Aluminum sulfate precipitation, floc filtration, if necessary, with pulverized active carbon, lime dosage, chlorination.
Comprehensive: Micro-screening, aluminum sulfate precipitation, floc filtration, active carbon filtration (or slow sand filtration), lime dosage, chlorination.
Complicated: Denitrification, aluminum sulfate precipitation when adding flocculation aids, active carbon filtration, ozonation.

Lake McIlwaine scores as follows:

Complex of hydrographic criteria   mean = 3
Complex of trophic criteria   mean = 3
Complex of hygenic criteria   mean = 1

Sum of complexes = 7/3 = 2.7 (rounded off to 3)

This value corresponds to the 'class of utilization'.

Table A.8 Classification of Lake McIlwaine[a], based on quality criteria

		Mean value	Rounded off
Group 1 of	1. Morphometry	2.8	
characteristic	2. Drainage basin	3.0	
features:	3. Load	3.0	
QUALITY CLASS based on hydrographic and territorial criteria:		2.9	3
Group 2 of	1. Oxygen balance	3.0	
characteristic	2. Nutrient concentration	2.3	
features	3. Bioproduction	3.5	
QUALITY CLASS based on trophic criteria:		2.9	3
Group 3 of	1. Salinity	1.0	
characteristic	2. Special criteria	1.9	
features:	3. Hygienically-relevant criteria	1.0	
QUALITY CLASS based on salinity, special and hygienically-relevant criteria:		1.3	1
TOTAL OF THREE CLASSES:		2.7	3(2.7)
Most unfavorable hygienically relevant criterion	Utilization concerned	Class value	Utilization class for respective utilization
pH	Drinking water	3.5	3[b]

[a]A man-made lake (Hunyanipoort Dam) southwest of Harare in the Republic of Zimbabwe with a full supply volume of $250 \times 10^6 \, m^3$ and a mean depth of 9.4 m. It is the city's primary water supply reservoir, a source of irrigation water, a recreational site, and an important fishery. It has been rehabilitated by advanced waste treatment in combination with crop irrigation. All numbers in this example were derived from data by J.A. Thornton and Nduku (1982).
[b]Alum dosing in water treatment, complicated by sharp fluctuations in pH.

4. If the class value of a hygienically relevant criterion is lower than the 'total class' value, this lower value determines the class of use, provided the water use has hygenic requirements. In Lake McIlwaine, the pH range of 6.5 to 9.5 (which yields a score of 3.5) identifies this criterion as the most unfavorable hygienically relevant criterion. Thus, the 'total class' score of 3 determines the class of use for the lake as 3, which suggests complicated treatment is required to produce 'acceptable' drinking water from this waterbody (see Table A.7). This value, in fact, describes the situation in regard to Lake McIlwaine (J.A. Thornton, 1982; McKendrick, 1982).

An example of this approach applied to Lake McIlwaine is summarized in Table A.8.

In additional to a tabular compilation, four-piece signs can be used to graphically display the class values of the three criteria complexed. The quality class derived from hydrographic, trophic and special criteria, as well as the use class, should be plotted from left to right. An example of this approach is shown in Figure A.6.

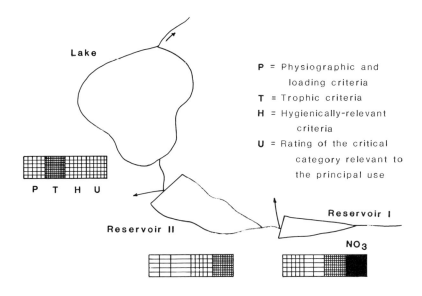

Presentation of Results: For mapping purposes, the results can be expressed on the basis of colors or signatures:

Class	In color	In black and white
1	Blue	
2	Green	
3a	Yellow	
3b	Brown	
4	Red	
5	Black	

Figure A.6 Graphical representation of quality class criteria applied to a water body

ANNEX 2

CASE STUDIES OF EUTROPHICATION CONTROL MEASURES

Lake/reservoir system	Location	Reference
(1). *Aeration/Destratification*:		
– Arbuckle Lake; Ham's Lake	Oklahoma, USA.	Toetz (1977; 1981)
– Farm pond	Oregon, USA	Malueg et al. (1973)
– Fischkaltersee	upper Bavaria, Federal Republic of Germany	Steinberg (1983)
– Ham's Lake	Oklahoma, USA.	Strecker et al. (1977)
– Klopeiner See; Kraiger See; Piburger See	Austria	Pechlaner (1976)
– Larson Lake; Mirror Lake	Wisconsin, USA	S.A. Smith et al. (1975)
– Occoquan Reservoir	Virginia, USA	Eunpu (1973)
– Spruce Run Reservoir	New Jersey, USA	Whipple et al. (1975)
– Various lakes and reservoirs	North America; Europe; Asia	Lorenzen & Fast (1977); Pastorok et al.(1981;1982)
– Worthersee Ossiacher See Millstädter See Weißeusee	Austria Austria	Sampl (1975) Sampl (1975)
(2). *Biomanipulation*:		
– Bautzen Reservoir	German Democratic Republic	Benndorf et al. (1984a)
– Farm ponds	Nebraska, USA	Hergenrader (1983)
– Lake Trummen;	Sweden	Andersson et al. (1978)

	Lake/reservoir system	Location	Reference
(2).	Biomanipulation (*continued*)		
	Lake Bysjön		
	– Lilla Stockelidsvatten	southwest Sweden	Henrikson *et al.* (1980)
	– Various lakes and reservoirs	Argentina; Guyana; India; Poland; Sudan; Sweden; USA; USSR; Zimbabwe	Schuytema (1977)
(3).	*Covering bottom sediments*:		
	– Cox Hollow Lake	Wisconsin, USA	Engel (1984)
	– Marion Millpond	Wisconsin, USA	Engel & Nichols (1984)
	– Several lakes and reservoirs	United States; Canada	Cooke (1980a)
(4).	*Dilution/Flushing*:		
	– Green Lake; Moses Lake	Washington, USA	E.B. Welch (1981a; 1981b)
	– Snake Lake	Wisconsin, USA	Born *et al.* (1973)
(5).	*Harvesting of macrophytes*:		
	– Laguna Lake	Philippine Islands	T.C. Rey (personal communication, Laguna Lake Development Authority, Manila)
	– Lake Sallie	Minnesota, USA	Peterson *et al.* (1974)
	– Several lakes and reservoirs	Michigan, Minnesota, Wisconsin, USA	Nichols (1974)
(6).	*Hypolimnetic injection of nutrient effluents*:		
	– Precambrian lake	Experimental Lakes Area, Canada	Schindler *et al.* (1980)
(7).	*Lake drawdown*:		
	– Lake Apopka	Florida, USA	Fox *et al.* (1977)
	– Various lakes and reservoirs	United States	Cooke (1980b)
(8).	*Nutrient inactivation*:		
	– Beerenplaat	The Netherlands	Harelaar & Rook (1978)
	– East Twin Lake; West Twin Lake	Ohio, USA	Cooke & Kennedy (1978); Cooke *et al.* (1978; 1980); Kennedy & Cooke (1982)

	Lake/reservoir system	Location	Reference
(8).	*Nutrient inactivation:* (continued)		
	– Horseshoe Lake	Wisconsin, USA	Peterson et al. (1973)
	– Medical Lake	Washington, USA	Soltero et al. (1981)
	– Stone Lake	Michigan, USA	Theis & DePinto (1976)
	– Various lakes	Europe; Australia; North America	Cooke & Kennedy (1981)
	– Lake Jabel	Mecklenburg, German Democratic Republic	Uhlmann & Klapper (1985)
	– Suesser See	Halle District, German Democratic Republic	Uhlmann & Klapper (1985)
	– Talsperre Haltern	Federal Republic of Germany	Kotter & Patsch (1980)
	– Tegeler See	Federal Republic of Germany	Hasselbarth (1979)
	– Wahnbachtalsperre	Federal Republic of Germany	Bernhardt et al. (1985)
(9).	*Sediment removal (Dredging):*		
	– Beverinsee	Berlin, German Democratic Republic	Uhlmann & Klapper (1985)
	– Lilly Lake	Wisconsin, USA	Dunst et al. (1984)
	– Lilly Lake; Lake Herman; Lake Trummen; Spring ponds Steinmetz Lake	Wisconsin, USA; South Dakota, USA; Sweden; Wisconsin, USA; New York, USA	Peterson (1981, 1982)
	– Lake Stubenberg	Austria	Stundl (1978)
	– Lake Trehörningen	Sweden	Ryding (1982)
	– Lake Trummen	Sweden	Björk (1972)
	– 64 lakes; Lake Trummen	USA; Sweden;	Peterson (1981)
(10).	*Phosphorus removal at rivermouth (Pre-reservoirs):*		
	– Wahnbach Reservoir; various pre-dams	Federal Republic of Germany	Bernhardt (1981, 1983)
	– 25 pre-dams	German Democratic Republic	Benndorf & Pütz (1987); Uhlmann & Klapper (1985)

Lake/reservoir system	Location	Reference
(11). *Wastewater diversion/Seepage trenches:*		
– Lake Fuschl	Austria	Haslauer et al. (1984)
– Lake Gjersjoen	Norway	Faafeng & Nilssen (1981)
– Kerspetalsperre	Federal Republic of Germany	Gräu (1977)
– lower Madison lakes	Wisconsin, USA	Sonzogni & Lee (1974a); Sonzogni et al. (1975)
– Mauensee	Switzerland	Gächter (1976)
– Lake McIlwaine	Zimbabwe	J.A. Thornton & Nduku (1982)
– Lake Minnetonka	Minnesota, USA	V.H. Smith & Shapiro (1981)
– Lake Norrviken	Sweden	Ahlgren (1978)
– Lake Ossiacher	Austria	Fricker (1980a)
– Lake Øyesjön	Sweden	Cullen & Forsberg (1988)
– Lake Sammamish	Washington, USA	E.B.Welch (1977); E.B.Welch et al. (1980; 1986)
– Schliersee; Tegernsee	Federal Republic of Germany	Hamm (1976)
– Stechlinsee	Potsdam District, German Democratic Republic	Uhlmann & Klapper (1985)
– Lake Washington	Washington, USA	Edmondson (1972); Edmondson & Lehman (1981)
– Lake Waubesa	Wisconsin, USA	Sonzogni et al. (1975)
– Lake Wegonsa	Wisconsin, USA	Sonzogni et al. (1975)
– Various lakes	Austria	Sampl et al. (1982)
– Lake Vesijarn	Finland	Keto (1982)
– Worthersee; Ossiacher See; Millstadter See; Weißensee	Austria	Sampl (1975)
– 4 lakes	Sweden	Ryding (1981b)
(12). *Wastewater treatment for phosphorus removal (including phosphate detergent restrictions:*		
– Lake Åsvalltjärn	Sweden	Holmgren (1985)
– Lake Boren; Lake Ekoln	Sweden	Forsberg et al. (1978)
– Lake Burrinjuck	Australia	NCDC (1985)
– Lake Constance	Federal Republic of Germany	Fricker (1980a)
– Finger Lakes	New York, USA	Troutman et al. (1982); Oglesby & Schaffner

Lake/reservoir system	Location	Reference
		(1978)
– Görväln Bay	Sweden	Ericsson et al. (1984)
– Gravenhurst Bay	Canada	Michalski et al. (1975); Dillon et al. (1978)
– Greifensee	Federal Republic of Germany	Stabel (1984)
– Haley Pond	USA	Bailey et al. (1979)
– Kootenay Lake	Canada	Parker (1976)
– Little Otter Lake	Canada	Michalski & Conroy (1973)
– Lake Mjøsa	Norway	Holtan (1979, 1980); Baalsrud (1982)
– Lake Ringsjön	Sweden	Ryding (1984)
– Lower St. Regis	New York, USA	Fuhs et al. (1977)
– Saginaw Bay	Michigan, USA	Bierman et al. (1984)
– Shagawa Lake	Minnesota, USA	Larsen et al. (1979); Larsen & Malueg (1980); Kibby & Hernandez (1976)
– Stockholm Archipelago	Sweden	Forsberg (1985)
– Lake Vättern	Sweden	Olsen & Willen (1980)
– Walensee	Switzerland	Zimmerman (1984)
– Zurichsee	Switzerland	Fricker (1980a); Schanz & Thomas (1981); Zimmerman (1984)
– Various lakes	Sweden	Ryding (1981b); Ryding & Forsberg (1976); Forsberg et al. (1975)

(13). *Multiple control measures*:

– Lake Balaton	Austria; Hungary	Somlyódy (1983); Somlyódy & van Straten (1986); Duckstein et al. (1982)
– North American Great Lakes	USA; Canada	PLUARG (1978a); Johnson et al., (1978); Sonzogni et al. (1980)

LITERATURE CITED

Ackerfors, H. & Enell, H. (1989). Environmental impact by Swedish fish farming with special regard to nutrient loads. *Ambio.* In Press.

ADCP (Agriculture Development and Coordination Programmes). 1979. *Aquaculture development in China.* Technical Report 79/10, ADCP, Fishery and Agricultural Organization (FAO), Rome, Italy. 65 p.

Ahlgren, I. 1978. Response of Lake Norrviken to reduced nutrient loading. *Verh. Internat. Verein. Limnol.* **20**:846–50.

Allen, G.H. & Hepher, B. 1978. Recycling of wastes through agriculture and constraints to wider application. In: Pillay, T.V. & Dill, W. (eds.), *Advances in Aquaculture*, FAO Technical Conference on Aquaculture, 1976, Farnam/Fishery News Book Press.

Almazan, G. & Boyd, C.E. 1978. Plankton production and *tilapia* yield in ponds. *Aquaculture* **15**: 75–7.

American Public Health Association, American Water Works Association and Water Pollution Control Federation. 1985. *Standard Methods for the Examination of Water and Wastewater*, 16th Edition. American Public Health Association, Washington, D.C.

Andersson, G., Berggen, H., Cronberg, G. & Gelin, G. 1978. Effects of planktivorous and benthivorous fish on organisms and chemistry in eutrophic lakes. *Hydrobiol.* **59**: 9–15.

Appler, H.N. & Jauncey, K. 1983. The utilization of a filamentous green alga (*Cladophora glomerata*) as a protein source for *Sarotherodon* (Tilapia) *niloticus* fingerlings. *Aquaculture* **30** : 21–30.

Ashton, P.J. 1981. Nitrogen fixation and the nitrogen budget of a eutrophic impoundment. *Water Res.* **15**: 823–33.

Ayles, G.B. & Barica, J. 1977. An empirical method for predicting trout survival in Canadian prairie lakes. *Aquaculture* **12**: 181–85.

Azov, Y., Shelef, G., Moraine, G. & Levy, A. 1980. Controlling algal genera in high rate wastewater oxidation ponds. In: Shelef, G. & Soeder, C. (eds.), *Algae Biomass: Production and Use*, Elsevier Press, Amsterdam, The Netherlands. p. 245–53.

Baalsrud, K. 1982. The rehabilitation of Norway's largest lake. *Wat. Sci. Tech.* **14**: 21–30.

Bailey, J.H., Scott, M., Courtemanch, D. & Dennis, J. 1979. Response of Haley Pond, Maine, to changes in effluent load. *Jour. Water Pollut. Cont. Fed.* **51**:728–34.

Baker, A.L. & Brook, A.J. 1971. Optical density profiles as an aid to the study of microstratified phytoplankton populations in lakes. *Arch. Hydrobiol.* **69**: 214–33.

Baker, L.A., Brezonik, P.L. & Kratzer, C.R. 1985. Nutrient loading models for Florida lakes. In: *Lake and Reservoir Management – Practical Applications*, Proceedings, Fourth Annual Conference and International Symposium, held at McAfee, New Jersey, USA. October 16–19, 1984 (North American Lake Management Society, Merrifield, Virginia, USA.). p. 253–58.

Balon, E.K. & Coche, A.G. 1974. *Lake Kariba: A Man-made Tropical Ecosystem in Central Africa.* Monographiae Biologicae, Volume 24, Junk Publishers, The Hague, The Netherlands. 767 p.

Bannink, B.A., van der Meulen, J.H.M. & Peeters, J.C.H. 1980. Hydrobiological consequences of the addition of phosphate precipitants to inlet water of lakes. *Hydrobiol. Bull.* **14**: 73–89.

Barica, J., Kling, H. & Gibson, J. 1980. Experimental manipulation of algal bloom composition by nitrogen addition. *Can. Jour. Fish. Aquat. Sci.* **37**: 1175–83.

Barnes, R.K. & Mann, K.H. (eds.). 1980. *Fundamentals of Aquatic Ecosystems*. Blackwell Scientific Publications, Oxford, United Kingdom.

Barthelmes, D. 1981. *Hydrobiologische Grundlagen der Biennenfischerei* (In German: Hydrobiological Fundamentals of Inland Fisheries). VEB Gustav Fischer Verlag, Jena, German Democratic Republic and G. Fischer, Stuttgart, Federal Republic of Germany. 252 p.

Barthelmes, D. 1983. High productivity as a principal management goal. Fisheries management of ponds, lakes and rivers. In: *UNEP/Unesco international postgraduate training course in ecological approaches to resource development, land management and impact assessment*, Volume VI/B, Berlin, German Democratic Republic. (Division of Ecological Sciences, Unesco, Paris.) p. 4–44.

Barthelmes, D. & Kliebs, K. 1978. Wirkungen von Silberkarpfen auf das Plankton in Flachgewassern nach Untersuchungen in Karpfenteichen. *Int. Revue ges. Hydrobiol.* 63: 411–19.

Baumol, W.J. & Oates, W.E. 1975. *The Theory of Environment Policy*. Prentice-Hall Publishers, Englewood Cliffs, New Jersey, USA. 272 p.

Beck, M.B. 1981. Hard or soft environmental systems? *Ecol. Modelling* 11: 233–51.

Beck, M.B. 1982. Identifying models of environmental system's behaviour. *Math. Modelling* 3: 467–80.

Beck, M.D. & van Straten, G. (eds). 1983. *Uncertainty and Forecasting of Water Quality*. Springer-Verlag, Berlin, Federal Republic of Germany.

Belsare, D.K. 1984. Fish production in tropical waters as protein and food source in developing countries. In: *UNEP/Unesco international postgraduate training course in ecological approaches to resource development, land management and impact assessment*, Volume VI/B, Berlin, German Democratic Republic. (Division of Ecological Sciences, Unesco, Paris.) p. 81–162.

Benemann, J.R. 1981. Energy from fresh and brackish water aquatic plants. In: Klass, D.L. (ed.), *Biomass as a Non-fossil Fuel Source*, Symposium Series No. 144, American Chemical Society, Washington, D.C. p. 99–121.

Benndorf, J. 1979. A contribution to the phosphorus loading concept. *Int. Revue ges. Hydrobiol.* 64: 177–88.

Benndorf, J. & Pütz, K. 1987. Control of eutrophication of lakes and reservoirs by means of pre–dams. *Water Res.* 21: 829–38.

Benndorf, J. & Recknagel, F. 1982. Problems of application of the ecological model SALMO to lakes and reservoirs having various trophic states. *Ecol. Modelling* 17: 129–45.

Benndorf, J., Recknagel, F. & Schultz, H. 1984a. Food–web manipulation for a man-made lake (Bautzen Reservoir). 1. The management problem and decision-making. In: *UNEP/Unesco international postgraduate training course in ecological approaches to resources development, land management and impact assessment in developing countries*, Volume VI/3, Dresden, German Democratic Republic (Division of Ecological Sciences, Unesco, Paris). p. 20–41.

Benndorf, J., Uhlmann, D. & Putz, K. 1981. Strategies for water quality management in reservoirs in the German Democratic Republic. *WHO Water Quality Bulletin* 6: 68–73 (World Health Organization, Canada Centre for Inland Waters, Burlington, Ontario, Canada.)

Benndorf, J., Kneschke, H., Kossatz, K. & Penz, E. 1984b. Manipulation of the pelagic food web by stocking with predacious fishes. *Int. Revue ges. Hydrobiol.* 69: 407–28.

Benndorf, J., Pütz, K., Krinitz, H. & Henke, W. 1975. Die Function der Vorsperren zum schutz der Talsperren vor Eutrophierung. *Wasserwirtschaft–Wassertechnik* 25: 19–25.

Bernhardt, H. 1978. Die hypolimnische Belüftung der Wahnbachtalsperre. *GWF* 119: 177–82.

Bernhardt, H. 1981. Reducing nutrient inflows. In: Rast, W. & J.J. Kerekes (compilers), Proceedings, *International Workshop on the Control of Eutrophication*, International Institute for Applied Systems Analysis (IIASA), A–2361 Laxenburg, Austria, October 12–15, 1981. p. 43–51.

Bernhardt, H. 1983. Input control of nutrients by chemical and biological methods. *Water Supply* 1: 187–206.

Bernhardt, H. & Wilhelms, A. 1984. Phosphateliminierung durch Aluminumsalz–Fallung in kleinen Fließgewässern. *Vom Wasser* 63: 299–323.

Bernhardt, H., Clasen, J., Hoyer, O. & Wilhelms, A. 1985. Oligotrophierung stehender Gewasser durch chemische Nahrstoft – Eliminierung aus den Zuflussen am Beispiel der Wahnbachtalsperre. *Arch. Hydrobiol. Suppl.* 70: 481–533.

Beveridge, M.C.M. 1984. *Cage and pen fish farming: Carrying capacity and environmental impacts*. Fisheries Technical Paper No. 255, Fishery and Agricultural Organization (FAO),

Rome, Italy. 131 p.

Bierman, V.J. 1976. Mathematical model of selective enhancement of blue green algae by nutrient enrichment. In: Canale, R.P. (ed.), *Modeling Biochemical Processes in Aquatic Ecosystems*, Ann Arbor Science Publishers, Ann Arbor, Michigan, USA. p. 1–31.

Bierman, V.J., Verhoff, F.H., Poulson, T.L. & Tenney, M.W. 1973. Multi–nutrient dynamic models of algal growth and species competition in eutrophic lakes. In: Middlebrooks, E.J., D.H. Falkenburg and T.E. Maloney, (eds.), *Modeling the Eutrophication Process*, Ann Arbor Science Publishers, Ann Arbor, Michigan, USA. p. 89–109.

Bierman, V.J., Dolan, D.M. & Kaspryzyk, R. 1984. Retrospective analysis of the response of Saginaw Bay, Lake Huron, to reductions in phosphorus loadings. *Environ. Sci. Technol.* **18**:23–31.

Biswas, A.K. 1976. Mathematical modeling and water-resources decision-making. In: Biswas, A.K. (ed.), *Systems Approach to Water Management*, McGraw–Hill, Inc., New York. p. 398–414.

Björk, S. 1972. Swedish lake restoration programme gets results. *Ambio* **1**: 153–65

Bogardi, I., David, L. & Duckstein, L. 1983. *Trade–off between cost and effectiveness of control of nutrient loading into a waterbody*. Report RR-83-19, International Institute for Applied Systems Analysis (IIASA), A-2361 Laxenburg, Austria. 36 p.

Born, S.M., Wirth, T.L., Peterson, J.O., Wall, J.P. & Stephenson, D.A. 1973. *Dilutional pumping at Snake Lake, Wisconsin*. Technical Bulletin No. 66, Wisconsin Department of Natural Resources, Madison, Wisconsin, USA. 32 p.

Boström, S., Jansson, M. & Forsberg, C. 1982. Phosphorus release from lake sediments. *Arch. Hydrobiol. Beih. Ergebn. Limnol.* **18**: 5–39.

Boyd, C.E. 1982. *Water Quality Management for Pond Fish Culture*. Elsevier Press, Amsterdam, The Netherlands. 318 p.

Breck, J.E., Prentki, R.T. & Loucks, O.L. (eds.). 1979. *Aquatic Plants, Lake Management and Ecosystem Consequences of Lake Harvesting*. Conference Proceedings, Center for Biotic Systems, University of Wisconsin, Madison, Wisconsin, USA., February 14–16, 1979.

Breck, J.E. 1981. Relationship between primary production of waters and fish yield (In Russian). In: Winberg, G.G. (ed.), *The Basis of Freshwater Ecosystem Investigation*, Zoological Institute, Academy of Sciences, Leningrad, USSR.

Brezonik, P.L. 1969. Eutrophication: The process and its modeling potential. In: *Modeling the Eutrophication Process*, Workshop Proceedings, St. Petersburg, Florida, November 19–21, 1969 (Department of Environmental Engineering, University of Florida, Gainesville, Florida and Federal Water Quality Administration, U.S. Department of Interior, Washington, D.C.). p. 68–110.

Brylinski, M. & Mann, K.H. 1973. An analysis of factors governing productivity in lakes and reservoirs. *Limnol. Oceanogr.* **18**: 1–14.

Bulon, V.V. & Winberg, G.G. 1981. Relationship between primary production of waters and fish yield (In Russian). In: Winberg, G.G. (ed.), *The Basis of Freshwater Ecosystem Investigation*, Zoological Institute, Academy of Sciences, Leningrad, USSR.

Cale, W.G. & McKown, M.P. 1986. A cost analysis technique for research management and design. *Environ. Management* **10**: 89–96.

Canale, R.P. 1976. *Modeling Biochemical Processes in Aquatic Ecosystems*. Ann Arbor Science Publishers, Ann Arbor, Michigan, USA. 389 p.

Canfield, D.E. & Bachmann, R.W. 1981. Prediction of total phosphorus concentrations, chlorophyll *a* and Secchi depths in natural and artificial lakes. *Can. Jour. Fish. Aquat. Sci.* **38**: 414–23.

Canfield, D.E. & Jones, J.R. 1984. Assessing the trophic status of lakes with aquatic macrophytes. In: *Lake and Reservoir Management*, Report No. EPA-440/5-84-001, U.S. Environmental Protection Agency, Office of Water Regulations and Standards, Washington, D.C. p. 446–51.

Canfield, D.E., Langeland, K.A., Linda, S.B. & Haller, W.T. 1985. Relations between water transparency and maximum depth of macrophyte colonization in lakes. *Jour. Aquat. Plant Management.* **23**: 25–28.

Canfield, D.C., Langeland, K.A., Maceina, M.J., Haller, W.T., Shireman, J.V. & Jones, J.R. 1983. Trophic state classification of lakes with aquatic macrophytes. *Can. Jour. Fish Aquatic Sci.* **40**: 1713–18.

Canfield, D.E., Shireman, J.V., Coole, D.E., Haller, W.T., Watkins, C.E. & Maceina, M.J.

1984. Prediction of chlorophyll *a* concentrations in Florida lakes: Importance of aquatic macrophytes. *Can. Jour. Fish. Aquat. Sci.* **41**: 497–501.

Carignan, R. 1981. An empirical model to estimate the relative importance of roots in phosphorus uptake by aquatic macrophytes. *Can. Jour. Fish. Aquat. Sci.* **39**: 243–47.

Carignan, R. & Kalff, J. 1980. Phosphorus sources for aquatic weeds: Water or sediments? *Science* **207**: 987–89

Carlson, R.E. 1977. A trophic state index for lakes. *Limnol. Oceanogr.* **23**: 361–69.

Chapra, S.C. & Reckhow, K.H. 1979. Expressing the phosphorus loading concept in probabilistic terms. *Jour. Fish. Res. Bd. Can.* **36**: 225–29.

Chapra, S.C. & Reckhow, K.H. 1983. *Engineering Approaches for Lake Management. Vol. 2. Mechanistic Modeling.* Ann Arbor Science, Butterworth Publishing Company, Woburn, Massachusetts, USA. 492 p.

Chapra, S.C. & Robertson, A. 1977. Great lakes eutrophication: The effect of point source control of total phosphorus. *Science* **196**: 1448–50.

Chiaudani, G. & Vighi, M. 1982. Multistep approach to identification of limiting nutrients in northern Adriatic eutrophied coastal waters. *Water Res.* **16**: 1161–66.

Chiaudani, G. & Virglis, M. 1974. The N:P ratio and tests with *Selenastrum* to predict eutrophication in lakes. *Water Res.* **8**: 1063–69.

Clasen. J. & Bernhardt, H. 1980. *OECD Eutrophication Programme. Shallow Lakes and Reservoir Project.* Final Report. Water Research Centre, Medmenham Laboratory, Medmenham, Marlow, Bucks SL7 2HD, United Kingdom. 289 p.

Coche, A.G. 1982. *Aquaculture in fresh waters: A list of reference books 1953–1981.* Fisheries Circular No. 724, Fishery and Agricultural Organization (FAO), Rome, Italy. 19 p.

Cochran, W.G. 1963 *Sampling Techniques,* 2nd Edition. Wiley & Sons, Inc., New York. 413 p.

Conn, W.D. 1985. *Review of techniques for weighing, or valuing in benefits associated with environmental quality improvements.* Technical Report, Social and Economic Considerations Committee, Great Lakes Science Advisory Board, International Joint Commission, Great Lakes Regional Office, Windsor, Ontario, Canada. 25 p.

Cooke, G.D. 1980a. Covering bottom sediments as a lake restoration technique. *Water Resour. Bull.* **16**: 921–26.

Cooke, G.D. 1980b. Lake level drawdown as a macrophyte control technique. *Water Resour. Bull.* **16**: 317–22.

Cooke. G.D. & Kennedy, R.H. 1978. Effects of hypolimnetic application of aluminum sulfate to a eutrophic lake. *Verh. Internat. Verein. Limnol.* **20**: 486–9.

Cooke. G.D. & Kennedy, R.H. 1981. *Precipitation and inactivation of phosphorus as a lake restoration technique.* Report No. EPA600/3-81-012, U.S. Environmental Protection Agency, Environmental Research Laboratory, Corvallis, Oregon, USA. 41 p.

Cooke, G.D., Heath, R.T., Kennedy, R.H. & McComas, M.R. 1978. *Effects of diversion and alum application on two eutrophic lakes.* Report No. EPA–600/3–78–033, U.S. Environmental Protection Agency, Environmental Research Laboratory, Corvallis, Oregon, USA. 102 p.

Cooke. G.D., R.T. Heath, R.H. Kennedy, & McComas, M.R. 1980. Change in lake trophic state and internal phosphorus release after aluminum application. *Water Resour. Bull.* **18**: 699–705.

Cooke, G.D., Welch, E.G., Peterson, S.A. & Newroth, P.R. 1986. *Lake and Reservoir Restoration.* Butterworth Publishing Company, Stoneham, Massachusetts, USA. 392 p.

Cooper, S., Smith, D.W., & Bence, J.R. 1985. Prey selection by freshwater predators with different foraging strategies. *Can. Jour. Fish. Aquat. Sci.* **42**: 1720–32.

Coote, D.R., MacDonald, E.M. & De Haan, R. 1979. Relationships between agricultural land and water quality. In: Loehr, R.C., Haith, D.A., Walter, M.F. & Martin, C.S. (eds.), *Best Management Practices for Agriculture and Silviculture,* Ann Arbor Science Publishers, Ann Arbor, Michigan, USA. p. 79–92.

Cowen, W.F. & Lee, G.F. 1976. *Algal nutrient availability and limitation in Lake Ontario during IFYGL Part I.* Report No. EPA–600/3–76–094a, U.S. Environmental Protection Agency, Environmental Research Laboratory, Duluth, Minnesota, USA.

Cowen, W.F., Sirisinha, K. & Lee, G.F. 1978. Nitrogen and phosphorus in Lake Ontario tributary waters. *Water, Air & Soil Pollut.* **10**:343–50.

Cullen, P. & Forsberg, C. 1988. Experiences with reducing point sources of phosphorus to

lakes. *Hydrobiol.* **170**: 321–36.
Davies, B.R. & Walmsley, R.D. 1985. *Perspectives in Southern Hemisphere Limnology.* Developments in Hydrobiology, Volume 28 (reprinted from Hydrobiol., Volume 125, 245 p.), Junk Publishers, The Hague, The Netherlands. 263 p.
DeBruijn, P.J., Lijklema, L. & van Straten, G. 1980. Transport of heat and nutrients across a thermocline. *Prog. Water Tech.* **12**: 751–65.
Deevey, E.S. 1940. Limnological studies in Connecticut. V. A contribution to regional limnology. *Amer. Jour. Sci.* **238**: 717–41.
DePinto, J.V., Edzwald, J.K., Switzenbaum, M.S. & Young, T.C. 1980. *Phosphorus removal in lower Great Lakes municipal treatment plants.* Report No. EPA–600/2–80–117, U.S. Environmental Protection Agency, Municipal Environmental Research Laboratory, Cincinnati, Ohio, USA. 147 p.
DeRooij, N.M. 1980. A chemical model to describe nutrient dynamics in lakes. In: Barica, J. & L.R. Murs (eds.), *Hypertrophic Ecosystems*, Developments in Hydrobiology, Volume 2, Junk Publishers, The Hague, The Netherlands p. 139–49.
DiGiano, F.A., Lijklema, L. & van Straten, G. 1978. Wind induced dispersion and algal growth in shallow lakes. *Ecol. Modelling* **4**: 237–52.
Dillon, P.J. & Kirchner, W.B. 1975. The effects of geology and land use on the export of phosphorus from watersheds. *Water Res.* **9**: 135–48.
Dillon, P.J. & Rigler, F.H. 1974. The chlorophyll–phosphorus relationship in lakes. *Limnol. Oceanogr.* **19**: 767–73.
Dillon, P.J. & Rigler, F.H. 1975. A simple method for predicting the capacity of a lake for development based on lake trophic status. *Jour. Fish. Res. Bd. Can.* **32**: 1519–31.
Dillon, P.J., Nicholls, K.H. & Robinson, G.W. 1978. Phosphorus removal at Gravenhurst Bay, Ontario: An 8-year study on water quality changes. *Verh. Internat. Verein. Limnol.* **20**: 263–71.
Dinges, R. 1973. *Ecology of Daphnia in stabilization ponds.* Technical Report, Texas State Department of Health, Division of Wastewater Technology and Surveillance, Austin, Texas, USA. 155 p.
DiToro, D.M. 1976. Combining chemical equilibrium and phytoplankton models – A general methodology. In: Canale, R.P. (ed.), *Modeling Biochemical Processes in Aquatic Ecosystems*, Ann Arbor Science Publishers, Ann Arbor, Michigan, USA. p. 233–55.
DiToro, D.M. 1980. Applicability of cellular equilibrium and Monod theory to phytoplankton growth kinetics. *Ecol. Modelling* **8**: 201–18.
Downing, J.A. & Anderson, M.R. 1985. Estimating the standing biomass of aquatic macrophytes. *Can. Jour. Fish. Aquat. Sci.* **42**: 1860–69.
Droop, M.R. 1973. Some thoughts on nutrient limitation in algae. *Jour. Phycol.* **9**: 264–77.
Duarte, C.M. & Kalff, J. 1986. Littoral slope as a predictor of the maximum biomass of submerged macrophyte communities. *Limnol. Oceanogr.* **31**: 1072–80.
Duckstein, L., Bogardi, I. & David, L. 1982. Dual objective control of nutrient loading into a lake. *Water Resour. Bull.* **18**: 21–26.
Dugdale, R.C. & Dugdale, V.A. 1961. Sources of phosphorus and nitrogen for lakes on Afognak Island. *Limnol. Oceanogr.* **6**: 13–23.
Duncan, N. & Rzóska, J. 1978. *Land Use Impacts on Lake and Reservoir Ecosystems.* Workshop Proceedings, Programme on Man and the Biosphere (Unesco, Paris), Activity 5 (MAB–5), Warsaw, Poland, May 26–June 2, 1978. 294 p.
Dunseth, R. 1977. *Polyculture of channel catfish, silver carp, and three all male tilapias.* Ph.D. Thesis, Auburn University, Auburn, Alabama, USA.
Dunst, R.C., Born, S.M., Hormork, P.D.U., Smith, S.A., Nichols, S.A., Peterson, J.O., Knauer, S.R., Sens, S.L., Winter, D.R. & Wirth, T.L. 1974. *Survey of lake rehabilitation techniques and experiences.* Technical Bulletin No. 75, Wisconsin Department of Natural Resources, Madison, Wisconsin, USA. 179 p.
Dunst, R.C., Vennie, J.G., Corey, R.B. & Peterson, A.E. 1984. *Effect of dredging Lilly Lake, Wisconsin.* Project Summary, Report No. EPA–600/S3–84–097, U.S. Environmental Protection Agency, Environmental Research Laboratory, Corvallis, Oregon, USA. 3 p.
Eberly, W.R. 1959. The metalimnetic oxygen maximum in Myers Lake. *Invest. Indiana Lakes Streams* **5**: 1–46.
Edmondson, W.T. 1970. Phosphorus, nitrogen and algae in Lake Washington after diversion

of sewage. *Science* **169**: 690–91.
Edmondson, W.T. 1972. Nutrients and phytoplankton in Lake Washington. In: Likens, G.E. (ed.), *Nutrients and Eutrophication: The Limiting Nutrient Controversy*, Special Symposium, Volume I, *Limnol. Oceanogr.* p.172–93.
Edmondson, W.T. 1985. *Recovery of Lake Washington from eutrophication.* In: Proceedings, International Congress on Lakes Pollution and Recovery, European Water Pollution Control Association, Rome, Italy, April 15–18, 1985. p. 228–34.
Edmondson, W.T. & Lehman, J.H. 1981. The effect of changes in the nutrient income on the condition of Lake Washington. *Limnol. Oceanogr.* **26**: 1–29.
Edwards, P. 1980. *Food Potential of Aquatic Macrophytes.* Studies and Reviews No. 5, International Center for Living Aquatic Resources Management, Manila, The Philippines. 51 p.
Effler, S.W., Field, S.D., Meyer, M.A. & Sze, P. 1981. Response of Onandaga Lake to restorative efforts. *Amer. Soc. Civ. Engr. Env. Engr. Div.*, **107**: 191–210.
Ekedahl, G., Röndell, B. & Wilson, A.L. 1982. Analytical errors. In: Suess, M.S. (ed.), *Examination of Water for Pollution Control. A Reference Handbook*, Volume 1. Pergamon Press, Oxford, United Kingdom. p. 266–315.
Enell, M. 1984. *Environmental impact of cage fish farming.* Nordisk Jordbrugsforstares Forening (NJF), NJF Utredning Report No. 38.
Engel, S. 1984. Evaluating stationary blankets and removable screens for macrophyte control in lakes. *Jour. Aquat. Plant Management* **22**: 43–8.
Engel, S. & Nichols, S.A. 1984. Lake sediment alteration for macrophyte control. *Jour. Aquat. Plant Management* **22**: 38–41.
Ericsson, P., Hajdu, S. & Willen, E. 1984. Vattenkvaliteten Görväln, en dynamisk Mälarfjärd. Vattenkemi och växplankton i ett fyrtioårigt perspektiv. *Vatten* **40**:193–211.
Eunpu, F.F. 1973. Control of reservoir eutrophication. *Jour. Amer. Water Works Assoc.* **65**: 268–74.
Faafeng, B.A. & Nilssen, J.P. 1981. A twenty-year study of eutrophication in a deep softwater lake. *Verh. Internat. Verein. Limnol.* **21**: 412–24
FAO (Fishery and Agricultural Organization). 1979. *Handbook of utilization of aquatic plants.* FAO Fisheries Technical Paper No. 187, FAO, Rome, Italy. 176 p.
FAO (Fishery and Agricultural Organization). 1983. *Freshwater aquaculture development in China.* FAO Fisheries Technical Paper No. 215, FAO, Rome, Italy. 124 p.
Fedra, K., Beck, M.B. and van Straten, G. 1981. Uncertainty and arbitrariness in ecosystems modelling: A lake modelling example. *Ecol. Modelling* **13**: 87–110.
Fee, E.J. 1976. The vertical and seasonal distribution of chlorophyll in lakes of the Experimental Lakes Area, northwestern Ontario: Implications for primary productivity. *Limnol. Oceanogr.* **21**: 767–83.
Fleming, R.H. 1940. *The composition of plankton and units for reporting population and production.* Proceedings. 6th Pacific Scientific Congress California **9**: 535–40.
Flett, R.J., Schindler, D.W., Hamilton, R.D. & Campbell, N.E. 1980. Nitrogen fixation in Canadian Precambrian Shield lakes. *Can. Jour. Fish. Aquat. Sci.* **37**: 494–505.
Forsberg, C. 1985. *Lake recovery in Sweden.* In: Proceedings, International Congress on Lakes Recovery, European Water Pollution Control Association, Italy, April 15–18, 1985. p. 352–81
Forsberg, C. & Ryding, S.-O. 1980. Eutrophication parameters and trophic state indices in 30 Swedish waste–receiving lakes. *Arch. Hydrobiol.* **89**: 189–207.
Forsberg, C. & Ryding, S.-O. 1981. Swedish experience of nutrient removal from wastewater. In: *International Symposium on Inland Waters and Lake Restoration*, Report No. EPA–440/5–81–010, U.S. Environmental Protection Agency, Office of Water Regulations and Standards, Washington, D.C. p. 298–303.
Forsberg, C., Ryding, S.-O. & Claesson, A. 1975. Recovery of polluted lakes. A Swedish research programme on the effects of advanced waste water treatment and sewage diversion. *Water Res.* **9**: 51–9.
Forsberg, C., Ryding, S.-O. & Claesson, A. 1978. Improved water quality in Lake Boren and Lake Eköln after nutrient reduction. *Verh. Internat. Verein. Limnol.* **20**: 825–32.
Fox, J.L., Brezonik, P.L. & Keirn, M.A. 1977. *Lake drawdown as a method of improving water quality.* Report No. EPA–600/3–77–005, U.S. Environmental Protection Agency,

Environmental Research Laboratory, Corvallis, Oregon, USA. 94 p.
Fricker, H. 1980a. *OECD Eutrophication Programme. Regional Project*. Alpine Lakes. Swiss Federal Board for Environmental Protection (Bundesamt für Umweltschutz), Bern, Switzerland. 234 p.
Fricker, H. 1980b. Methods of assessing nutrient loading. In: *Restoration of Lakes and Inland Waters*, Report No. EPA-440/5-81-010, U.S. Environmental Protection Agency, Office of Water Regulations and Standards, Washington, D.C. p. 56–60.
Frisk, T. 1981. New modifications of phosphorus models. *Aqua Fennica* 11: 7–17.
Fruh, G.E., Stewart, K.M., Lee, G.F. & Rohlich, G.A. 1966. Measurement of eutrophication and trends. *Jour. Water Pollut. Cont. Fed.* 38: 1237–58.
Fuhs, G.W., Demmerle, S.D., Canelli, E. & Chen, M. 1972. Characteristics of phosphorus limited plankton algae (with reflections on the limiting–nutrient concept). In: Likens, G.E. (ed.), *Nutrients and Eutrophication: The Limiting Nutrient Controversy*, Special Symposium, Volume I, Limnol. Oceanogr. p. 113–33.
Fuhs, G.W., Allen, S.P., Hetling, L.J. & Tofflemire, T.J. 1977. *Restoration of Lower St. Regis Lake (Franklin County, New York)*. Ecological Research Series, Report No. EPA-600/3-77-021, U.S. Environmental Protection Agency, Environmental Research Laboratory, Corvallis, Oregon, USA. 107 p.
Gächter, R. 1976. Die Tiefenwasserableitung ein Weg zur Sanierung von Seen. *Schweiz. Zeitschr. Hydrol.* 38: 1–28.
Gaudet, J.J. & Denny, P. 1981. *Ecology of Aquatic and Wetland Vegetation in Africa*. Junk Publishers, The Hague, The Netherlands.
Gerloff, G.C. & Skoog, F. 1954. Cell contents of nitrogen and phosphorus as a measure of their availability for growth of Microcystis aeruginosa. *Ecology* 35: 348–53.
Goldman, J.C. 1979. Outdoor algal mass cultures. I. Applications. *Water Res.* 13: 1–19.
Golterman, H.L. (ed.). 1971. *Methods for Chemical Analysis of Fresh Waters*. International Biological Programme (IBP) Handbook Series, No. 8, Blackwell Scientific Publications, Oxford, United Kingdom. 188 p.
Golterman, H.L. 1973. Natural phosphate sources in relation to phosphate budgets: A contribution to the understanding of eutrophication. *Water Res.* 7: 3–17.
Golterman, H.L. 1975. *Physiological Limnology. An Approach to the Physiology of Lake Ecosystems*. Elsevier Scientific Publishing Co., New York. p. 366–402.
Golterman, H.L., Sly, P.G. & Thomas, R.I. 1983. *Study of the relationship between water quality and sediment transport. A guide for the collection and interpretation of sediment water quality data*. Technical Paper in Hydrology No. 26, Unesco, Paris. 231 p.
Gonzalez, H.B. 1984. Scientists and Congress. *Science* 224: 127–9.
Gordon, M.S., Chapman, D.J., Kawasaki, L.Y., Tarifino–Silva, E. & Yu, D.P. 1982. Aquacultural approaches to recycling of dissolved nutrients in secondarily treated domestic wastewaters. IV: Conclusions, design and operational considerations for artificial food chains. *Water Res.* 16: 67–71.
Gräu, A. 1977. Untersuchungen zur Ermittlung der Leistungsfahigkeit einer Hangversickerung als spezielles Verfahren einer Bodenpassage. *Gewasserschutz Wasser–Abwasser*, Volume 24.
Gupta, A.B. & Roy, D.C. 1975. Utilisation of algae by some common fishes of two ponds. *Nova Hedwigia* 26: 509–16.
Hagedorn. H. 1981. Hemmung des Algenwachstums durch Lichtreduktion. Neue Technologien in der Trinkwasserversorgung *DVGS–Schriftenreihe Wasser* 102: 411–13.
Haith, D.A. 1982. Development and testing of watershed loading functions for nonpoint sources. In: Vogt, W.G. & Mickle, M.H. (eds.), *Modeling and Simulation*, Volume 13, School of Engineering, University of Pittsburg, Pittsburgh, USA. p. 1463–67.
Haith, D.A. & Tubbs, L.J. 1981. Watershed loading functions for nonpoint sources. *Jour. Amer. Soc. Civ. Engr., Env. Engr. Div.* 107: 121–37.
Haith, D.A., Tubbs, L.J. & Pickering, N.B. 1984. *Simulation of Pollution by Soil Erosion and Soil Nutrient Loss*. Center for Agricultural Publishing and Documentation (Pudoc), Wageningen, The Netherlands.
Hamm, A. 1976. Untersuchungen zur Nährstoffbilanz am Tegernsee und Schliersee nach der Abwasserfernhaltung – zugleich ein Beitrag uber die diffusen Nährstoffquellen im Einzugsgebiet bayerischer Alpen-und Voralpenseen. *Zeitschr. f. Wasser- und Abwasser-Forsch.* 9: 110–49.

Hanson, J.M. & Leggett, W.C. 1982. Empirical prediction of fish biomass and yield. *Can Jour. Fish. Aquat. Sci.* **39**: 257–63.
Harelaar, A. & Rook, J.J. 1978. Algenbekämpfung im Beerenplaat-Speicherbecken durch Zugatz von Eisen II – Sulfat. *BVGW-Schriftenreihe Wasser, Volume 16*.
Hart, R.C. & Allanson, B.R. 1984. *Limnological criteria for management of water quality in the southern hemisphere.* Report No. 93, South African National Scientific Programmes, Foundation for Research and Development, Pretoria, Republic of South Africa 181 p.
Haslauer, J., Moog, D. & Horvath, F.J. 1982. The effect of sewage removal on lake water quality. *Jour. Water Pollut. Cont. Fed* **54**: 193–97.
Hasler, A.D. 1947. Eutrophication of lakes by domestic drainage. *Ecology* **28**: 383–95.
Hasselbarth, U. 1979. Phosphoreliminierung aus den Zuflussen berliner Seen. *Zeitschr. f. Wasser- und Abwasser-Forsch.* **12**: 133–47.
Healey, F.P. 1973. Inorganic nutrient uptake and deficiency in algae. *CRC Crit. Rev. Microbiol.* **3**: 69–113.
Healey, F.P. & Hendzel, L.L. 1975. Effect of phosphorus deficiency on two algae growing in chemostats. *Jour. Phycol.* **11**: 303–09.
Healey, F.P. & Hendzel, L.L. 1976. Physiological changes during the course of blooms of *Aphanizomenon flos-aquae. Jour. Fish. Res. Bd. Can.* **33**: 36–41.
Healey, F.P. & Hendzel, L.L. 1980. Physiological indicators of nutrient deficiency in lake phytoplankton. *Can. Jour. Fish. Aquat. Sci.* **37**: 442–53.
Hecky, R.E., Fee, E.J., Kling, H.J. & Rudd, J.W.M. 1981. Relationship between primary production and fish production in Lake Tanganyika. *Trans. Amer. Fish. Soc.* **110**: 336–45.
Henderson, D.W. 1974. *Social indicators: A rationale and research framework.* Technical Report, Economic Council of Canada, Ottawa, Ontario, Canada. 90 p.
Henderson, S. 1983. *An evaluation of filter feeding fishes for removing excessive nutrients and algae from wastewater.* Report No. EPA–600/S2–83–019, U.S. Environmental Protection Agency, Environmental Research Laboratory, Corvallis, Oregon, USA.
Hendry, C.D., Brezonik, P.L. & Edgerton, E.S. 1981. Atmospheric deposition of nitrogen and phosphorus in Florida. In: Eisenreich, S.J. (ed.), *Atmospheric Pollutants in Natural Waters*, Ann Arbor Science Publishers, Ann Arbor, Michigan, USA. p. 199–215.
Henrikson, L., Nyman, H.G., Oscarson, H.G. & Stenson, J.A.E. 1980. Trophic changes, without changes in the external nutrient loading. *Hydrobiologia* **68**: 257–63.
Henry, R. & Tundisi, J.G. 1982. Evidence of limitation by molybdenum and nitrogen on the growth of the phytoplankton community of the Lobo Reservoir (Sao Paulo, Brasil). *Rev. Hydrobiol. Tropicale* **15**: 201–08.
Henry, R., Hino, K., Tundisi, J.G. & Ribiero, J.S.B. 1985. Responses of phytoplankton in Lake Jacaretinga to enrichment with nitrogen and phosphorus in concentrations similar to those of the River Solimoes (Amazon, Brazil). *Arch. Hydrobiol.* **103**: 453–77
Hergenrader, G.L. 1983. *Enhancement of water quality in Nebraska farm ponds by control of eutrophication through biomanipulation.* Completion Report, Project No. A–027–NEB, Water Resources Center, University of Nebraska, Lincoln, Nebraska, USA. 30 p.
Heyman, U., Ryding, S.-O. & Forsberg, C. 1984. Frequency distributions of water quality variables. Relationships between mean and maximum values. *Water Res.* **18**: 787–94.
Hill, G. & Rai, H. 1982. A preliminary characterization of the tropical lakes of the Central Amazon by comparison with polar and temperate systems. *Arch. Hydrobiol.* **96**: 97–111.
Holmgren, S. 1985. Phytoplankton in a polluted subarctic lake before and after nutrient reduction. *Water Res.* **19**: 63–71.
Holtan, H. 1979. The Lake Mjøsa story. *Arch. Hydrobiol. Beih. Ergebn. Limnol.* **13**: 242–58.
Holtan, H. 1980. The case of Lake Mjøsa. *Prog. Wat. Tech.* **12**: 103–120.
Hornberger, G.M. & Spear, R.C. 1980. Eutrophication in Peel Inlet. I. Problem-defining behaviour and a mathematical model for the phosphorus scenario. *Water Res.* **14**: 29–42.
Horne, A.J. & Viner, A.B. 1971. Nitrogen fixation and its significance in tropical Lake George, Uganda. *Nature.* **232**: 417–18.
Hrbaček, J. 1966. A morphological study of some backwaters and fish ponds in relation to the representative plankton sample. *Hydrobiol. Studies* I: 221–66.
Hrbaček, J. 1969. Relationships between some environmental parameters and the fish yield as a basis for a predictive model. *Verh. Internat. Verein. Limnol.* **17**: 1069–84.
Huber, W.C. & Heaney, J.P. 1980. Operational models for stormwater quality management.

Literature cited

In: Overcash, M.R. & Davidson, J.M. (eds.), *Environmental Impact of Nonpoint Source Pollution*, Ann Arbor Science Publishers, Ann Arbor, Michigan, USA. p. 397–444.
Hudson, N. 1971. *Soil Conservation.* Cornell University Press, Ithaca, New York, USA.
Hutchinson, G.E. 1957. *A Treatise on Limnology.* John Wiley & Sons, Inc., New York. Volume 1.
Hutchinson, G.E. 1961. The paradox of the plankton. *Amer. Natur.* **95**: 137–45.
Ichimura, S., Nagasawa, S. & Tanaka, T. 1968. On the oxygen and chlorophyll maximum found in the metalimnion of a mesotrophic lake. *Bot. Mag. Tokyo* **81**: 1–10.
Imboden, D.M. 1982. Modellvorstellungen uber den Phosphor–Kreislauf in stehenden Gewässern. *Zeitschr. f. Wasser- und Abwasser-Forsch.* **15**: 89–95.
Imboden, D.M. & R. Gächter. 1978. A dynamic lake model for trophic state prediction. *Ecol. Modelling* **4**: 77–98.
Jacoby, J.M., Lynch, D.D., Welch, E.B. & Perkins, M.A. 1982. Internal phosphorus loading in a shallow, eutrophic lake. *Water Res.* **16**: 911–19.
James, D.H. & Lee, G.F. 1974. A model of inorganic carbon limitation in natural waters. *Water, Air & Soil Pollut.* **3**: 315–20.
Janus, L.L. & Vollenweider, R.A. 1981. *The OECD Cooperative Programme on Eutrophication. Canadian contribution.* Scientific Series No. 131, Canada Centre for Inland Waters, Burlington, Ontario, Canada. 371 p.
Jarvis, M.J.F. 1982. Avifauna of Lake McIlwaine. In: Thornton, J.A. (ed.), *Lake McIlwaine, the Eutrophication and Recovery of a Tropical Man-Made Lake.* Monographiae Biologicae, Volume 49, Junk Publishers, The Hague, The Netherlands. p. 188–94.
Jayangoudar, I.S. & Ganapati, S.V. 1965. Algae of importance and their control in water supplies. *Hydrobiol.* **22**: 317–30.
Jayaraman, K.S. 1981. From rank weeds to riches. *Nature* **291**: 183.
Jenkins, R.M. 1967. *The influence of some environmental factors on standing crop and harvest of fishes in U.S. reservoirs.* In: Proceedings, Reservoir Fishery Resources Symposium, Southern Division, American Fisheries Society. p. 298–321.
Jenkins, R.M. & Morais, D.J. 1971. Reservoir sport fishing effort and harvest in relation to environmental variables. In: Hall, G.E. (ed.), *Reservoir Fisheries and Limnology*, Special Publication No. 8, American Fisheries Society, Washington, D.C.. p. 371–84.
Johnson, M.G. & Berg, N. 1979. A framework for nonpoint pollution in the Great Lakes Basin. *Jour. Soil and Water Conserv.* **34**: 68–73.
Johnson, M.G., Conneau, J.C., Heidtke, T.M., Sonzogni, W.C. and Stahlbaum, B.W. 1978. *Management information base and overview modelling.* Report prepared for Pollution From Land Use Activities Reference Group (PLUARG), International Joint Commission, Great Lakes Regional Office, Windsor, Ontario, Canada. 90 p.
Jolánkai, G. 1984. An approach to solve the missing link problem of non-point source modelling. In: Jolankai, G. & Roberts, G. (eds.), *Land Use Impacts on Aquatic Systems*, Workshop Proceedings, Programme on Man and the Biosphere (Unesco, Paris), Activity 5 (MAB-5), Budapest, Hungary, October 10–14, 1983. p. 271–83.
Jones, J.R. 1977. Chemical characteristics of some Missouri reservoirs. *Trans. Mo. Acad. Sci.* **10/11**:58–71.
Jones, J.R. & Bachmann, R.W. 1976. Prediction of phosphorus and chlorophyll levels in lakes. *Jour. Water Pollut. Cont. Fed.* **48**: 2176–82.
Jones, J.R. & Bachmann, R.W. 1978. Trophic status of Iowa lakes in relation to origin and glacial geology. *Hydrobiol.* **57**: 267–73.
Jones, J.R. & Hoyer, M.V. 1982. Sportfish harvest predicted by summer chlorophyll a concentration in midwestern lakes and reservoirs. *Trans. Amer. Fish. Soc.* **111**: 176–79.
Jones, J.R. & Novak, J.T. 1981. Limnological characteristics of Lake of the Ozarks, Missouri. *Verh. Int. Ver. Limnol.* **21**: 919–25.
Jones, J.R., Borofka, B.P. & Bachmann, R.W. 1976. Factors affecting nutrient loads in some Iowa streams. *Water Res.* **10**: 117–22.
Jones, R.A. & Lee, G.F. 1979. Phosphorus removal at the Dansbury, Connecticut sewage treatment plant on water quality in Lake Lillinonah. *Water, Air & Soil Pollut.* **16**: 511–31.
Jones, R.A. & Lee, G.F. 1982. Recent advances in assessing impact of phosphorus loads on eutrophication–related water quality. *Water Res.* **16**: 503–15.
Jones, R.A. & Lee, G.F. 1984. Application of OECD eutrophication modelling approach to

South African dams (reservoirs). *Water SA* **10**: 109–14.

Jones, R.A., Rast, W. & Lee, G.F. 1979. Relationship between mean and maximum chlorophyll *a* concentrations in lakes. *Environ. Sci. Technol.* **13**: 869–70.

Joó, O. 1986. Role of the Zala River in the eutrophication of Lake Balaton. In: Somlyódy, L. & van Straten, G. (eds.), *Modeling and Managing Shallow Lake Eutrophication, with Application to Lake Balaton*, Springer-Verlag, New York. p. 341–65.

Kalbe, L. 1976. Restoration of highly eutrophic shallow lakes and possible uses of lake sediments. In: EUTROSYM '76, Proceedings, International Symposium, *Eutrophication and Rehabilitation of Surface Waters*, Volume 5, Karl-Marx-Stadt, Berlin, German Democratic Republic, September 20–25, 1976. p. 59–71.

Kalčeva, R., Outrara, J.V., Schindler, Z. & Straškraba, M. 1982. An optimization model for the economic control of reservoir eutrophication. *Ecol. Modelling* **17**: 121–28.

Kelly, C.A., Rudd, J.W.M., Cook, R.B. & Schindler, D.W. 1982. The potential importance of bacterial processes in regulating rate of lake acidification. *Limnol. Oceanogr.* **27**: 868–82.

Kennedy, R.H. & Cooke, G.D. 1982. Control of lake phosphorus with aluminum sulfate: Dose determination and application techniques. *Water Resour. Bull.* **18**: 389–95.

Kennedy, R.H., Thornton, K.W. & Ford, D.E. 1985. Characterization of the reservoir system. In: Gunnison, D. (ed.), *Microbial Processes in Reservoirs*, Junk Publishers, Dordrecht, The Netherlands. p. 27–38.

Kerekes, J.J. 1975. Phosphorus supply in undisturbed lakes in Kejimkujik National Park, Nova Scotia (Canada). *Verh. Inernat. Verein. Limnol.* **19**: 349–57.

Kerekes, J.J. 1982. The application of phosphorus load – trophic response relationships to reservoirs. *Cdn. Water Resour. Jour.* **7**: 349–54.

Kerekes, J.J., Freedman, B, Howell, G. & Clifford, P. 1984. Comparison of the characteristics of an acidic eutrophic and an acidic oligotrophic lake near Halifax, N.S. *Water Pollut. Res. Jour. Can.* **19**: 1–10.

Keto, J. 1982. The recovery of Lake Vesijärvi following sewage diversion. *Hydrobiologia* **86**: 195–99.

Kibby, H. & Hernandez, D.J. 1976. *Environmental impacts of advanced wastewater treatment at Ely, Minnesota*. Ecological Research Series, No. EPA–600/3–76–092, U.S. Environmental Protection Agency, Environmental Research Laboratory, Corvallis, Oregon, USA. 30 p.

Kibler, D.F. (ed.). 1982. *Urban Storm Hydrology*. Water Resources Monograph No. 7, American Geophysical Union, Washington, D.C. 271 p.

Kilham, P. 1982a. The effect of hippopotamuses on potassium and phosphate ion concentrations in an African lake. *Amer. Midland Naturalist* **108**: 202–05.

Kilham, P. 1982b. Acid precipitation: Its role in the alkalization of a lake in Michigan. *Limnol. Oceanogr.* **27**: 856–67.

Kilham, S.S. & Kilham, P. 1978. Natural community bioassays: Predictions of results based on nutrient physiology and competition. *Verh. Internat. Verein. Limnol.* **20**: 68–74.

Kimmel, B.L. & Groeger, A.W. 1984. Factors controlling primary production in lakes and reservoirs: A perspective. In: *Lake and Reservoir Management*, Report No. EPA–440/5–84–001, U.S. Environmental Protection Agency, Office of Water Regulations and Standards, Washington, D.C. 20460. p. 277–81.

Kirchner, W.B. & Dillon, P.J. 1975. An empirical method for estimating the retention of phosphorus in lakes. *Water Resour. Res.* **11**: 182–83.

Kitchell, J.F., O'Neill, R.V., Webb, D., Gallepp, G., Bartell, S., Koonce, J.F. & Ausmus, B.S. 1979. Consumer regulation of nutrient cycling. *Bioscience* **29**: 28–34.

Klapper, H. 1980. Experience with lakes and reservoir restoration techniques in the German Democratic Republic. *Hydrobiol.* **72**: 31–41.

Koschel, R., Benndorf, J., Proft, G. & Recknagel, F. 1983. Calcite precipitation as a natural control mechanism of eutrophication. *Arch. Hydrobiol.* **98**: 380–408.

Kotter, K. & Patsch, B. 1980. Verminderung der Algenentwicklung in der Talsperre Haltern durch Phosphatfallung. *GWF* **121**: 496–98.

Krutilla, J.W. & Fisher, A.C. 1975. *The economics of natural resources*. Technical Report, Resources for the Future, Washington, D.C. 292 p.

Kudelska, D., Cydziak, D. & Soszka, H. 1981. Proprzycja systemu oceny jakosci jezior (In Polish: Design of lake quality evaluation systems). *Wiadomosci ekologiczne* **27**: 149–73.

LaBaugh, J.W. 1985. Uncertainty in phosphorus retention, Williams Fork Reservoir, Colorado.

Water Resour. Res. **21**: 1684–92.
LaBaugh, J.W. & Winter, T.C. 1984. The impact of uncertainties in hydrologic measurements on phosphorus budgets and empirical models for two Colorado reservoirs. *Limnol. Oceanogr.* **29**: 322–39.
Lam, D.C.L. & Jaquet, M. 1976. Computation of physical transport and regeneration of phosphorus in Lake Erie. *Jour. Fish. Res. Bd. Can.* **33**: 550–63.
Landner. L. 1976. *Eutrophication of lakes. Causes, effects and means for control, with emphasis on lake rehabilitation.* Report prepared for World Health Organization, Regional Office for Europe, Copenhagen, Denmark. 78 p.
Landner, L. 1986. *Eutrophisation des lacs et reservoirs en climat chaud.* Technical Report, Swedish Environmental Research Group, Götgatan, Stockholm, Sweden.
Lang, D.S. & Brown, E.J. 1981. Phosphorus-limited growth of a green alga and a blue–green alga. *Appl. Environ. Microbiol.* **42**: 1002–09.
Larsen, D.P. & Malueg, K.W. 1980. Whatever became of Shagawa Lake? In: *Restoration of Lakes and Inland Waters*, Report No. EPA–440/5–81–010, U.S. Environmental Protection Agency, Office of Water Regulations and Standards, Washington, D.C. p. 67–72.
Larsen, D.P. & Mercier, H.T. 1976. Phosphorus retention capacity of lakes. *Jour. Fish. Res. Bd. Can.* **33**: 1742–50.
Larsen, D.P., Sickle, J.V., Malueg, K.W. & Smith, P.D. 1979. The effect of wastewater phosphorus removal on Shagawa Lake, Minnesota: Phosphorus supplies, lake phosphorus and chlorophyll *a*. *Water Res.* **13**: 1259–72.
Lean, D.R.S. & Pick, F.R. 1981. Photosynthetic response of lake plankton to nutrient enrichment: A test for nutrient limitation. *Limnol. Oceanogr.* **26**: 1001–19.
Lee, G.F. 1971. Eutrophication. In: *Encyclopedia of Chemical Technology*, Supplemental Volume, John Wiley & Sons, Inc., New York. p. 315–38.
Lee, G.F. & Jones, R.A. 1980. *Study programme for development of information for use of OECD eutrophication modeling in water quality management.* Manuscript, Quality Control in Reservoir Committee, American Water Works Association. 35 p.
Lee, G.F. & Jones, R.A. 1981a. Application of the OECD eutrophication modeling approach to estuaries. In: Neilson, B.J. and Cronin, L.E. (eds.), *Estuaries and Nutrients*, Humana Press, Clifton, New Jersey, USA. p. 549–68.
Lee, G.F. & Jones, R.A. 1981b. *Effect of eutrophication on fisheries.* Manuscript, prepared for American Fisheries Society Washington, D.C.
Lee, G.F., Bentley, E. & Amundson, R. 1975. Effect of marshes on water quality. In: *Coupling of Land and Water Systems*, Ecological Studies 10, Springer-Verlag, New York. p. 105–27.
Lee, G.F., Jones, R.A. & Rast, W. 1980. Availability of phosphorus to phytoplankton and its implications for phosphorus management strategies. In: Loehr, R.C., Martin, C. and Rast, W. (eds.), *Phosphorus Management Strategies for Lakes*, Ann Arbor Science Publishers, Inc., Ann Arbor, Michigan, USA. p. 259–308.
Lee, G.F., Sonzogni, W.C. & Spear, R.D. 1977. Significance of oxic vs. anoxic conditions for Lake Mendota sediment phosphorus release. In: Golterman, H.L. (ed.), *Interactions Between Sediments and Freshwater*, Junk Publishers, The Hague, The Netherlands. p. 294–306
Leidy, G.R. & Jenkins, R.M. 1977. *The development of fishery compartments and population rate coefficients for use in reservoir and ecosystem modeling.* Contract Report No. Y–77–1, U.S. Army Corps of Engineers, Waterways Experiment Station, Vicksburg, Mississippi, USA. p. A1–A8.
Leonardson, L. & Ripl, W. 1980. Control of undesirable algae and induction of algal successions in hypertrophic ecosystems. In: Barica, J. & Mur, L.R. (eds.), *Hypertrophic Ecosystems*, Developments in Hydrobiology, Volume 2, Junk Publishers, The Hague, The Netherlands. p. 57–65.
Leslie, A.J., Nall, L.E. & Van Dyke, J.M. 1983. Effects of vegetation control by grass carp on selected water quality variables in four Florida lakes. *Trans. Amer. Fish. Soc.* **112**: 777–87.
Lester, J.N. & Kirk P.W.W. (eds) (1986). *Management Strategies for Phosphorus in the Environment*. Selter Ltd, London, United Kingdom.
Lewis, W.M. 1983. Temperature, heat and mixing in Lake Valencia, Venezuela. *Limnol. Oceanogr.* **28**: 273–86.
Liang, Y., Melack, J.M. & Wang, J. 1981. Primary production and fish yields in Chinese ponds and lakes. *Trans. Amer. Fish. Soc.* **110**: 346–50.

Likens, G.E. & Bormann, F.H. 1974. Linkages between terrestrial and aquatic ecosystems. *Bioscience* **24**: 447–56.

Likens, G.E., Bormann, F.H., Pierce, R.S., Eaton, J.S. & Johnson, N.M. 1977. *Biogeochemistry of a Forested Ecosystem*. Springer-Verlag, New York. 146 p.

Lindmark, G. 1979. *Phosphorus as a growth-controlling factor for phytoplankton in Lago Paranoa, Brasilia*. Ph.D. Thesis, Part IV, Institute of Limnology, Lund University, Sweden. 49 p.

Lindmark. G. & Shapiro, J. 1982. Effects of environmental stresses on the relationship between *Plectonema boryanum* and Cyhanophage LPP-1. Experiments and experiences in biomanipulation. Interim Report No. 19, Limnological Research Center, University of Minnesota, Minneapolis, Minnesota, USA. p. 129–54.

Little, E.C.S. (ed.). 1968. *Handbook of utilization of aquatic plants*. FAO Fisheries Technical Paper No. 187, FAO, Rome, Italy. 176 p.

Little, I.M.D. 1957. *A Critique of Welfare Economics*. Clarendon Press, Oxford, United Kingdom. 302 p.

Loehr, R.C. 1974. Characteristics and comparative magnitude of non-point sources. *Jour. Water Pollut. Cont. Fed.* **46**:1849–72.

Loehr, R.C., Martin, C.S. & Rast, W. (eds.). 1980. *Phosphorus Management Strategies for Lakes*. Ann Arbor Science, Ann Arbor, Michigan, USA. 490 p.

Löffler, H. 1982. Limnological aspects of shallow lakes. In: Proceedings, SCOPE/UNEP International Scientific Workshop, *Ecosystem Dynamics on Wetlands and Shallow Water Bodies*, Volume 1, Centre of International Projects, GKNT, Moscow, USSR. pp. 37–62.

Lorenzen, M. & Fast, A. 1977. *A guide to aeration/circulation techniques for lake management*. Ecological Research Series, No. EPA-600/3-77-004, US Environmental Protection Agency, Environmental Research Laboratory, Corvallis, Oregon, USA. 126 p.

Lorenzen, M. & Mitchell, R. 1973. Theoretical effects of artificial destratification on algal production in impoundments. *Environ. Sci. Technol.* **7**:939–44.

Loucks, D.P. 1976. Surface-water quantity management models. In Biswas, A.K. (ed.), *Systems Approach to Water Management*, McGraw-Hill, Inc., New York. p. 156–218.

Loucks, D.P., Stedinger J.R. & Haith, D.A. 1981. *Water Resource Systems Planning and Analysis*. Prentice-Hall Publishers, Englewood Cliffs, New Jersey, USA. 559 p.

Lum, L.W.K., Armstrong, N.E. & Fruh, E.G. 1981. Incorporation of chemical equilibrium, physico-chemical reactions and the simulation of inorganic qualities in a reservoir ecological model. *Ecol. Modelling* **14**:95–124.

Lund, J.W.G. 1969. Phytoplankton. In: *Eutrophication: Causes, Consequences, Correctives*, Symposium Proceedings, National Academy of Sciences, Washington, DC. p. 306–30.

Lund, J.W.G.,Jaworski, G.H.M. & Bucka, H. 1971. A technique for bioassays of freshwater, with special reference to algal ecology. *Acta Hydrobiologica* **13**:235–49.

Maceina, M.J., Shireman, J.V., Langeland, K.A. & Canfield, D.E. 1984. Prediction of submersed plant biomass by use of a recording fathometer. *Jour Aquat. Plant. Management* **22**:35–38.

Mackenthun, K.M., Ingram, W.M. & Porges, R. 1964. *Limnological aspects of recreational lakes*. Publication No. 1167. Public Health Service, US Department of Health, Education and Welfare, Washington, DC. 176 p.

Malmqvist, P.A. 1982. *Pollution from urban stormwater*. Technical Report, National Swedish Environment Protection Board, Stockholm, Sweden.

Maloney, T.E. & Miller, W.E. 1975. Algal assays: Development and application. In: Special Technical Publication No. 573, American Society of Testing Materials (ASTM), Philadelphia, Pennsylvania, USA. p. 344–55.

Maloney, T.E., Miller, W.E. & Shiroyama, T. 1972. Algal responses to nutrient additions in natural waters. I. Laboratory assays. In: Likens, G.E. (ed.), *Nutrients and Eutrophication: The Limiting Nutrients Controversy*, Special Symposium, Volume I, *Limnol. Oceanogr.* p. 134–40.

Malueg, K.W., Tilstra, J.R., Schults, D.W. & Powers, C.F. 1973. Effect of induced aeration on stratification and eutrophication processes in an Oregon farm pond. In: Ackermann, W.C., White, G.F. and Worthington, E.B. (eds.), *Man-Made Lakes: Their Problems and Environmental Effects*, Geographical Monograph Series, Volume 17. p. 578–87.

Mancy, K.H. & Allen, H.E. 1982. Design of measurement systems. In: Suess, M.J. (ed.). *Examination of Water for Pollution Control. A Reference Handbook*, Volume 1, Pergamon

Literature cited

Press, Oxford, United Kingdom. p. 1–22.
Matuszek, J.E. 1978. Empirical predictions of fish yields of large North American lakes. *Trans. Amer. Fish. Soc.* **107**:385–94.
McConnell, W.J. 1963. Primary production and fish harvest in a small desert impoundment. *Trans. Amer. Fish. Soc.* **92**:1–12.
McConnell, W.J., Lewis, S & Olson, S.J. 1977. Gross photosynthesis as an estimator of potential fish production. *Trans. Amer. Fish. Soc.* **106**:417–23.
McCoy, G.A. 1983. Nutrient limitation in two arctic lakes, Alaska. *Can. Jour. Fish. Aquat. Sci.* **40**:1195–1202.
McCuen, R.H. 1982. *A Guide to Hydrologic Analysis Using SCS Methods.* Prentice-Hall Publishers, Englewood Cliffs, New Jersey, USA. 145 p.
McKendrick, J. 1982. Water supply and sewage treatment in relation to water quality in Lake McIlwaine. In: Thornton, J.A. (ed.), *Lake McIlwaine, the Eutrophication and Recovery of a Tropical, Man-Made Lake.* Monograhiae Biologicae, Volume 49, Junk Publishers, The Hague, The Netherlands. p. 202–17.
McNabb, C.D. 1976. The potential of submersed vascular plants for reclamation of wastewater in temperate zone ponds. In: Tourbier, J. and Pierson, R.W. (eds.), *Biological Control of Water Pollution,* University of Pennsylvania Press, Philadelphia, Pennsylvania, USA. p. 123–32.
Meadley, G.R.W. 1970. The control of algae. *Jour. Dept. Agric. West Aust.* **11**:252–53.
Melack, J. 1976. Primary production and fish yields in tropical lakes. *Trans. Amer. Fish. Soc.* **105**:575–80.
Melack, J.M., Kilham, P. & Fisher, T.R. 1982. Responses of phytoplankton to experimental fertilization with ammonium and phosphate in an African soda lake. *Oecologia* **52**:321–26.
Michalski, M.F.P. & Conroy, N. 1973. *The 'oligotrophication' of Little Otter Lake, Parry Sound District.* Proceedings, 16th Conference, Great Lakes Research. p. 934–48.
Michalski, M.F.P., Nicholls, K.H. & Johnson, M.G. 1975. Phosphorus removal and water quality improvements in Gravenhurst Bay, Ontario. *Verh. Internat. Verein. Limnol.* **19**: 644–59.
Miller, W.E., Greene, J.C. & Shiroyama, T. 1978. *The Selenastrum capricornutum Printz algal assay bottle test.* Technical Report No. EPA-600/9-78-018, US Environmental Protection Agency, Environmental Research Laboratory, Corvallis, Oregon, USA. 125 p.
Mitchell, D.S. (ed.). 1974. *Aquatic vegetation and its use and control.* Technical Report, UNESCO, Paris. 135 p.
Mitchell, D.S. & Marshall, B.E. 1974. Hydrobiological observations on three Rhodesian reservoirs. *Freshwater Biol.* **4**:6 61–72.
Mitra, E. & Banerjee, A.C. 1976. Utilisation of higher aquatic plants in fishery waters. In: Varshney, C.K. and Rzóska, J.R. (eds.), *Aquatic Weeds in Southeast Asia,* Junk Publishers, The Hague, The Netherlands. p. 375–81.
MMBW (Melbourne and Metropolitan Board of Works). 1976. *Werribee Farm.* Technical Report, Melbourne and Metrolpolitan Board of Works, Melbourne, Austarlia. 20 p.
Moll, R.A., Brahce, M.Z. & Peterson, T.P. 1984. Phytoplankton dynamics within the subsurface chlorophyll maximum of Lake Michigan. *Jour. Plank. Res.* **6**:751–66.
Monaghan, M. Ltd. 1977. *Evaluation of remedial measures to control nonpoint sources of water pollution in the Great Lakes Basin.* Report prepared by Macklin Monaghan Ltd. for Pollution From Land Use Activities Reference Group (PLUARG), Task Group A, International Joint Commission, Great Lakes Regional Office, Windsor, Ontario, Canada. 159 p.
Monteith, T.J., Sullivan, R.A.C., Heidtke, T.M & Sonzogni, W.C. 1981. *Watershed. A management technique for choosing among point and nonpoint control strategies.* Handbook prepared for Great Lakes National Programmes Office, US Environmental Protection Agency, Chicago, Illinois, USA. 107 p.
Morris, I. (ed.). 1980. *The Physiological Ecology of Phytoplankton.* Studies in Ecology, Volume 7, University of California Press, Berkeley, California, USA. 625 p.
Moss, B. 1969. Limitation of algal growth in some central African lakes. *Limnol. Oceanogr.* **14**:591–601.
Moyle, J.B. 1956. Relationship between the chemistry of Minnesota surface water and wildlife management. *Jour. Wildlife Management* **20**:303–20.
Mueller, D.K. 1982. Mass balance model estimation of phosphorus concentrations in reservoirs.

Water Resour. Bull. **18**:377–82.
Musgrave, G.W. & Holtan, H.N. 1984. Infiltration. In: Chow, V.T. (ed.) *Handbook of Applied Hydrology*, McGraw-Hill Inc., New York. Chapter 12.
Nalbandov, J.P., Sapoznikov, V.V. & Cernjakova, A.M. 1979. Rascet biochimiceskogo potreblenija kisloroda na meridial' nom razreze v Indijskom okeane. In: Sbornik, *Chimiko-Okeanograficeskie issledovanija morej i okeanov*, Nauka, Moscow, USSR. p. 46–52.
Nash, C.E. & Brown, C.M. 1979. A theoretical comparison of waste treatment processing ponds and fish production ponds receiving animals wastes. In: Roger, S.V., Pullin, S. and Shehadeh, Z.H. (eds.), *Integrated Agricultural Farming Systems*, Proceedings, ICLARM-SEARCA Conference on Integrated Agriculture — Aquaculture Systems. p 87–97.
National Academy of Sciences. 1969. *Eutrophication: Causes, Consequences, Correctives*, Symposium Proceedings, National Academy of Sciences, Washington, D.C. 661 p.
Naumann, E. 1919. Några synpunkter angående limnoplanktons ekologi med särskild hänsyn till fytoplankton. *Svensk Botanisk Tidskrift* **13**:129–63.
NCDC (National Capital Development Commission). 1985. *Data from water quality monitoring programme*. National Capital Development Commission, Canberra, Australia.
Nichols, S.A. 1974. *Mechanical and habitat manipulation for aquatic plant management*. Technical Bulletin No. 77, Wisconsin Department of Natural Resources, Madison, Wisconsin, USA. 34 p.
NIWR (National Institute for Water Research). 1985. *The limnology of Hartbeespoort Dam*. Report. No. 110, South African National Scientific Programmes, Foundation for Research and Development, Pretoria, Repulic of South Africa. 269 p.
Novotny, V. & Chesters, G. 1981. *Handbook of Nonpoint Pollution*. Van Nostrand Reinhold Publishers, New York. 555 p.
Nürnberg, G.K. 1984. The prediction of internal phosphorus load in lakes with anoxic hypolimnia. *Limnol. Oceanogr.* **29**:111–24.
Odum, E.P. 1971. *Fundamentals of Ecology*, 3rd Edition. W.B. Saunders Co., Philadelphia, Pennsylvania, USA. p. 106–39.
OECD (Organization for Economic Cooperation and Development). 1974. *Environmental damage costs*. Technical Report, OECD, Paris. 332 p.
OECD (Organization for Economic Coopertion and Development). 1982. *Eutrophication of Waters.. Monitoring, Assessment and Control*. Final Report. OECD Cooperative Programme on Monitoring of Inland Waters (Eutrophication Control), Environment Directorate, OECD, Paris. 154 p.
Oglesby, R.T. 1977. Relationships of fish yield to lake phytoplankton standing crop, production and morphoedaphic factors. *Jour. Fish. Res. Bd. Can.* **34**:2271–79.
Oglesby, R.T. & Schaffner, W.R. 1978. Phosphorus loadings to lakes and some of their responses. Part 2. Regression models of summer phytoplankton standing crops, winter total P and transparency of New York lakes with known phosphorus loadings. *Limnol. Oceanogr.* **23**:135–45.
Ogrosky, H.O. & Mockus, V. 1964. Hydrology of agricultural lands. In: V.T. Chow (ed.), *Handbook of Applied Hydrology*, McGraw-Hill Publishers, New York. Chapter 21.
Olsen, P. & Willén, E. 1980. Phytoplankton response to sewage reduction in Vättern, a large oligotrophic lake in central Sweden. *Arch. Hydrobiol.* **89**:171–88.
Olszewski, P. 1973. 15 Jahre Experiment auf dem Kortowosee. *Verh. Internat. Verein. Limnol* **18**:1792–97.
Omernik, P. 1976. *The influence of land use on stream nutrient levels*. Ecological Research Series, No. EPA-600/3-76-014, US Environmental Protection Agency, Environmental Research Laboratory, Corvallis, Oregon, USA. 106 p.
Omernik, J.M. 1977. *Nonpoint sources-stream nutrient level relationships*. Ecological Research Series, No. EPA-600/3-77-105, US Environmental Protection Agency, Environmental Research Laboratory, Corvallis, Oregon, USA. 151 p.
Opuszynski, K. 1979. Silver carp in carp ponds. IV. Influence on ecosystem. *Ekol. Polsk.* **27**, 117–33.
Oskam, G. 1978. Light and zooplankton as algae regulating factors in eutrophic Biesbosch reservoirs. *Verh. Internat. Verein. Limnol.* **20**:1612–18.
Oswald, W.J. 1976. *Removal of algae in natural bodies of water*. Report No. EPA-600/3-76-056, US Environmental Protection Agency, Environmental Research Laboratory, Corvallis,

Oregon, USA.
Oswald, W.J. & Golueke, C.G. 1968. Harvesting and processing of waste-grown microalgae. In: Jackson, D. (ed.), *Algae, Man and the Environment*, Syracuse University Press, Syracuse, New York, USA. p. 371–89.
Paerl, H.W. & Ustach, J.F. 1982. Blue-green algal scums: An explanation for their occurrence during freshwater blooms. *Limnol. Oceanogr.* **27**:212–17.
Palmer, M.D. 1975. Coastal region residence time estimates from concentration gradients. *Jour. Great Lakes Res.* **1**:130–41.
Pantulu, V.R. 1982. Effects of water resource development on wetlands in the Mekong Basin. In: Proceedings, SCOPE/UNEP International Scientific Workshop, *Ecosystem Dynamics in Wetlands and Shallow Water Bodies*, Volume 1, Centre of International Projects, GKNT, Moscow, USSR. p. 249–60.
Park, R.A. 1979. The aquatic ecosystem model CLEANER. In: Jørgensen, S.E. (ed.), *State-of-the-Art in Ecological Modelling*, Denmark, 1978. p. 579–602.
Park, R.A., O'Neill, R.V., Bloomfield, J.A., Shugart, H.H., Booth, R.S., Goldstein, R.A., Mankin, J.B., Koonce, J.F., Scavia, D., Adams, M.S., Clesceri, L.S., Colon, E.M., Dettman, E.H., Hoopes, J.A., Huff, D.D., Katz, S., Kitchell, J.F., Kohberger, R.C., La Row, E.J., McNaught, D.C., Peterson, J.L., Titus, J.F., Weiler, R.P., Williamson, J.M. & Zahorcak, C.S. 1974. A generalized model for simulating lake ecosystems. *Simulation* **23**:33–50.
Parker, R.A. 1976. *Phosphate reduction and response of plankton populations in Kootenay Lake*. Ecological Research Series, No. EPA-600/3-76-063, US Environmental Protection Agency, Environmental Research Laboratory, Corvallis, Oregon, USA. 63 p.
Pastorok, R.A., Lorenzen, M.W. & Ginn, T.C. 1982. *Artificial aeration and oxygenation of reservoirs: A review of theory, technique and experience.* Technical Report E-82-3, Environmental and Water Quality Operational Studies, US Army Corps of Engineers, Waterways Experiment Station, Vicksburg, Mississippi, USA. 192 p.
Pastorok, R.A., Ginn, T.C. & Lorenzen, M.W. 1981. *Evaluation of aeration/circulation as a lake restoration technique*. Ecological Research Series, No. EPA-600/3-81-014, US. Environmental Protection Agency, Environmental Research Laboratory, Corvallis, Oregon, USA. 58 p.
Patalas, K. 1968. Landschaft und Klima als Faktoren der Messenproducktion von Algen (In German: Landscape and climate as factors of algae mass production). *Fortschr. Wasserchem.* **8**:21–31.
Pechlaner, R. 1976. *Ziele, Wege und Erfolge der See-Restaurierung*. Projekt Life 2000, Tagung Salzburg, Austria.
Peterson, S.A. 1981. *Sediment removal as a lake restoration technique*. Ecological Research Series, No. EPA-600/3-81-013, US Environmental Protection Agency, Environmental Research Laboratory, Corvallis, Oregon, USA. 55 p.
Peterson, S.A. 1982. Lake restoration by sediment removal. *Water Resour. Bull.* **18**:423–35.
Peterson, S.A., Smith, W.L. & Malueg, K.W. 1974. Full-scale harvest of aquatic plants: Nutrient removal from a eutrophic lake. *Jour. Water Pollut. Cont. Fed.* **46**:697–707.
Peterson, S.A., Wall, J.P., Wirth, T.L. & Born, S.M. 1973. *Eutrophication control: Nutrient inactivation by chemical precipitation at Horseshoe Lake, Wisconsin*. Technical Bulletin No. 62, Wisconsin Department of Natural Resources, Madison, Wisconsin, USA. 19 p.
Phosphorus Management Strategies Task Force. 1980. *Phosphorus management for the Great Lakes*. Final Report to the International Joint Commission, Great Lakes Regional Office, Windsor, Ontario, Canada. 125 p.
Pick, F.R., Nalewajko, C. & Lean, D.R.S. 1984. The origin of a metalimnetic chrysophyte peak. *Limnol. Oceanogr.* **29**:125–34.
Pineau, M., Villeneuve, J.P. and Campbell, P.G.C. 1985. Cost of alternative phosphorus control strategies for stream systems. *Environ. Technol. Letters* **6**:231–36.
Platt, T. (ed.). 1981. *Physiological Bases of Phytoplankton Ecology*. Bulletin 210, Canadian Bulletin Fisheries and Aquatic Science. 346 p.
PLUARG (Pollution From Land Use Activities Reference Group). 1987a. *Environmental management strategy for the Great Lakes ecosystem*. Final Report, Pollution From Land Use Activities Reference Group to the International Joint Commission, Great Lakes Regional Office, Windsor, Ontario, Canada. 115 p.
PLUARG. 1978b. *Reports of the United States Public Consultation Panels to the Pollution*

From Land Use Activities Reference Group. PLUARG, International Joint Commission, Great Lakes Regional Office, Windsor, Ontario, Canada. 148 p.
PLUARG. 1078c. Reports of the Canadian Public Consultation Panels to the Pollution From Land Use Activities Reference Group. PLUARG, International Joint Commission, Great Lakes Regional Office, Windsor, Ontario, Canada. 86 p.
Pokorny, J. & Rejmankova, E. 1983. Oxygen regime in fish ponds with duckweeds (*Lemnaeae*) and *Ceratophyllum*. *Aquat. Bot.* **17**:125–37.
Potter, J.S. 1971. Pollution in agriculture. *Jour. Agric. S. Aust.* **74**:160–73.
Powers, C.F., Schultz, D.W., Malueg, K.W., Brice, R.M. & Schuldt, M.D. 1972. Algal responses to nutrient additions in natural waters. II. Field experiments. In: Likens, G.E. (ed.). *Nutrients and Eutrophication: The Limiting Nutrients Controversy*, Special Symposium, Volume I, Limnol. Oceanogr. p. 141–56.
Prochazkova, L. 1975. Nitrogen and phosphorus budgets. Slapy Reservoir. In: Hasler, A. (ed.). *Coupling of Land Water Systems, Ecol. Studies* **10**:65–73.
Provasoli, L. 1969. Algal nutrition and eutrophication. In: *Eutrophication: Causes, Consequences and Correctives*, Symposium Proceedings, National Academy of Sciences, Washington, DC. p. 574–93.
Pütz, K. Benndorf, H., Glasebach, H. & Kummer, G. 1983. Die Massenentwicklung der Geißelalge *Synura uvella* in den Trinkwassertalsperren Klingenberg und Lehrmuhle — ihre Auswirkungen auf die Trinkwasserversorgung und ihre Bekampfung. *Wasserwirtschaft — Wassertechnik* **33**:135–39.
Rainey, R.H. 1967. Natural displacement of pollution from the Great Lakes. *Science* **155**: 1242–43.
Rast, W. 1981a *Practical Lake Management for Eutrophication Control*. Technical Manual prepared for Unesco/OECD/IIASA International Workshop on Control of Eutrophication, International Institute for Applied Systems Analysis, Laxenburg, Austria, October 12–15, 1981. 175 p.
Rast, W. 1981b. Quantification of the input of pollutants from land use activities in the Great Lakes Basin. In: Steenvoorden, J.H.A.M. and Rast, W. (eds.). *Impact of Non-Point Sources on Water Quality in Watersheds and Lakes*, Workshop Proceedings, Programme on Man and the Biosphere (Unesco, Paris), Activity 5 (MAB-5), Amsterdam, The Netherlands, May 11–14, 1981. pp. 98–141.
Rast, W. & Kerekes, J.J. (compilers). 1982. *International Workshop on the Control of Eutrophication*. Proceedings, Unesco/OECD/IIASA Workshop on Control of Eutrophication, International Institute for Applied Systems Analysis, A-2361 Laxenburg, Austria, October 12–15, 1981. 107 p.
Rast, W. & Lee, G.F. 1978. *Summary analysis of the North American (US portion) OECD Eutrophication Project: Nutrient loading-lake response relationships and trophic state indices*. Ecological Research Series, No. EPA-600/3-78-008, US Environmental Protection Agency, Environmental Research Laboratory, Corvallis, Oregon, USA. 454 p.
Rast, W. & Lee, G.F. 1983. Nutrient loading estimates for lakes. *Amer. Soc. Civil. Engr., Env. Engr. Div.*, **109**:502–05.
Rast, W., Jones, R.A. & Lee, G.F. 1983. Predictive capability of US. OECD phosphorus loading — eutrophication response models. *Jour. Water Pollut. Cont. Fed.* **55**:990–1003.
Rawson, D.S. 1939. Some physical and chemical factors in the metabolism of lakes. *Amer. Assoc. Adv. Sci.* **10**:9–26.
Rawson, D.S. 1955. Morphometry as a dominant factor in the productivity of large lakes. *Verh. Internat. Verein. Limnol.* **12**:164–75.
Rawson, D.S. 1960. A limnological comparison of twelve large lakes in northern Saskatchewan. *Limnol. Oceanogr.* **5**: 195–211.
Reckhow, K.H. 1978. Lake quality discriminant analysis. *Water Resour. Bull.* **14**:856–67.
Reckhow, K.H. 1979. Uncertainty analysis applied to Vollenweider's phosphorus loading criteria. *Jour. Water Pollut. Cont. Fed.* **51**:2123–28.
Reckhow, K.H. 1983. A method for the reduction of lake model prediction error. *Water Res.* **17**:911–16.
Reckhow, K.H. & Simpson, J.T. 1980. A procedure using modeling and error analysis for the prediction of lake phosphorus concentration from land use information. *Can Jour. Fish. Aquat. Sci.* **37**:1439–48.

Literature cited

Reckhow, K.H., Chapra, S.C. 1983. *Engineering Approaches for Lake Management. Vol. I. Data Analysis and Empirical Models.* Ann Arbor Science, Butterworth Publishing Co. Woburn, Massachusetts, USA. 340 p.

Reckhow, K.H., Beaulac, M.N. & Simpson, J.T. 1980. Modeling phosphorus loading and lake response under uncertainty: A manual and compilation of export coefficients. Report No. EPA-400/5-80-011, US Environmental Protection Agency, Office of Water Regulations, Criteria and Standards Division, Washington, DC. 214 p.

Recknagel, F. & Benndorf, J. 1982. Validation of the ecological simulation model SALMO. *Int. Revue ges. Hydrobiol.* **67**:113–25.

Redfield, A.C. 1934. On the proportions of organic derivatives in sea water and their relation to the composition of plankton. James Johnstone Memorial Volume, Liverpool, United Kingdom. p. 177–92.

Redfield, A.C., Ketchum, B.H. & Richards, F.A. 1963. The influence of organisms on the composition of sea water. In: Hill, M.N. (ed.). *The Sea*, Volume 2, Wiley-Interscience, New York. p. 26–77.

Reedy, K.R. & Tuckes, J.C. 1983. Productivity and nutrient uptake of water hyacinth (*Eichhornia crassipes*). I. Effect of nitrogen source. *Jour. Econ. Bot.* **37**:237–47.

Rhee, C.-Y. 1978. Effects of N : P atomic ratios and nitrate limitation on algal growth, cell composition, and nitrate uptake. *Limnol. Oceanogr.* **23**:10–25.

Richardson, J.L. 1975. Morphology and lacustrine productivity. *Limnol. Oceanogr.* **20**: 661–67.

Ripl. W. 1976. Biochemical oxidation of polluted lake sediments with nitrate — A new lake restoration method. *Ambio* **5**:132–25.

Robarts, R.D. 1982. Primary production of Lake McIlwaine. In: Thornton, J.A., *Lake McIlwaine, the Eutrophication and Recovery of a Tropical African Man-Made Lake.* Monographiae Biologicae, Volume 49, Junk Publishers, The Hague, The Netherlands. p. 110–17.

Robards, R.D. & Zohary, T. 1984. *Microcystis aeruginosa* and underwater light attenuation in a hypertrophic lake (Hartbeesport Dam, South Africa). *Jour. Ecol.* **72**:1001–17.

Rodhe, R.D. 1949. Die Bekämpfung einer Wasserblüte von *Microcystis* und die gleichzeitige Forderung einer neuen Hockproduktion von *Pediastrum* im See Norrviken bei Stockholm. *Verh. Internat. Verein. Limnol.* **10**:373–86.

Rodhe, W. 1969. Crystallization of Eutrophication concepts in northern Europe. In: *Eutrophication: Causes, Consequences, Correctives*, Symposium Proceedings, National Academy of Sciences, Washington, DC. p. 50–64.

Roger, S.V., Pullin, S. & Shehadeh, Z.H. (eds.) 1979. Proceedings, ICLARM-SEARCA Conference on Integrated Agriculture — Aquaculture Systems.

Rowe, D.B. 1982. Water pollution: Perspectives and control. In: Thornton, J.A., *Lake McIlwaine, the Eutrophication and Recovery of a Tropical African Man-Made Lake.* Monographiae Biologicae, Volume 49, Junk Publishers, The Hague, The Netherlands. p. 195–201.

Rudolf, G. & Uhlmann, D. 1968. Fallmittelzugahe in Gewässer (In German: Addition of precipitants to waterbodies). *Fortschritee Wasserchemie Grenzgebiete* **8**:267–78.

Ryder, R.A. 1982. The morphoedaphic index — use, abuse and fundamental concepts. *Trans. Amer. Fish. Soc.* **111**:154–64.

Ryder, R.A., Kerr, S.R., Loftus, K.H. & Regier, H.A. 1974. The morphoedahic index, a fish yield estimator — review and evaluation. *Jour. Fish. Res. Bd. Can.* **31**:663–68.

Ryding, S.-O. 1980. *Monitoring of Inland Waters: OECD Eutrophication Programme – The Nordic Project.* Publication 1980:2, Nordic Cooperative Organization for Applied Research (NORDFORSK), Helsinki, Finland, 207 p.

Ryding, S.-O. 1981a. Optimization of monitoring programmes by means of general correlations between standard water quality variables *Verh. Internat. Verein. Limnol.* **21**: 791–98.

Ryding, S.-O. 1981b. Reversibility of man-induced eutrophication. Experiences of a lake recovery in Sweden. *Int. Revue ges. Hydrobiol.* **66**:449–503.

Ryding, S.-O. 1982. Lake Trehörningen Restoration Project. Changes in water quality after sediment dredging. *Hydrobiology* **92**:549–58.

Ryding, S.-O. 1983. Mass-balance studies to, within, and from drainage basins. A new approach *for water monitoring programmes* (In Swedish). Publication 1983:1, Nordic Cooperative Organization for Applied Research (NORDFORSK), Helsinki, Finland. p. 149–62.

Ryding, S.-O. 1984. *Lake Ringsjön. The metamorphosis of a Swedish lake ecosystem during one century.* Technical Report, Institute of Limnology, Uppsala University, Uppsala, Sweden.

Ryding, S.-O. 1985. Chemical and microbiological processes as regulators of the exchange of substances between sediments and water in shallow, eutrophic lakes. *Int. Revue. ges. Hydrobiol.* **70**:657–702.

Ryding, S.-O. 1986. Identification and quantification of nonpoint source pollution as a base for effective lake management. In: Workshop Proceedings, *Land Use Impacts on Aquatic Ecosystems*, Programme on Man and Biosphere (Unesco, Paris), Activity 5 (MAB-5), Toulouse, France, April 28–May 2, 1986.

Ryding, S.-O. & Forsberg, C. 1976. Six polluted lakes: A preliminary evaluation of the treatment and recovery process. *Ambio* **5**:151–56.

Ryding, S.-O. & Forsberg, C. 1977. Sediments as a nutrient source in shallow, polluted lakes. In: Golterman, H.L. (ed.), *Interactions Between Sediments and Fresh Waters*, Junk Publishers, The Hague, The Netherlands. p. 227–34.

Ryding, S.-O. & Forsberg, C. 1979. Short-term load-reponse relationships in shallow, polluted lakes. In: Barica, J. & Mur, L.R. (eds.). *Hypertrophic Ecosystems*, Developments in Hydrobiology, Volume 2, Junk Publishers, The Hague, The Netherlands. p. 95–103.

Saffermann, R.S. 1973. Phycoviruses. In: Carr, N.G. & Whitton, B.A. (eds.), *The Biology of Blue-Green Algae*, Blackwell Scientific Publications, Oxford, United Kingdom. p. 214–37.

Sakamoto, M. 1966. Primary production by the phytoplankton community in some Japanese lakes and its dependence on lake depth. *Arch. Hydrobiol.* **62**:1–28.

Salas, H. 1982. *Calculation of sedimentation coefficient, K_s, for eutrophication study of tropical lakes.* Technical Report 299, Organizacion Panamericana de la Salud (Pan American Health Organization), Centro Panamericano de Ingeniería Sanitaria y Ciencias del Ambiente (CEPIS), Lima, Peru. 9 p.

Salas, H. 1983. *Desarrolo de metodologias simplificadas para la evaluacion de euroficacion en lagos calidos, antes lagos tropicales.* Resumen del Segundo Encuentro del Proyecto Regional, Organzacion Panamericana de la Salud (Pan American Health Organization), Centro Panamericano de Ingeniería Sanitaria y Ciencias del Ambiente (CEPIS), Lima, Peru. 33 p.

Sampl, H. 1975. Maßuahmen zur Sanierung und Regenerierung der Kartner Seen. *Natur und Land* **4**: 101–08.

Sampl. H. Gusinde, R.E. and Tomek, H. 1982. *Seenreinhaltung in Österreich* (In German: Lake protection in Austia). Wien, Bundesministerium f. Land u. Forstwirtschaft (Vienna, Austria). 256 p.

Sartor, J.D., Boyd, G.B. & Agardy, F.J. 1974. Water pollution aspects of street surface contaminants. *Jour. Water Pollut. Cont. Fed.* **46**:456–67.

Satterland, D.R. 1972. *Wildland Watershed Management.* Ronald Press, New York, 370 p.

Sawyer, C.N. 1966. Basic concept of eutrophication. *Jour. Water Pollut. Cont. Fed.* **38**:737–44.

Scavia, D. & Robertson, A. 1979. *Perspectives on Lake Ecosystem Modeling.* Ann Arbor Science Publishers, Ann Arbor, Michigan, USA. 326 p.

Schanz, F. & Thomas, E.A. 1981. Reversal of eutrophication in Lake Zurich. *Water Qual. Bull.* **6**:108–12.

Schäperclaus, W. 1979. *Fischkrankheiten* (In German: Diseases of Fish), 4th Edition. Akademie Verlag, Berlin. 1,089 p.

Schindler, D.W. 1971a. A hypothesis to explain differences and similarities among lakes in the Experimental Lakes Area. *Jour. Fish. Res. Bd. Can.* **28**:295–301.

Schindler, D.W. 1971b. Carbon, nitrogen and phosphorus, and the eutrophication of freshwater lakes. *Jour. Phycol.* **7**:321–29.

Schindler, D.W. 1977. Evolution of phosphorus limitation in lakes. *Science* **195**:260–62.

Schindler, D.W. 1985. The coupling of elemental cycles by organisms: Evidence from whole-lake biochemical perturbations. In: Stumm, W. (ed.), *Chemical Processes in Lakes*, John Wiley & Sons, New York. p. 225–50.

Schindler, D.W. & Fee, E.J. 1974. Experimental Lakes Area: Whole-lake experiments in eutrophication. *Jour. Fish. Res. Bd. Can.* **31**:937–53.

Schindler, D.W., Milles, K.W., Malley, D.F., Findlay, D.L., Shearer, J.A., Davies, I.J., Turner,

M.A., Linsey, G.A. & Cruikshank, D.R. 1985. Long-term ecosystem stress: The effects of years of experimental acidification on a small lake. *Science* **228**:1395–1401.

Schindler, D.W., Ruszcynski, T. & Fee, E.J. 1980. Hypolimnion injection of nutrient effluents as a method for reducing eutrophication. *Can. Jour. Fish. Aquat. Sci.* **37**:320–27.

Schlesinger, D.A. & Regier, H.A. 1982., Climatic and morphoedaphic indices of fish yields from natural lakes. *Trans. Amer. Fish. Soc.* **111**:141–50.

Schroeder, G.L. 1975. Some effects of stocking fish in waste treatment ponds. *Water Res.* **9**:591–93.

Schuytema, G.S. 1977. *Biological control of aquatic nuisances — A review.* Ecological Research Series, No. EPA-600/3-77-084, US Environmental Protection Agency, Environmental Research Laboratory, Corvallis, Oregon, USA. 90 p.

Schwab, G.O., Frevert, R.K., Edminster, T.W., & Barnes, K.K. 1981. *Soil and Water Conservation Engineering*, 3rd Edition. John Wiley & Sons, Inc., New York.

Scott, W.E., Barlow, D.J. & Hauman, J.H. 1981. Studies on the ecology, growth and physiology of toxic *Microcystis aeruginosa* in South Africa. In: Carmichael, W.W. (eds.). *The Water Environment, Algal Toxins and Health*, Plenum Press, New York. p. 49–69.

Seidel, K. 1976. Macrophytes and water purification. In: Tourbier, J. and Pierson, R.W. (eds.). *Biological Control of Water Pollution*, University of Pennsylvania Press, Philadelphia, Pennsylvania, USA. p. 109–21.

Serruya, C. 1975. Nitrogen and phosphorus balances and load biomass relationships in Lake Kinneret (Israel). *Verh. Internat. Verein. Limnol.* **19**:1357–69.

Shanahan, P. & Harleman, D.R.F. 1982. *Linked hydrodynamics and biogeochemical models of water quality in shallow lakes.* Report No. 268, Ralph M. Parsons Laboratory, Aquatic Science and Environmental Engineering, Massachusetts Institute of Technology, Cambridge, Massachusetts, USA.

Shanahan, P. & Harleman, D.R.F. 1984. Transport in lake water quality modeling. *Jour. Amer. Soc. Civil Engr., Env. Engr. Div.* **110**:42–57.

Shapiro, J. 1973. Blue-green algae: Why they become dominant. *Science* **179**:382–84.

Shapiro, J. 1979. The need for more biology in lake restoration. In: *National Conference on Lake Restoration*, Report No. EPA-440/5-79-001, US Environmental Protection Agency, Office of Water Planning and Standards, Washington, DC. p. 161–67.

Shapiro, J.V. & Wright, D.I. 1984. Lake restoration by biomanipulation. *Freshwater Biol.* **14**:371–83.

Shapiro, J., Lamarra, V. & Lynch, M. 1975. Biomanipulation: an ecosystem approach to lake restoration. In: Brezonik, P.L. and Fox, J.L. (eds.), *Water Quality Management Through Biological Control*, Symposium Proceedings, University of Florida, Department of Environmental Engineering, Gainesville, Florida, USA.

Sharpley, A.N. & Syers, J.K. 1981. Amounts and relative significance of runoff types in the transport of nitrogen into a stream draining an agricultural watershed. *Water, Air & Soil Pollut.* **15**:299–308.

Shelef, G. & Soeder, C.J. (eds.). 1980. *Algae Biomass: Production and Use.* Elsevier Press, Amsterdam, The Netherlands. 852 p.

Sinden, J.A. & Worrell, A.C. 1979. *Unpriced Values.* John Wiley & Sons, Inc., New York. 511 p.

Skimin, W.E., Powers, E.C. & Jarecki, E.A. 1978. *An evaluation of alternatives and costs for nonpoint source controls in the United States Great Lakes Basin.* Report prepared for Pollution From Land Use Activities Reference Group (PLUARG), Task Group A, International Joint Commission, Great Lakes Regional Office, Windsor, Ontario, Canada. 351 p.

Sly, P.G. 1978. Sedimentary processes in lakes. In: Lerman, A. (ed.). *Lakes: Chemistry, Geology and Physics*, Springer-Verlag, New York. p. 65–84.

Sly, P.G. (ed.). 1982. *Sediment/Freshwater Interaction.* Developments in Hydrobiology, Volume 9 (reprinted from Hydrobiol., Volume 91/92, 1982), Junk Publishers, The Hague, The Netherlands. 700 p.

Smart, M.M., Jones, J.R. & Sebaugh, J.L. 1985. Stream–watershed relations in the Missouri Ozark Plateau Province. *Jour. Environ. Qual.* **14**:77–82.

Smith, D.W. 1985. Biological control of excessive phytoplankton growth and the enhancement of aquacultural production. *Can. Jour. Fish. Aquat. Sci.* **42**:1940–45.

Smith, R.E.H. & Kalff, J. 1981. The effect of phosphorus limitation on algal growth rates: Evidence from alkaline phosphatase. *Can. Jour. Fish. Aquat. Sci.* **38**:1421–27.
Smith, R.E.H. & Kalff, J. 1982. Size-dependent phosphorus uptake kinetics and cell quota in phytoplankton. *Jour. Phycol.* **18**:275–84.
Smith, R.V. & Stewart, D.A. 1977. Statistical models of river loadings of nitrogen and phosphorus in the Lough Neagh system. *Water Res.* **11**:633–36.
Smith, R.V., Stevens, R.J., Foy, R.H. & Gibson, C.E. 1982. Upward trend in nitrate concentrations in rivers discharging into Lough Neagh for the period 1969–79. *Water Res.* **16**:183–88.
Smith, S.A., Knauer, D.R. & Wirth, T.L., 1975. *Aeration as a lake management technique.* Technical Bulletin No. 87, Wisconsin Department of Natural Resources, Madison, Wisconsin, USA. 40 p.
Smith, V.H. 1982. The nitrogen and phosphorus dependency of algal biomass in lakes: An empirical and theoretical analysis. *Limnol. Oceanogr.* **27**:1101–12.
Smith, V.H. 1983. Low nitrogen to phosphorus ratios favor dominance by blue-green algae in lake phytoplankton. *Science* **221**:669–70.
Smith, V.H. 1985. Predictive models for the biomass of blue-green algae in lakes. *Water Resour. Bull.* **21**:433–39.
Smith, V.H. & Shapiro, J. 1981. Chlorophyll–phosphorus relations in individual lakes. Their importance to lake restoration strategies. *Environ. Sci. Technol.* **15**:444–51.
Smith, V.H., Rigler, F.H., Choulik, O., Diamond, M., Griesbach, S. & Skraba, D. 1984. *A manual for eutrophication control in the subarctic.* Environmental Studies No. 33, Report prepared for Arctic Land Use Research Programmes, Environmental Protection Branch, Department of Indian and Northern Affairs, Ottawa, Canada. 32 p.
Snedecor, G.W. & Cochran, W.G. 1967. *Statistical Methods.* The Iowa State University Press, Ames, Iowa, USA. 593 p.
Snodgrass, W.J. & O'Melia, C.R. 1975. Predictive model for phosphorus in lakes. *Environ. Sci. Technol.* **9**:937–44.
Soltero, R.A., Nicholls, D.G., Gasperino, A.F. & Beckwith, M.A. 1981. Lake restoration: Medical Lake, Washington, *Jour. Freshwater Ecol.* **1**:155–65.
Somlyódy, L. 1983. Lake eutrophication management models. In: Somlyódy, L., Herodek, S. and Fischer, J. (eds.), *Eutrophication of shallow lakes: Modeling and management. The Lake Balaton study.* Report No. CP-83-S3, International Institute for Applied Systems Analysis (IIASA), A-2361 Laxenburg, Austria. p. 207–54.
Somlyódy, L. & van Straten, G. 1986. *Modelling and Managing Shallow Lake Eutrophication, with Application to Lake Balaton.* Springer-Verlag, Berlin, Federal Republic of Germany.
Somlyódy, L. & Wets, J.-B. 1985. *Stochastic optimization models for lake eutrophication management.* Report CP-85-16, International Institute for Systems Analysis (IIASA), A-2361 Laxenburg, Austria.
Sonzogni, W.C. & Lee, G.F. 1974a. Diversion of wastewaters from Madison lakes. *Jour. Env. Engr. Div., Amer. Soc. Civil Engr.* **100**:153–70.
Sonzogni, W.C. & Lee, G.F. 1974b. Nutrient sources for Lake Mendota–1972. *Trans. Wisc. Acad. Sci. Arts Lett.* **62**:133–64.
Sonzogni, W.C., Fitzgerald, G.P. & Lee, G.F. 1975. Effects of wastewater diversion on the lower Madison lakes. *Jour. Water Pollut. Cont. Fed.* **47**:535–42.
Sonzogni, W.C., Chapra, S.C., Armstrong, D.E. & Logan, T.J. 1982. Bioavailability of phosphorus inputs to lakes. *Jour. Environ. Qual.* **11**:555–63.
Sonzogni, W.C., Uttormark, P.C. & Lee, G.F. 1976. A phosphorus residence time model: Theory and application. *Water Res.* **10**:429–35.
Sonzogni, W.C., Jeffs, D.M., Konrad, J.C., Robinson, J.B., Chesters, G., Coote, D.R. & Ostry, R.C. 1980. Pollution from land runoff. *Environ. Sci. Technol.* **14**:148–53.
Sperling, L. 1962. The nutritive value of new sources of protein from waterbodies and wastewater (In German). *Wiss. Z. Karl-Marx-Univ., Leipzig, Math.-Nat. R.* **11**:207–11.
Spet, G.I. 1972. A comparison of effectivity in utilizing a unit area for fish cultivation or for agricultural purposes (In Russian). *Gidrobiologicheskie Zhurnal* **8**:115–29.
Sreenivasan, A. 1980. Fish production in some hypertrophic ecosystems in South India. In: Barica, J. & Mur L.R. (eds.), *Hypertrophic Ecosystems.* Developments in Hydrobiology, Volume 2, Junk Publishers, The Hague, The Netherlands.

Sridharan, N. & Lee, G.F. 1977. Algal nutrient limitation in Lake Ontario and tributary waters. *Water Res.* **11**:849–58.
Srisuwantach, V. 1978. Mean depth and fish yield in seven reservoirs of Thailand. *Thai Fish. Gazette* **31**:133–35.
Srisuwantach, V.& Soungchomphan, R. 1981. Morphoedaphic index and primary production in five reservoirs of Thailand, with emphasis on their relationship in fish yield. *Thai Fish. Gazette* **34**:57–69.
Stabel, H.H. 1984. Impact of sedimentation on the phosphorus content of the euphotic zone of Lake Constance. *Verh. Internat. Verein. Limnol.* **22**:964–69.
Steele, J.A. 1973. Reservoir algal productivity. In: James, A. (ed.). *The Use of Mathematical Models in Water Pollution*, Symposium Proceedings, University of Newcastle upon Tyne, Newcastle, United Kingdom. p. 107–35.
Steinberg,C. 1983. Effects of artificial destratification on the phytoplankton populations in a small lake. *Jour Plankton Res.* **5**:855–64.
Stefan, H., Skoglund, T. & Megard, R.O. 1976. Wind control of algal growth in eutrophic lakes. *Amer. Soc. Civil. Engr., Env. Engr. Div.*, **102**:1201–13.
Stěpánek, M. 1980. Cascade reservoirs as a method for improving the trophic state downstream. In: Barica, J. and Mur, L.R. (eds.), *Hypertrophic Ecosystems*, Developments in Hydrobiology, Volume 2, Junk Publishers, The Hague, The Netherlands. p. 323–27.
Stewart, K.M. and Rohlich, G.A. 1967. *Eutrophication, A review*. Publication No. 34, State Water Quality Control Board, Sacramento, California, USA. 188p.
Stockner, J.G., Shortreed, K.R.S. & Stephens, K. 1980. *The British Columbia Lake Fertilization Programmes: Limnological results from the first two years of nutrient enrichment.* Report No. 24, Canadian Fisheries and Marine Service, Ottawa, Ontario, Canada. 91 p.
Stråskraba, M. 1980. The effects of physical variables on freshwater production: Analyses based on models. In: LeCren, E.D. and Lowe-McConnell, R.H. (eds.), *The Functioning of Freshwater Ecosystems*, IBP Handbook No. 22, International Biological Programmes, Cambridge University Press, Cambridge, United Kingdom.
Stråskraba, M. 1982. The application of predictive mathematical models of reservoir ecology and water quality. *Can. Water Resour. Jour.* **7**:283–318.
Stråskraba, M. 1986. Ecotechnological measures against eutrophication. *Limnologica* (Berlin) **17**:237–49.
Stråskraba, M. & Gnauck, A.H. 1983., Aquatische Ekosysteme. In: *Modellierung und Simulation*, VEB, Gustav Fischer Verlag, Jena. 279 p.
Stråskraba, M. & Gnauck, A.H. 1985. *Freshwater Ecosystems. Modelling and Simulation.* Developments in Environmental Modelling, Volume 8, Elsevier Scientific Publishers, Amsterdam, The Netherlands. 300 p.
Strecker, R.G., Steichen, J.M., Garton, J.E. & Rice, C.E. 1977. Improving lake water quality by destratification. *Trans. Amer. Soc. Agric. Engr.* **20**:713–20.
Stumm, W. & Morgan, J.J. 1970. *Aquatic Chemistry.* John Wiley & Sons, Inc., New York. p. 429–36.
Stundl, K. 1978. Attempts to reduce eutrophication in an artificial bathing lake. *Verh. Internat. Verein. Limnol.* **20**:1878–83.
Suess, M.S. (ed.). 1982. *Examination of Water for Pollution Control. A Reference Handbook*, Volume 1. Pergamon Press, Oxford, United Kingdom.
Swartzman, G.L. & Bentley, R. 1979. A review and comparison of phytoplankton simulation models. *Jour. Int. Soc. Ecol. Modelling* **1**:30–81.
Sykora, J.B., Roche, B., Volk, R. & Kay, G. 1980. Endotoxicity, algae and *Limnulus amoeboecyte* lysate test in drinking water. *Water Res.* **14**:829–39.
Talling, J.F. & Talling, I.B. 1965. The chemical composition of African lake water. *Int. Revue ges. Hydrobiol.* **50**:421–63.
Taylor, W.D., Lambou, L.R., Williams, L.R. & Hern, S.C. 1980. *Trophic state of lakes and reservoirs.* Technical Report No. E-80-3, US Army Corps of Engineers, Waterways Experiment Station, Vicksburg, Mississippi, USA. 15 p.
Technical Standard. 1982. *Nützung und schütz der Gewässer. Stehende Binnengewässer. Klassifizierung* (In German: Utilization and protection of waterbodies. Standing inland waters. Classification). Technical Standard 27885/01, Berlin, German Democratic Republic, April 30, 1982. 16 p.

Theis, T.L. & DePinto, J.V. 1976. *Studies on the reclamation of Stone Lake, Michigan.* Ecological Research Series, No. EPA-600/3-76-106, US Environmental Protection Agency, Enviromental Research Laboratory, Corvallis, Oregon, USA. 85 p.

Thienemann, A. 1918. Untersuchungen uber die Beziehungen zwischen dem Sauerstoffgehalt des Wassers und der Zusammensetzung der Fauna in norddeutschen Seen. *Arch. Hydrobiol.* **12**:1–65.

Thomann, R.V., Di Toro, D.M., Winfield, R.P. & O'Connor, D.J. 1975. *Mathematical modeling in Lake Ontario. I. Model development and verification.* Ecological Research Series, No. EPA-660/3-75-005, US Environmental Protection Agency, Large Lakes Research Station, Grosse Ile, Michigan, USA. 177 p.

Thornburn, G. 1986. The social and institutional dimensions of developing phosphorus management straties. In: Lester, J.M. and Kirk P.W.W. (eds.). *Management Strategies for Phosphorus in the Environment*, Selter Ltd., London, United Kingdom. p. 417–24.

Thornton, J.A. 1979. *P-loading to lakes: Similarities between temperate and tropical lakes.* Technical Report, National Institute of Water Research (NIWR), Pretoria, Republic of South Africa, (presented at SIL-UNEP Workshop on African Limnology, Nairobi, Kenya). 16 p.

Thornton, J.A. 1980. A comparison of the summer phosphorus loadings to three Zimbabwean water-supply reservoirs of varying trophic states. *Water SA* **6**:163–70.

Thornton, J.A. 1982. *Lake McIlwaine: The Eutrophication and Recovery of a Tropical African Man-made Lake.* Monographiae Biologicae, Volume 49, Junk Publishers, The Hague, The Netherlands. 251 p.

Thornton, J.A. 1985. Nutrients in Africa lake ecosystems: Do we know all? *Jour. Limnol. Soc. S. Afr.* **12**:6–21.

Thornton, J.A. 1987. Aspects of eutrophication management in tropical/sub-tropical regions: A review. *Jour. Limnol. Soc. S. Afr.* **13**:25–43

Thornton, J.A. & Nduku, W.K. 1982. Water chemistry and nutrient budgets. In: Thornton, J.A., *Lake McIlwaine, the Eutrophication and Recovery of a Tropical African Man-Made Lake.* Monographiae Biologica, Volume 49, Junk Publishers, The Hague, The Netherlands. p. 43–59.

Thornton, J.A. & Walmsley, R.D. 1982. Applicability of phosphorus budget models to southern African man-made lakes. *Hydrobiology* **89**:237–45.

Thornton, J.A., Cochrane, K.L., Jarvis, A.C., Zohary, T., Robarts, R.D. & Chutter, F.M. 1986. An evaluation of management aspects of a hypertrophic African impoundment. *Water Res.* **20**:413–19.

Thornton, K.W. 1984. Regional comparisons of lakes and reservoirs: Geology, climatology and morphology. In: *Lake and Reservoir Management*, Report No. EPA-440/5-84-001, US Enviromental Protection Agency, Office of Water Regulations and Standards, Washington, DC. pp. 261–65.

Thornton, K.W., Kennedy, R.H., Magoun, A.D. & Saul, G.E. 1982. Reservoir water quality sampling design. *Water Resour. Bull.* **18**:471–80.

Thornton, K.W., Kennedy, R.H., Carroll, J.H., Walker, W.W., Gunkel, R.C. & Ashby, S. 1981. Reservoir sedimentation and water quality – an heuristic model. In: Stefan, H. (ed.), *Symposium on Surface Water Impoundments*, Volume I, American Society of Civil Engineers, New York. p. 654–61.

Tilman, D. 1976. Ecological competition between algae: Experimental confirmation of resource-based competition theory. *Science* **192**:463–65.

Tilman, D., Kilham, S.S. & Kilham, P. 1976. Morphometric changes in *Asterionella formosa* colonies under phosphate and silicate limitation. *Limnol. Oceanogr.* **21**: 883–84.

Tilman, D. Kilham, S.S. & Kilham, P. 1982. Phytoplankton community ecology: The role of limiting nutrients. *Ann. Rev. Ecol. Syst.* **13**, 349–72.

Timmons, D.R., Hold, R.F. & Latterell, J.J. 1970. Leaching of crop residues as a source of nutrients in surface runoff waters. *Water Resour. Res.* **6**:1367–75.

Toerien, D.F., Hyman, K.L. & Bruwer, M.J. 1975. A preliminary trophic status classification of some South African impoundments. *Water SA* **1**:15–23.

Toetz, D.W. 1977. Effects of lake mixing with an axial flow pump on water chemistry and phytoplankton. *Hydrobiologia* **55**:129–38.

Toetz, D.W. 1981. Effects of whole lake mixing on water quality and phytoplankton. *Water*

Res. **15**:1205–10.
Toews, D.R. & Griffith, J.S. 1979. Empirical estimates of potential fish yield for the Lake Bangweulu System, Zambia, Central Africa. *Trans. Amer. Fish. Soc.* **108**:241–52.
Torrey, M.S. & Lee, G.F. 1976. Nitrogen fixation in Lake Mendota, Madison, Wisconsin. *Limnol. Oceanogr.* **21**:365–79.
Troutman, N.M., McCulloch, C.E. & Oglesby, R.T. 1982. Statistical determination of data requirements for assessment of lake restoration programmes. *Can. Jour. Fish. Aquat. Sci.* **39**:607–10.
Turner, R.E. 1977. Intertidal vegetation and commercial yields of penaeid shrimp. *Trans. Amer. Fish. Sco.* **106**:411–16.
Uhlmann,D. 1958. Die biologische Selbstreinigung in Abwasserteichen (In German: Biological self-purification in sewage ponds). *Verh. Internat. Verein. Limnol.* **13**:617–23.
Uhlmann, D. 1971. Influence of dilution, sinking and grazing rate on phytoplankton populations of hyper-fertilized ponds and micro-ecosystems. *Mitt. Internat. Verein. Limnol.* **19**:100–24.
Uhlmann, D. 1979. *Hydrobiology – A Text for Engineers and Scientists.* John Wiley & Sons, Inc., New York. 313 p.
Uhlmann, D. 1980. Stability and multiple steady states of hyper-eutrophic ecosystems. In: Barica, J. and Mur L.R. (eds.), *Hypertrophic Ecosystems*, Developments in Hydrobiology, Volume 2, Junk Publishers, The Hague, The Netherlands. p. 235–47.
Uhlmann, D. 1984. Evaluation of strategies for controlling eutrophication of lakes and reservoirs. *Int. Reveue ges. Hydrobiol.* **67**:821–35.
Uhlmann, D. & Hŕbaček, J. 1976. Kriterion der eutrophie stehender Gewässer (In German: Trophic criteria of standing waters). *Limnologica* **10**:245–53.
Uhlmann, D & Klapper, H. 1985. *Protection and restoration of lakes and reservoirs in the German Democratic Republic. Case studies.* In: Proceedings, International Congress on Lakes Pollution and Recovery, European Water Pollution Control Association, Rome, Italy, April 15–18, 1985. p. 262–71.
United States Environment Protection Agency. 1971. *Algal assay procedure bottle test.* Technical Report, US Environmental Protection Agency, National Eutrophication Research Programmes, Corvallis, Oregon, USA. 82 p.
United States Environmental Protection Agency. 1974. *Marine algal assay procedure: Bottle test.* Technical Report, US Environmental Agency, Eutrophication Research Laboratory, Corvallis, Oregon, USA. 43 p.
United States Environmental Protection Agency. 1980. *Clean Lakes Programmes guidance manual.* Report No. EPA-440/5-81-003, US Environmental Protection Agency, Office of Water Regulations and Standards, Washington, DC. 103 p.
United States Environmental Protection Agency, 1988. *The lake and reservoir restoration guidance manual.* US Environmental Protection Agency, Report No. 440/5-88-002, Criteria and Standards Division, Nonpoint Source Branch, Washington, D.C.
Uttormark,P.D., Chapin, J.D. & Green, K.M. 1974. *Estimating nutrient loading of lakes from non-point sources.* Ecological Research Series, No. EPA-660-/3-74-020, US Environmental Protection Agency, Environmental Research Laboratory, Corvallis, Oregon, USA. 112 p.
Vallentyne, J.R. & Thomas, N.A. 1978. *Fifth year review of the Canada–United States Great Lakes Water Quality Agreement.* Final Report of Task Group III (Phosphorus loadings) to US and Canadian Governments, International Joint Commission, Great Lakes Regional Office, Windsor, Ontario, Canada. 84 p.
van Staten, G. 1981. Analysis of model and parameter uncertainty in simple phytoplankton models for Lake Balaton. In: Dubois, D.M. (ed.), *Progress in Ecological Engineering and Management by Mathematical Modelling* (CEBEDOC, Liege, Belgium). p. 107–34.
van Straten, G. 1983. Lake eutrophication models. In: Somlyódy, L. (ed.), *Eutrophication of Shallow Lakes. Modelling and Management. The Lake Balaton study*, International Institute for Applied Systems Analysis (IIASA), A-2361 Laxenburg, Austria. pp. 175–206.
van Straten, G. 1986. Lake eutrophication models. In: Somlyódy, L. & van Straten, G. (eds.), *Modeling and Managing Shallow Lake Eutrophication, with Application to Lake Balaton*, Springer-Verlag, New York, p. 35–68.
Varshney, C.K. & Rzóska, J.R. (eds). 1976. *Aquatic Weeds in Southeast Asia.* Junk Publishers, The Hague, The Netherlands. 396 p.

Vass, K.K. & Zutshi, D.P. 1983. Energy flow, trophic evolution and management of a Kashmir Himalayan lake. *Arch. Hydrobiol.* **97**:39–59.
Verhagen, J.H.G. 1976. The construction of models of simulation: A guideline for investigations into and control of water quality in reservoirs. *Hydrobiol. Bull.* **10**:172–78.
Verhoff, F.H. & Heffner, M.R. 1979. The rate of availability in river waters. *Environ. Sci. Technol.* **13**:844–49.
Verhoff, F.H., Melfi, D.A. & Yaksich, S.M. 1979. Storm travel distance calculations for total phosphorus and suspended materials in rivers. *Water Resour. Res.* **15**:1354–60.
Verhoff, F.H., Yaksich, S.M. & Melfi, D.A. 1980. River nutrient and chemical transport estimation. *Jour. Amer. Soc. Civil Engr., Env. Engr. Div.* **106**:591–608.
Vincent, W.F., Wurtsbaugh, W., Vincent, C.I. & Richerson, P.J. 1984. Seasonal dynamics of nutrient limitation in a tropical, high altitude lake (Lake Titicaca, Peru-Brazil): Application of physiological bioassays. *Limnol. Oceanogr.* **29**:540–52.
Viner, A.B. 1975.The supply of minerals to tropical rivers and lakes (Uganda). In: Hasler, A. (ed.), *Coupling of Land and Water Systems*, Ecological Studies, Volume 10, Springer-Verlag, Berlin, Federal Republic of Germany. p. 227–61.
Vitousek, P.M., Gosz, J.R. Grier, C.C., Melillo, J.M., Reiners, W.A. & Todd, R.L. 1979. Nitrate losses from disturbed ecosystems. *Science* **204**:469–74.
Vollenweider, R.A. 1968. *Scientific fundamentals of the eutrophication of lakes and flowing waters, with particular reference to nitrogen and phosphorus as factors in eutrophication.* Technical Report DAS/CSI/68.27, Environmental Directorate, Organization for Economic Cooperation and Development (OECD), Paris. 154 p.
Vollenweider, R.A. 1969. Möglichkeiten und Grenzen elementarer Modelle der Stoffbilanz von Seen (In German: Possibilities and limits of elementary models concerning the budget of substances in lakes). *Arch. Hydrobiol.* **66**:1–36.
Vollenweider, R.A. 1974. *A manual on methods for measuring primary production in aquatic environments*, 2nd Edition. IBP Handbook No. 12, International Biological Programmes, Blackwell Scientific Publications, Oxford, United Kingdom. 225 p.
Vollenweider, R.A. 1975. Input–output models with special reference to phosphorus loading concept in limnology. *Schweiz. Zeit. Hydrol.* **37**:53–84.
Vollenweider, R.A. 1976a. Advances in defining critical loading levels for phosphorus in lake eutrophication. *Mem. Ist. Ital. Idrobiol.* **33**:53–83.
Vollenweider, R.A. 1976b. Rotsee, a source, not a sink for phosphorus? A comment to and a plea for nutrient balance studies. *Hydrologie* **38**:29–34.
Vollenweider, R.A. & Janus, L.L. 1982. *Statistical models for predicting hypolimnetic oxygen depletion rates. Mem. Ist. Ital. Idrobiol.* **40**:1–24
Vollenweider, R.A., Munawar, M. & Stadelmann, P. 1974. A comparative review of phytoplankton and primary production in the Laurentian Great Lakes. *Jour. Fish. Res. Bd. Can.* **31**:739–62.
Walker, W.W. 1981. *Empirical methods for predicting eutrophication in impoundments. Report 1. Data base development.* Technical Report E-81-9, Environmental & Water Quality Operational Studies, US Army Corps of Engineers, Watersways Experiment Station, Vicksburg, Mississippi, USA. 153 p.
Walker, W.W. 1982. *Empirical methods for predicting eutrophication in impoundments. Report 2. Model testing.* Technical Report E-81-9, Environmental & Water Quality Operational Studies, US Army Corps of Engineers, Waterways Experiment Station, Vicksburg, Mississippi, USA. 229 p.
Walker, W.W. 1985. *Empirical methods for predicting eutrophication in impoundments. Report 3. Model refinements.* Technical Report E-81-9, Environmental & Water Quality refinements. Technical Report E-81-9, Environmental & Water Quality Operational Studies, US Army Corps of Engineers, Waterways Experiment Station, Vicksburg, Mississippi, USA. 297 p.
Walmsley, R.D. & Toerien, D.F. 1977. The summer conditions of three eastern Transvaal reservoirs and some considerations regarding the assessment of trophic status. *Jour. Limnol. Soc. S. Afr.* **3**:37–41.
Walmsley, R.D. & Butty, M. 1980. *Guidelines for the control of eutrophication in South Africa.* Technical Report, Water Research Commission and National Institute for Water Research, Pretoria, Republic of South Africa. 27 p.
Walmsley, R.D. & Thornton, J.A. 1984. An evaluation of OECD-type phosphorus eutrophic-

ation models for predicting the trophic status of southern African man-made lakes. *S. Afr. Jour. Sci.* **80**:257–59.
Wanieslista, M.P. 1979. *Stormwater Management*. Ann Arbor Science Publishers, Ann Arbor, Michigan, USA. 383 p.
Wanielista, M.P., Yousef, Y.A. & McLellon, W.M. 1977. Non-point source effects on water quality. *Jour Water Pollut. Cont. Fed.* **49**:441–51.
Watson, N.H.F., Thomson, K.P.B. & Elder, F.C. 1975. Sub-thermocline biomass concentration detected by transmissometer in Lake Superior. *Verh. Internat. Verein. Limnol.* **19**:682–88.
Weber, C.A. 1907. Aufbau und Vegetation der Moore Norddeutschlands. *Bot. Jahrb.* **40**. Beibl. **90**:19–32.
Weise, G. & Jorga, W. 1981. Aquatic macrophytes – a potential resource. *WHO Water Quality Bulletin* **6**:104–07 (World Health Organization, Canada Centre for Inland Waters, Burlington, Ontario, Canada).
Welch, E.B. 1977. *Nutrient diversion: Resulting lake trophic state and phosphorus dynamics*. Ecological Research Series, No. EPA-600/3-77-003, US Environmental Protection Agency, Environmental Research Laboratory, Corvallis, Oregon, USA. 91 p.
Welch, E.B. 1980. *Ecological Effects of Waste Water*. Cambridge University Press, Cambridge, United Kingdom. 337 p.
Welch, E.B. 1981a. *The dilution/flushing technique in lake restoration*. Report No. EPA-600/3/-81-016, US Environmental Protection Agency, Environmental Research Laboratory, Corvallis, Oregon, USA. 13 p.
Welch, E.B. 1981b. The dilution/flushing technique in lake restoration. *Water Resour. Bull.* **17**:558–64.
Welch, E.B., Rock, C.A., Howe, R.C. & Perkins, M.A. 1980. Lake Sammamish response to wastewater diversion and increasing urban runoff. *Water Res.* **14**:821–28.
Welch, E.B., Spyridakis, D.E. Shuster, J.I. & Horner, R.R. 1986. Declining lake sediment phosphorus release and oxygen deficit following wastewater diversion. *Jour. Water Pollut. Cont. Fed.* **58**:92–96.
Welch, P.S. 1952. *Limnology*, 2nd Edition, McGraw-Hill, Inc., New York. 538 p.
Wells, S.A. & Gordon, J.A. 1982. Geometric variations in reservoir water quality. *Water Resour. Bull.* **18**:661–70.
Westerdahl, H.E., Ford, W.B., Harris, J. & Lee, C.R. 1981. *Evaluation of techniques to estimate annual water quality loadings to lakes*. Technical Report E-81-1, Environmental & Water Quality Operational Studies, US Army Corps of Engineers, Waterways Experiment Station, Vicksburg, Mississippi, USA. 61 p.
Wetzel, R.G. 1975. *Limnology*. W.B. Saunders Co., Philadelphia, Pennsylvania, USA. 743 p.
Wetzel, R.G. & Hough, R.A. 1973. Productivity and role of aquatic macrophytes in lakes. An assessment. *Pol. Arch. Hydrobiol.* **20**:9–19.
Wetzel, R.G. & Likens, G.E. 1979. *Limnological Analysis*. W.B. Saunders Co., Philadelphia, Pennsylvania, USA. 357 p.
Whipple, W., Hunter, J.V., Trama, F.B. & Tuffey, T.J. 1975. *Oxidation of lake and impoundment hypolimnia*. Final Report, OWRT Project B-050-NJ, Water Resources Research Institute, Rutgers University, New Brunswick, New Jersey, USA. 87 p.
Whitfield, P. 1982. Selecting a method for estimation substance loadings. *Water Resour. Bull.* **18**:203–10.
Wijeyaratne, W.M.S. & Costa, H.H. 1981. Stocking rate estimations of *Tilapaia mossambica* fingerlings for some inland reservoirs of Sri Lanka. *Int. Revue ges. Hydrobiol.* **66**:327–33.
Williams, W.D. & Wan, H.F. 1972. Some distinctive features of Australian inland waters. *Water Res.* **6**:829–36.
Wilson, A.L. 1982. Design of sampling programmes. In: Suess, M.J. (ed.), *Examination of Water for Pollution Control. A Reference Handbook*, Volume 1, Pergamon Press, Oxford, United Kingdom. p. 23–77.
Winter, T.C. 1981. Uncertainty in estimating the water balance in lakes. *Water Resour. Bull.* **17**:82–115.
Wischmeier, W.H. & Smith, D.D. 1978. *Predicting Rainfall Erosion Losses – A Guide to Conservation Planning*. Agricultural Handbook No. 537, US Government Printing Office, Washington DC.
Wolny, P. & Grygierek, E. 1970. *Intensification of fish pond production*. Proceedings,

IBP/Unesco Symposium on Productivity Problems of Freshwater, May, 1970. pp. 1–13.
Wolverton, B. & McDonald, R.C. 1976. Don't waste waterweeds. *New Scientist* **71**:318–20.
Wood, G. 1975. *An assessment of eutrophication in Australian inland waters*. Technical Paper No. 15, Australian Water Research Council, Canberra, Australia.
Yaksich, S.M. & Verhoff, F.H. 1983. Sampling strategy for river pollutant transport. *Amer. Soc. Civil. Engr., Env. Engr. Div.* **109**:219–31.
Young, T.C. DePinto, J.V., Flint,S.E., Switzenbaum, M.S. & Edzwald, J.K. 1982. Algal availability of phosphorus in municipal wastewater. *Jour. Water Pollut. Cont. Fed.* **54**: 1505–16.
Zaret, T.M., Devol., A.H. & Santos, A.D. 1981. Nutrient addition experiment in Lago Jacaretinga, central Amazon Basin, Brazil. *Verh. Internat. Verein. Limnol.* **21**:721–24.
Zevenboom, W., de Vaate, A.B. & Mur, L.R. 1981. Assessment of factors limiting growth rate of *Oscillatoria agardhii* in hypertrophic Lake Wolderwijd, 1978, by use of physiological indicators. *Limnol. Oceanogr.* **27**:39–52.
Zimmerman, U. 1984. *Phosphorkonzentration und phytoplankton biomasse am vesipiel des Systemes Walensee-Zurichsee der Jahre 1974–1984*. Arbeitsgemeinschaft Wasserwerke Bodensee-Rhein AWBR. Jahresbericht 1984, Karlsruhe: 121–44.
Zutshi, D.P. & Wanganeo, A. 1984. The phytoplankton and primary production of a high altitude subtropical lake. *Verh. Internat. Verein. Limnol.* **22**:1168–72.

INDEX

Acidification of lakes, reversal of, 230
Algae, as nutrient removers, 213
  filamentous, 215
  (see also under phytoplankton)
Algal bioassays, 47, 48, 58–60, 120
  bioassay, disadvantages of, 59
  blooms, 6, 55, 56, 113, 116, 155, 166
  cultural techniques, 214
  cultural technique problems, 214
  density, 53
  growth in bioreactors, 174
  growth in dynamic models, 108
  growth in limiting nutrient concept, 53, 55, 58
  harvesting, 215
  physiology, 60–63
  primary productivity, 53
  sedimentation, 53
  species as eutrophication indicators, 48, 49
  (see also under phytoplankton)
Algicides, 21, 183
Analytical procedures, 148
Annual conditions, 6, 49, 57
  model predictions, 89
  nutrient load, 70, 89, 124, 145
  (see also under nutrient unit load)
  phosphorus input values, 74, 90, 94
Anthropogenic load (see under nutrient load)
Atmospheric nutrient inputs, 116
  precipitation, 68, 116
  unit loads, *126*
Aquaculture, 3, 9, 12, 24, 40
  fish of, 218–229
  macrophytes of, 215
  phytoplankton of, 215
Aquatic plants, as feed for cattle, 217
  humans, 217
  pigs, 217
  biomass determination, 105
  commercial use of, 217
  economic use of types of, 216, 217
  harvest of, 20, 21, 24, 34, 215, 216
  inputs (nutrient), 20
  in simple models, 103, 104

Baseflow, 88
  in watershed models, 88, 90, 92
Biogas, 214, 218
Biological control of phytoplankton
  biomass, 80–83, 183
  of macrophytes, 81–84
  response of lakes using models, 87, 103
  transformation in simulation models, *109*
Biologically available nutrients, 20, 54, 55, 57, 119, 120
  models to calculate, 88
Bioreactors, 174
Blue-green algae, 47, 52, 56, 57, 119, 135, 187, 189, 244
  as physiological indicators, 61
Boundary conditions, 16, 37, *38*, 46, *47*, 55, 56, 258, *260, 262*
  (see also under trophic categories)

C^{14}, use in bioassays, 60, 61
Cage fish farming, 83
Canalization of wastewaters, 179
Carbon, from macrophyte biomass, 218
  in limiting nutrient concept, 52
  : nitrogen: phosphorus ratios, 52, 54, 57
Catchment area physiography, *68–71*
CEPIS study, 49, 99
Chemical cycling, 14
  transformations in models, *109*
Chlorophyll, in lakes, 44
  in reservoirs, 45
  in trophic boundary levels, *259*
  mean levels, 258, *261*
  peak levels, 258, *259, 261*
  Secchi depth relationships, 104
Chlorophyll 'a' as measure phytophankton standing crop, 224
  empirical model predictions of, 94, 98,

309

Control of eutrophication

empirical model predictions of, 94, 98, 102
in trophic boundary levels, 37
macrophyte model predictions, and, 104, 106
sportfish yeild and, *225*
Classification of trophic categories, see under trophic categories
of waterbodys, see under waterbodys
Climate, effects on eutrophication, 65
temperate and tropical lake systems, 46 49
Control costs in optimization models, 210–212
of phosphorus, 195–212
of phosphorus target loads, 199, *208–210*
of relative point and non-point sources, 197–210
methods, In-lake, biological control, 183, 186, 187
chemical control, 183, 188
circulation, 182
flow augmentation, 181, 186
harvesting, 182, 189
hypolimnetic aeration, 181, 182, 185, 189, 190
lake level drawdown, 183
light control, 187
nutrient inactivation, 181
removal hypolimnetic waters, 182
sediment removal, 182, 183, 185
methods, of non-point source nutrients
agricultural, 192–195, *196, 197*
cropland, 195, *198*
summary of, *200–207*
urban, 191, *193, 194*
programmes, 9, 13, 27, 34, 90, 232
dynamic models of, 251–254
long-term, 22, 34, 49, 51, 169, 211
selection of, 32, 234–255
short-term, 34, 211
Cost, 13, 19, 22–30, 120
benefit analysis, 25
harvesting aquatic plants, of, 217
sampling, 158–164 (see also under sampling)
(see also under control)
Cultural eutrophication (see under eutrophication)

Data, average values, 165
desired statistical confidence level, 158
geometric mean values, 166
normal distribution of, 166
precision, 163
presentation, 164–67
weighting factors, 165

Diffuse nutrient loads (see under nutrient loads, non-point)
Domestic wastewater, affecting acidic lakes, 230
nutrient reuse, 229
Drainage basin characteristics, 65, 69
chemical cycling in, 19
manmade factors relating to, 71–74
land usage, 71
point sources, 71
mineral composition, 69
natural factors related to, 65–71
climate, 65
hydrology, 67
geology, 68
physiography, 68
surface area ratios, 42, *43*

Epilimnion, hypolimnion relationships in dynamic models, 108
samples for limiting nutrient concept, 55
Erosion, as control measure, 192
as nutrient source, 118, 135
Eutrophic, exchange processes, 111
mesotrophic boundary, 16, 259 (see also under trophic categories)
Eutrophication benefits of, 12, 24, 213–230
definition of cultural, 1, 37
ecological consequences of cultural, 2, 15
of natural, 2
general symptoms of, 38–49

Fathometer, 105
Fertilizer application, 17, 20, 23, 24, 48, 71, 89
as nutrient source, 117
control measures, 193–195
from fish pond sludge, 229
Filtration of phosphorus from tributaries, 175
through aluminium oxide, 178
Fish culture and nutrient excretion, 220, 222
and use of domestic effluents, 229
as nutrient reuse, 219
as short term improvement, 222
exclusion from parts of water column, 221
filter feeding, 221
kills, 2, 226
pond management, 226–229
fish types for stocking, 227
oxygen depletion, 226
turbidity, 226
yield estimation, 222–226
and chlorophyll 'a', phytophankton measurements, 224, *225*
and total phosphorus, *223*

Index

Morphoedaphic Index (MEI), 223
Fisheries and use of nutrients, 213, 218–229
   cold water, 226
Flocculation of phosphorus from tributaries, 175
Flushing rates, 33, 48, 79, 120, 186
   from dynamic models, 108
   nutrient load estimates, 143–146
Freeze-thaw cycles, 46, 70

Geology affecting degree of eutrophication, 68
   of lakes and reservoirs, 44
Government concerns, 31, 33, 35
   expenditure, 25
   local, 32
   role, 16, 18
Ground water seepage as nutrient source, 115, 118
Guide to contents of book, 8

Harvesting cultured algae, 214
   macrophytes, 21, 215
Health effects, 15, 22, 29, 40, 190, 227, 240
Hydraulic residence time, 42, 44, 80, 145, 246
Hydrodynamic, components in dynamic models, 108
   properties affecting sampling strategy, 150
Hydrographic criteria in classifications, 264 (see also under waterbody)
Hydrologic, components of dynamic models, 108
Hydrology affecting degree of eutrophication, 17, 67
   in classification criteria, 264
Hypolimnion, 40
   depth affecting nutrient load, 74–77
   models of oxygen depletion rates, 101
   oxygen depletion, 46, 226, 227

In-lake control measures, 21, 34 (see also under control)
   control methods, 180–919 (see also under control)
   nutrient inactivation, 21, 28
   phosphorus retention characteristics, 98
   sampling programmes, 45 (see also under sampling)
   total phosphorus concentrations, 98, 99
Inorganic turbidity, 234

Lake basin morphology, 74
   biological stability, 15
   comparisons with reservoirs, 42, *43*
   definition of, 42, 46
   temperate and tropical differences, 46–49

Lag periods, 33, 34
Land management and nutrient loads, 74
   management practices, 24
   use in control of eutrophication, 73, 173
Legislation of control programmes, 30–32, 195
   institutional concerns, 30
   regulatory concerns, 31, 171
Light intensity, calculation for, 80
Limiting growth, 48
Limiting nutrient concept, 7, 47, 49–63
   and algal growth, 49, 53
   assessment using C^{14}, 61
   carbon, 52
   history of, 50
   limitations of, 52, 56
   physiological indicators of, 60–63
   seasonal changes in, 53
Littoral zone, 40, 190
   and effect of carp, 226
   nutrient removal by macrophytes, 215
   slope, and biomass of macrophytes, 106

Macro-nutrients, 50
Macrophages, biological control of, 81
   harvesting of floating, 216
   harvesting of rooted, 215
   removal of nutrients from sediments, 215
   simple phosphorus models and, 103
   water transparency model predictions and, 105
Management assessment of
   alternative phosphorus control options, 241, *242*
   control of nitrogen, 240
   effectiveness of control programmes, 244
   further in-lake control measures, 243
   limiting nutrients, 239
   model applicability, 236
   problem and definition of goals, 239
   consultation with, 17
   decision sequences, *234, 235, 237*
   decision tree approach, 236 (see also under control methods)
   general considerations, 12, 231
   goals, 14–18, 239
   implementations of remedial programmes, *235*
   matching quality and desired use, 232
   models, 87, 90, 94
   of fish ponds, 226, *228*
   of response failure of
      biological factors, 250
      geological and chemical factors, 250
      geographic factors, 249
      hydrodynamic factors, 250
      limiting nutrients, 247

morphometric factors, 250
nutrient loading, 248
phosphorus-chlorophyll levels, 247
response time, 245
sampling programme, 249
option of doing nothing, 21
options general, 13, 19–22
programmes general, 16, 18
specific steps to follow, 238
strategies general, 231–255
Matrix development of control options, 28
Mesotrophic, definition of, 37 (see also under trophic categories)
Models, 5, 8, 14, 45, 85–113
and sediment effects, 111
and water transparency, 105
CLEAN, 122
CLEANER, 112
cooperative programme for monitoring inland waters, 5
for estimating nutrient loads, 137, 138–143
for macrophytes, 104
for management decisions, 85, 236
for prediction of hypolimnetic oxygen depletion rate, 101
fully transport orientated, 109
general, 85–87, 110
multi-section, 108, 109
optimization, static, for phosorous control, 210
dynamic, for phosphorous control, 211
SALMO, 113, 212, 252–254
Waterbody empirical, 94–107
and macrophytes, 103
for chlorophyll 'a', 98
for nitrogen loading, 98
for reservoirs, 102, 103
for phosphorus (simple), 95, 96, 97, 99, 100
specific chlorophyll-phosphorus, 98, 101
uptake by rooted macrophytes, 104
simulation, 107–113
chemical and biological transformations, 109
hydrodynamic components, 108
hydrologic components, 108
interface terms, 107, 110, 111
selection of a, 110, 112
Watershed, 88–94
empirical, 88–90
simulation, 90–94
rural, 91
urban, 92
Monitoring, post treatment, 254–255
programmes, 9, 19
Municipal wastewater treatments, 4, 17, 18, 20, 21, 119
and application of algal culture techniques, 214
cost of, 22
cost of phosphate control, 195
phosphate control, 170
unit loads, 121, 122

N_2-fixing algae, 47, 56, 118, 244
Nitrate as limiting nutrient, 48
Nitrogen as limiting nutrient, 47, 48
biologically available, 55
dimensionless concentrations, 57, 58
fixation, 118, 135
for aquatic growth, 20
levels and human health, 40
loading models, 98
major sources of, 115–120
phosphorous ratios, 54–63
as boundary values, 55, 56
physiological indicators, 60–63
rural sources, 117
total, in trophic boundary levels, 47
Nomograms, 241, 242
Nutrient export coefficients, 90
external loads, atmospheric inputs, 126
countryside & household, 127
farm wastewaters, 128
general, 122–131
indirect measurements, 125
rivermouth measurements, 123
rural cropland, 129
rural non-cropland, 129–130
rural urban inputs, 126, 127
wetland, 130
internal loads, groundwater, 131
nitrogen fixation, 135
phosphorus, 131–135
sediment release, 131
shoreline erosion, 135
waterfowl, 135
limitation (see under limiting nutrient concept)
loads, anthropogenic, 2, 264, 268, (see also under eutrophication cultural)
estimation of, 115–146
reliability, 146
simplified, 135–143
of phosphorus, 137–143
of nitrogen, 138
watershed model, 138, 139–143
with rapid flushing rates, 143–146
quantifying, 120–135
unit, 120
using models, 86
Nutrient non-point sources, 9, 14, 20, 21
as source aquatic plant nutrients, 71–74
control measures, 23, 28

definition of, 4
estimated by watershed models, 87
external sources, 116–118
  atmospheric, 116
  erosion, 118
  fertiliser, 117
  rural, 117
  septic tank, 117
  soil, 117
internal sources, 118
  ground water seepage, 118
  nitrogen fixation, 118
  recycling of nutrients, 118
remedial measures, 191–195
Nutrient point sources, 9, 14, 20, 21, 23, 115–*116*
  aquatic plants, 71
  control measures, 23
  cost of control, 195
  definition of, 4
  reduction of inputs, 20, 21
Nutrient recycling, 118
  reduction, 236
  removal by algae, 213–215
  reuse 213–230 (see under aquaculture, aquatic plants, fish)
  transport mechanisms in models, 108

Oligotrophic lakes, and hypolimnetic oxygen concentrations, 76
  definition of, 37
  geology of, 69
Open boundary systems (see under trophic categories)
Optimal models (see under models)
  production of algal cultures, 214
  trophic conditions, 15, 59, 260
Oxygen consumption of hypolimnion, 76

Parasitic disease, 22
Phosphate control by chemical precipitation, 170–*171*
  control in detergents, 171
  from watershed geology, 69
  in detergents, 21
  release from sediments, 78
Phosphorous as limiting nutrient, 47, 48, 51, 53
  biologically available, 55, 119
  chlorophyll models, 101
  control of external load, 169–180
  cycling 77–79
  cycling models, 104
  deficiency, physiological indications of, 60–63
  dimensionless concentrations, 57, *58*
Phosphorus elimination plant (PEP), 175–178
  empirical model predictions, 98
  for aquatic growth, 20
  from atmospheric precipitation, 68
  from geological sources, 69
  in influent waters, 178
  in reservoirs, 44, 47
  load and fish yields, *223, 224*
  loading models 96, 97, 99, *100*, 236
  major sources of, 115–120
  mass & atomic ratios, 55
  models and macrophytes, 103
  models applied to reservoirs, 102
  models based on multiple lake loading characteristics, 95
  precipitating chemicals in influent waters, 178
  ratios as boundary values, 56
  ratio to nitrogen, *54–63*
  reduction in tributaries, 174
  removal from effluent, 21, 28
  retention characteristics and model predictions, 98
  rural sources, 117
  sediment, release from, 131–134
  sediment role in cycling *77–79*
  total concentration predictions, 94
  total, indication of boundary trophic levels, 37, 46, 47, 55, 258, *260*
  total, in-lake concentrations, 16, 43, 44, 98
  unit loads, 122, *127*
Photosynthesis, and lake altitude, 65
  in limiting nutrient concept, 51, 60
  gross, and fish yields, 225
  regulation, 188
Phytoplankton, as human and animal foodstuffs, 214
  as nutrient removers, 213
  biological control of, 83
  blooms, 46 (see also under algae)
  control by filter-feeding fish, 221
  control by zooplankton, 221
  filamentous, 215
  flushing rate relationships, 80
  growth curves, 51
  light limitation of growth of, 48, 50
  reduction by artifical mixing, 81
  reservoir and lake comparisons, 44
Point sources (see under nutrients)
Policymaker role of, 11–14, 19, 29
Polyculture ponds, 227
Pre-reservoir in phosphorus elimination, 80, 174
Public attitudes, 24, 25, 31, 79
  awareness, 14, 29, 34, 99, 104
  feedback, 35
  role, 17, 33

Rainfall affecting nutrient inputs, 68
Regulatory concerns, 31–32

Reservoirs definition of, 3, 42–46
  phosphorus models, 102
  spatial patterns, 44, *45*
Resource competition by algae, 52
Rural sources of nitrogen, 117
  of phosphorus, 117
Sample depths for indentification of limiting nutrient, 55
  validity of, 148
Sampling costs, 158–164
  fixed constant, 159
  as function of programme variable, 159
  hypothetical of monitoring programme, *160–164*
  variable, 159
Strategies, 147–167
  for waterbodies with longitudinal water quality
  gradients, 155–158
  random sampling, 156
  sampling station, *157*
  what to sample, 147–150
    analytical procedures, 148
    primary parameters, *149*
    statistical characteristics, 150
Sampling strategies
  when to sample, 154–155
  where to sample, 150–154
    chlorophyll concentration, 152
    from hypolimnion, 153
    lateral heterogeneity, 152
    spatial consistency, 151, 153
    thermocline, 151
    vertical distribution, 151
Scenario analysis, 251–254
Secchi-depths, and macrophyte colonization, 105, *106*
  for indication of trophic boundary levels, 37, *262*
  model predictions of, 102, 103
Sediments as nutrient sink, 118
  as nutrient source, 104, 118
  biologically available nutrients, 119
  effects of carp, 226
  effects of macrophytes, 215
  effects on dynamic models, *111*, 184
  recycling, 84, 185
  release of phosphorus, 78, 132
  role of trophic status, 77–79
Seepage trenches, 179
Septic tanks as nutrient sources, 116, 117, 122, 127
Sewage and population size, 3
  cleansing by macrophytes, 215
Soil erosion, 27, 68
  nutrient contents of, 117
Spatial patterns, 44, *45*
Spiked tests, 59

Technical standard classification system, 257, 269
Thermal stratification, causes of, 106
  models of, 106
Treatment techniques, 169–212
  by diversion of wastewaters, 179
  canalization, 179
  direct reduction of phosphorus at source, 170–174
  in-lake eutrophication, 180–191
  non-point source nutrients in drainage basin, 191–195
  tributary influent waters, 174–179
Trophic categories, definition of, 37, *38*
  distribution, of *260*
  eutrophic, 37, *38, 41,* 42
  fixed boundary system, *38*, 258
  mesotrophic, 37
  oligotrophic, 37, *38, 41*
  open boundary, 37, *39*, 259
  Secchi depths, 37, *38*
  total phosphorus, 37, *38*

Uncertainty analysis, 94, 95

Water, hyacinth harvesting, 215, 216
  quality and relation to desired use, 3, 15, 232, *272, 273*
    detailed classification of, *263*
    transparency, 55, 105, 233
    usability, 233, *234*
Waterbody, factors affecting degree of eutrophication, 74–78
  models, 86
  morphology, 74–77, 264
  morphology and macrophyte relationship predictions, 106
  phosphorus models, 15
  shape affecting use of simple models, 103
  shape affecting water quality, 76
Waterbody classification, a detailed system, 262–275
  graphical representation, *275*
  classification, how to classify, example of, 269–275
  hydrographic criteria, 264, 265
  hygenically relevant criteria, 269, *270*
  morphometric crtieria, *266*, 268
  simple empirical system, 257–262
  Technical Standard, 257, 264, 269
  trophic criteria, *266*, 268
Watershed, effect of upstream waterbody, 70
  models, 86
  nitrogen flux, 70
  tributary effects, 71 (see also under drainage basin)

Zooplanktivorous fish, 200
Zooplankton, grazing, 53, 81, 187
  role in control of algal biomass, 221

314